Thomas W. Wieting
Reed College
1985

Geometric Quantization in Action

Mathematics and Its Applications

Volume 8

Norman E. Hurt

MRJ Incorporated, Fairfax, Virginia, U.S.A.

Geometric Quantization in Action

Applications of Harmonic Analysis in Quantum Statistical Mechanics and Quantum Field Theory

D. REIDEL PUBLISHING COMPANY

Dordrecht : Holland / Boston : U.S.A. / London : England

Library of Congress Cataloging in Publication Data

Hurt, Norman,
Geometric quantization in action.

(Mathematics and its applications; v. 8)
Bibliography: p.
Includes index.
1. Geometric quantization. 2. Quantum statistics.
3. Quantum field theory. 4. Harmonic analysis.
I. Title. II. Series: Mathematics and its applications
(D. Reidel Publishing Company); v. 8.
QC174.17.G46H87 1982 530.1′33 82-12370
ISBN 90-277-1426-6

Published by D. Reidel Publishing Company
P.O. Box 17, 3300 AA Dordrecht, Holland.

Sold and distributed in the U.S.A. and Canada
by Kluwer Boston Inc.,
190 Old Derby Street, Hingham, MA 02043, U.S.A.

In all other countries sold and distributed
by Kluwer Academic Publishers Group,
P.O. Box 322, 3300 AH Dordrecht, Holland.

D. Reidel Publishing Company is a member of the Kluwer Group.

Printed in The Netherlands.

Table of Contents

To my parents, and to
Susan, Michael and Jason

Editor's Preface

Approach your problems from the right end and begin with the answers. Then, one day, perhaps you will find the final question.

'The Hermit Clad in Crane Feathers' in R. Van Gulik's *The Chinese Maze Murders.*

It isn't that they can't see the solution. It is that they can't see the problem.

G. K. Chesterton, *The Scandal of Father Brown* 'The Point of a Pin'.

Growing specialization and diversification have brought a host of monographs and textbooks on increasingly specialized topics. However, the 'tree' of knowledge of mathematics and related fields does not grow only by putting forth new branches. It also happens, quite often in fact, that branches which were thought to be completely disparate are suddenly seen to be related.

Further, the kind and level of sophistication of mathematics applied in various sciences has changed drastically in recent years: measure theory is used (non-trivially) in regional and theoretical economics; algebraic geometry interacts with physics; the Minkowsky lemma, coding theory and the structure of water meet one another in packing and covering theory; quantum fields, crystal defects and mathematical programming profit from homotopy theory; Lie algebras are relevant to filtering; and prediction and electrical engineering can use Stein spaces.

This series of books, *Mathematics and Its Applications*, is devoted to such (new) interrelations as *exempla gratia*:

- a central concept which plays an important role in several different mathematical and/or scientific specialized areas;
- new applications of the results and ideas from one area of scientific endeavour into another;
- influences which the results, problems and concepts of one field of enquiry have and have had on the development of another.

With books on topics such as these which are stimulating rather than definitive, intriguing rather than encyclopaedic, we hope to contribute something towards better communication among the practitioners in diversified fields.

The present book is a good example of the synergetic effects which can occur when initially discrete areas of inquiry come into contact and when the tools or ideas ((geometric) quantization) more or less designed for one field are used in another (representation theory). There is something very healthy and promising about current applied mathematics if in one research level treatise words and concepts like (Epstein) zeta function, Selberg Trace formula, Schubert cells, Automorphic forms and Riemann-Roch occur next to Ising model, Toda lattice, De Sitter space and quantum (statistical) mechanics.

The unreasonable effectiveness of mathematics in science . . .

Eugene Wigner

Well, if you knows of a better 'ole, go to it.

Bruce Bairnsfather

What is now proved was once only imagined.

William Blake

As long as algebra and geometry proceeded along separate paths, their advance was slow and their applications limited.

But when these sciences joined company, they drew from each other fresh vitality and thenceforward marched on at a rapid pace towards perfection.

Joseph Louis Lagrange

Krimpen a/d IJssel
July, 1982

Michiel Hazewinkel

Preface

Although there were many ideas and goals in geometric quantization at the beginning, the major thrust of this volume is to present the single theme that geometric quantization provides the structure for the geometric realizations of the irreducible unitary representations of the groups involved in physics. Once this theme and the geometric realizations are established, we will present some examples of the use of this representation theory in two areas of physics – quantum statistical mechanics and quantum field theory.

We have reviewed in this volume the orbit technique of Kostant and Kirillov. However, the basic philosophy that we want the reader to appreciate is that for almost all the standard examples presented in elementary quantum theory – the harmonic oscillator, the hydrogen atom or Kepler's problem, the spinning particle, etc. – the geometric underpinnings are contained in classical results on cohomology of bundles over compact complex homogeneous spaces. The orbit theory becomes somewhat superfluous. And the heart of the results on independence of polarization and the calculation of the degeneracy of the eigenvalues is contained in the Hirzebruch–Riemann–Roch theorem. In fact, it is the beautiful results of Kodaira, Hirzebruch, Borel, Bott and others which should be studied by physicists.

The second point we hope the reader begins to appreciate is that all the geometry of quantum theory was known to the classical geometers – in particular Cartan. (Maybe Bob Hermann is right that *everything* is in Cartan.)

When we turn to the development of the representation theory aspects of quantum statistical mechanics we hope the reader begins to understand the history of modern spectral geometry. After Weyl, Jeans, Sommerfeld and others, the first major break-through in spectral geometry came from quantum statistical mechanics in the work of Fowler. His work foreshadowed all the later work by Minakshisundaram, Berger, McKean, Singer, Gilkey and company. In fact, the higher order terms in the spectral expansion were also known to Kirkwood in his work in statistical mechanics – predating Gilkey by twenty years. The physicist deWitt also developed much of the spectral theory expansions in his work in renormalization theory in quantum field theory in the 1950s. This brings us to the third major point – we hope the reader begins to see that quantum field theory is not mysterious at all. In fact, nearly all the machinations of quantum field theory are already present in quantum statistical mechanics.

Finally, one great theorem or theory looms over this entire tome – viz. the Selberg trace formula. This theory of Selberg contains nearly all the major results we use – e.g. the Plancherel theorem, the Riemann–Roch theorem, the Poisson summation formula, etc. The Selberg theory artfully combines physics – esp. scattering theory, representation theory, and geometry in one beautiful series of results. The Selberg theory allows us to actually perform calculations on noncompact but finite volume universes which are of interest to those working in general relativity. Thus we end our volume with an introduction to the ideas of temperatures of black holes.

But, we first turn our attention to the groups, group representations and homogeneous spaces which have played a fundamental role in physics. Certainly basic to understanding the most elementary classical systems the study of the rotation groups $SO(n)$, the Euclidean groups $E(n)$ and the generalized Lorentz groups $SO(n, 1)$ is required. In the next chapters the representation theory of these groups will be presented based on geometric quantization.

The author has many people to thank for help over the years during which the work leading to this volume was performed. The author especially wants to thank Marshall Stone, Chris Byrnes, Hans Fischer, Bob Hermann, Stuart Dowker, Enrico Onofri, John Rawnsley, K. Maurin, and Richard Cushman. Finally, the author wishes to thank Michiel Hazewinkel for the invitation to contribute to this series.

Chapter 0

Survey of Results

0.1. Introduction

Symmetry groups and their properties form the foundations of quantum mechanics and classical mechanics. Usually the study of symmetry groups is tied closely to one of these subjects or the other. It is the intent of this volume to introduce the reader to a more unified view of symmetry groups in physics – both quantum and classical. The vehicle to achieve this approach is the study of quantum statistical physics of systems with high degree of symmetry – in particular the study of the high temperature asymptotics of these systems. These asymptotics provide the natural connection with classical (statistical) mechanics.

The history of quantum statistical mechanics has been told many times. What interests us herein are several beautiful and deep results relating the quantum and classical systems which first were derived in the years between the discovery of Plank's radiation law and the early days of modern quantum theory. These results are basically theorems regarding the asymptotics of certain partial differential equations. However, the importance of these results, which we develop in this volume, is the relationship of the asymptotics to the underlying symmetry groups and geometry of the physical system. The message which the reader should take away is that quantum statistical mechanics is merely a part of, what is today called by mathematicians, spectral geometry. Spectral geometry in its simplest form considers the case of a compact Riemannian manifold (M, g) and a Laplace-Beltrami operator Δ on (M, g) with spectrum $\{\lambda_n\}$; the basic question of spectral geometry is whether the differential geometric properties of (M, g) can be determined by knowledge of $\{\lambda_n\}$–e.g. can we determine the volume of M, its dimension, its scalar curvature, its Ricci curvature, etc.? We will see that spectroscopists and statistical mechanicians were performing spectral geometry before it was even a discipline. Viz. while taking observations of the spectra of (simple) physical systems, data regarding the geometry of these systems were being determined.

Once the reader is familiar with this philosophy for studying statistical mechanics, the second aim of this volume is to show the reader that these results can be transferred to a very different area of quantum physics – viz. quantum field theory. These vastly different fields of study require many different subjects at the fingertips of the reader. Clearly space does not permit a thorough review of each of these areas. The physicist will be expected to bring with him certain mathematical trappings while the mathematician will be assumed to know some elementary physics. To begin to set the notation and language and to give the reader a flavor of results which will be studied in greater depth in later chapters we briefly present some examples.

0.2. Some elementary quantum systems

A quantum system is described by a partial differential operator, H, called the Hamiltonian, which acts on elements Ψ in a Hilbert space $\mathscr{H}$. The object of study is the Schrödinger equation $H\Psi = E\Psi$. For elementary physical systems there are interesting relationships between the Hamiltonians.

Example 0.2.1. The four dimensional isotropic harmonic oscillator is described by

$$-\frac{1}{2m}\left[\sum\frac{\partial^2}{\partial x_k^2} - m^2\omega^2\sum x_k^2\right]\Psi = E\Psi.$$

The geometry of the system is specified by its configuration space $M = R^4$ with elements $(x_1, \ldots, x_4)$. The line element of this space is

$$\mathrm{d}s^2 = \sum(\mathrm{d}x_i)^2.$$

If we change coordinates to $(R, \vartheta, \varphi, \psi)$ where

$$x_1 = R\cos\left(\frac{\varphi+\psi}{2}\right), \quad x_2 = R\sin\left(\frac{\vartheta}{2}\right)\cos\left(\frac{\varphi-\psi}{2}\right),$$

$$x_3 = R\cos\left(\frac{\vartheta}{2}\right)\sin\left(\frac{\varphi+\psi}{2}\right), \quad x_4 = R\sin\left(\frac{\vartheta}{2}\right)\sin\left(\frac{\varphi-\psi}{2}\right),$$

the line element becomes

$$\mathrm{d}s^2 = \mathrm{d}R^2 + \frac{R^2}{4}(\mathrm{d}\vartheta^2 + \mathrm{d}\varphi^2 + \mathrm{d}\psi^2 + 2\cos\vartheta\,\mathrm{d}\varphi\,\mathrm{d}\psi).$$

In these coordinates with $r = R^2$ the Schrödinger equation for the isotropic oscillator becomes

$$\frac{-2}{m}\left[\frac{1}{r}\frac{\partial}{\partial r}\left(r^2\frac{\partial}{\partial r}\right)+\frac{1}{r}\left\{\frac{1}{\sin\vartheta}\frac{\partial}{\partial\vartheta}\left(\sin\vartheta\frac{\partial}{\partial\vartheta}\right)+\right.\right. \tag{$*$}$$

$$\left.\left.+\frac{1}{\sin^2\vartheta}\left(\frac{\partial^2}{\partial\varphi^2}+\frac{\partial^2}{\partial\psi^2}-2\cos\vartheta\frac{\partial^2}{\partial\varphi\partial\psi}\right)\right\}-\frac{m^2\omega^2 r}{4}\right]\Psi=E\Psi.$$

Imposing the constraint $\partial\Psi/\partial\psi=0$ we obtain the Schrödinger equation for the hydrogen atom

$$\frac{-1}{2m}\left[\frac{1}{r^2}\frac{\partial}{\partial r}\left(r^2\frac{\partial}{\partial r}\right)+\frac{1}{r^2}\left\{\frac{1}{\sin\vartheta}\frac{\partial}{\partial\vartheta}\left(\sin\vartheta\frac{\partial}{\partial\vartheta}\right)+\right.\right.$$

$$\left.\left.+\frac{1}{\sin^2\vartheta}\frac{\partial^2}{\partial\varphi^2}\right\}-\frac{k}{r}\right]\Psi=E'\Psi,$$

where $k=-mE/2$, $E'=-m\omega^2/8$. Solving ($*$) by separation of variables $\Psi=\Phi_1(r)\Phi_{\text{rot}}(\vartheta,\varphi,\psi)$ we obtain the Schrödinger equation for the spherical rotator

$$\left[\frac{1}{\sin\vartheta}\frac{\partial}{\partial\vartheta}\left(\sin\vartheta\frac{\partial}{\partial\vartheta}\right)+\frac{1}{\sin^2\vartheta}\left(\frac{\partial^2}{\partial\varphi^2}+\frac{\partial^2}{\partial\psi^2}\right.\right.$$

$$\left.\left.-2\cos\vartheta\frac{\partial^2}{\partial\varphi\partial\psi}\right)+\lambda\right]\Phi_{\text{rot}}=0.$$

The solutions of the spherical rotator are known to be

$$\Phi_{\text{rot}}=\Theta_{JKM}(\vartheta)e^{iK\varphi}e^{iM\psi},$$

where

$$\Theta_{JKM}(\vartheta)=x^{|K-M|/2}(1-x)^{|K+M|/2}F(\alpha\beta\gamma x)$$

$$\lambda=J(J+1)$$

$$J=0,1,2,\ldots,K=J,J-1,\ldots,-J$$

$$M=J,J-1,\ldots,-J.$$

$F(\ldots)$ is the hypergeometric function with $x=(1-\cos\vartheta)/2$,

$$\alpha=-J+\tfrac{1}{2}|K-M|,\quad \beta=J+\tfrac{1}{2}|K+M|+\tfrac{1}{2}|K-M|,$$

$$\gamma=|K-M|+1.$$

The equation for Φ_1 is

$$\left[\frac{1}{r^2}\frac{\mathrm{d}}{\mathrm{d}r}\left(r^2\frac{\mathrm{d}}{\mathrm{d}r}\right)+\left(\frac{-m^2\omega^2}{4}+\frac{mE}{2r}-\frac{\lambda}{r^2}\right)\right]\Phi_1=0$$

which is the equation for radial motion of a particle with angular momentum J in a Coulomb field. The solution of this equation is known to be

$$\Phi_{nJ} = e^{-\rho/2}\rho^J L_{n+J}^{2J+1}(\rho), \qquad E = 2n\omega, n = 1, 2, \ldots$$

$\rho = m\omega r$, L_{n+J}^{2J+1} is the associated Laguerre polynomial. The wave functions for the isotropic harmonic oscillator are thus

$$\Psi_{\text{osc}} = \Phi_{nJ}(\rho)\Theta_{JKM}(\vartheta)e^{iK\varphi}e^{iM\psi}$$

and the wave functions for the hydrogen atom are those with $M = 0$ – i.e.

$$\Psi_{Hy} = \Phi_{nJ}(\rho)\Theta_{KJO}e^{iK\varphi}.$$

The number of linearly independent wave functions with the same energy level, i.e. the *degree of degeneracy*, is easily calculated for each of the cases just described:

$$d_{\text{osc}} = \sum_{j=0}^{n-1}(2j+1) = (2n-1)n(2n+1)/3,$$

$$d_{Hy} = \sum_{j=0}^{n-1}(2j+1) = n^2,$$

$$d_{\text{rot}} = (2j+1)^2.$$

The Schrödinger operators for the systems of interest are determined by the Laplace–Beltrami operators Δ on manifolds M. Recall that if (M, g) is a Riemannian manifold with local coordinates $(q_1, \ldots, q_n)$ and Riemannian metric

$$g_{ij}(q) = g\left(\frac{\partial}{\partial q_i}, \frac{\partial}{\partial q_j}\right)(q)$$

then the *Laplace–Beltrami operator* is

$$\Delta = -\sum_{i,j}\frac{1}{\sqrt{g}}\frac{\partial}{\partial q_i}\left(g^{ij}\sqrt{g}\frac{\partial}{\partial q_j}\right)$$

where $g = \det(g_{ij})$ and $(g^{ij}) = (g_{ij})^{-1}$.

EXAMPLE 0.2.2. Consider the coordinate system $(q^1, q^2, q^3) = (\vartheta, \varphi, \psi)$ with the metric

$$\begin{pmatrix} I & 0 & 0 \\ 0 & I & I\cos\vartheta \\ 0 & I\cos\vartheta & I \end{pmatrix}$$

The reader can check that Laplace–Beltrami operator in this case is

$$-\frac{1}{I}\left\{\frac{1}{\sin\vartheta}\frac{\partial}{\partial\vartheta}\left(\sin\vartheta\frac{\partial}{\partial\vartheta}\right)+\frac{1}{\sin^2\vartheta}\left[\frac{\partial^2}{\partial\varphi^2}+\frac{\partial^2}{\partial\psi^2}-\right.\right.$$
$$\left.\left.-2\cos\vartheta\frac{\partial}{\partial\varphi}\frac{\partial}{\partial\psi}\right]\right\}$$

which is the Schrödinger operator in the spherical rotator; cf. Exercise 0.9.

0.3. Examples of Group Representation in Physics

The degree of degeneracies of eigenstates are related to representations of the underlying symmetry groups. Before we examine this let us introduce some more notation. Consider the symmetry group of a triangle:

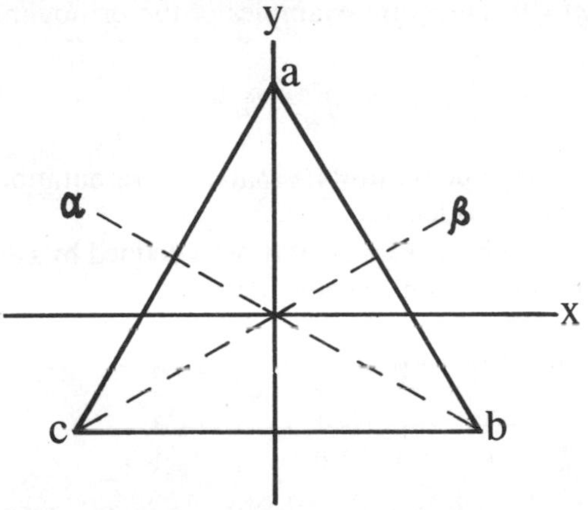

e = the identity map, g_1 = reflection in the yz plane; g_2 = reflection in the αz plane; g_3 = reflection in the βz plane; g_4 = clockwise rotation of 120°; g_5 = counterclockwise rotation of 120°. Consider now a representation of G – i.e. a map $T: G \to GL(V)$ where T satisfies $T(gh) = T(g)T(h)$ and $T(e) = I$. The dimension of V is called the *degree* of T. We normally are going to study representations with elements in the complex field C. Furthermore we will usually treat unitary representations of G; if T has matrix form $(T(g)_{ij})$, this means that $(T(g)^*_{ji}) = (T(g^{-1})_{ij})$. Two representations T and T' are called *equivalent* if there is an S in $GL(V)$ such that $T'(g) = = ST(g)S^{-1}$ for all g in G. A subspace W of V is called *invariant* under T if $T(g)w \in W$ for all g in G and w in W. A representation is called *reducible* if there is a proper sub-space W of V which is invariant under T; otherwise T is called *irreducible*. We will let $\hat{G}$ denote the set of inequivalent irreducible unitary representations of G.

The reader can check that the following maps are representations for the triangle group defined above:

$$
\begin{array}{ccccccc}
 & e & g_1 & g_2 & g_3 & g_4 & g_5 \\
T^1 & (1) & (1) & (1) & (1) & (1) & (1) \\
T^2 & (1) & (-1) & (-1) & (-1) & (1) & (1) \\
T^3 & \begin{pmatrix} 1 & 0 \\ 0 & 1 \end{pmatrix} & \begin{pmatrix} -1 & 0 \\ 0 & 1 \end{pmatrix} & \begin{pmatrix} \frac{1}{2} & -\frac{\sqrt{3}}{2} \\ -\frac{\sqrt{3}}{2} & -\frac{1}{2} \end{pmatrix} & \begin{pmatrix} \frac{1}{2} & \frac{\sqrt{3}}{2} \\ \frac{\sqrt{3}}{2} & -\frac{1}{2} \end{pmatrix} & \begin{pmatrix} -\frac{1}{2} & \frac{\sqrt{3}}{2} \\ -\frac{\sqrt{3}}{2} & -\frac{1}{2} \end{pmatrix} & \begin{pmatrix} -\frac{1}{2} & -\frac{\sqrt{3}}{2} \\ \frac{\sqrt{3}}{2} & -\frac{1}{2} \end{pmatrix}.
\end{array}
$$

These are in fact irreducible representations. If we let T^j_{mn} denote the mnth component of these matrix representations we can check that $\sum_g T^k_{11}(g)T^k_{11}(g) = 6$ for $k = 1, 2$ while $\sum_g T^3_{11}(g)T^3_{11}(g) = 3$. Also we see that $\sum_g T^1_{11}(g)T^2_{11}(g) = 0$. These are examples of the orthogonality relations

$$
\sum_g T^i_{mn}(g)\sqrt{\frac{l_i}{h}}T^{j*}_{m'n'}(g)\sqrt{\frac{l_i}{h}} = \delta_{ij}\delta_{mm'}\delta_{nn'}
$$

which hold for any irreducible unitary matrix representations T^j of degree l_j of a finite group G of order h.

The *character* of a matrix representation is defined by $\mathfrak{x}_k(g) = \sum_m T^k_{mm}(g)$. For the triangle group the character table is

$$
\begin{array}{ccccccc}
 & e & g_1 & g_2 & g_3 & g_4 & g_5 \\
\mathfrak{x}_1 & 1 & 1 & 1 & 1 & 1 & 1 \\
\mathfrak{x}_2 & 1 & -1 & -1 & -1 & 1 & 1 \\
\mathfrak{x}_3 & 2 & 0 & 0 & 0 & -1 & -1
\end{array}
$$

Using the orthogonality relations we have immediately that

$$
\sum_g \mathfrak{x}_i(g)\mathfrak{x}_j(g)^* = h\delta_{ij}.
$$

Consider now the action of G on the configuration space of a quantum mechanical system – i.e. a map $f: G \times M \to M$ where $f(g_1, f(g_2, m)) = = f(g_1 g_2, m)$ and $f(e, m) = m$. Usually we denote $f(g, m)$ as gm. Assume this action leaves the Hamiltonian $H(m)$ invariant – i.e. $H(gm) = H(m)$. If we consider the map $g \to (T(g)\Psi)(m) = \Psi(g^{-1}m)$ then one easily checks that this is a group representation. If we set $E_\lambda = \{\Psi \in \mathcal{H} | H\Psi = \lambda\Psi\}$ then for Ψ in E_λ we have $HT(g)\Psi = T(g)H\Psi = \lambda T(g)\Psi$. Thus $T(g)\Psi \in E_\lambda$ and E_λ is invariant under the representation T of G. Often the eigenspace is irreducible under T; in this case the dimension of the representation is equal to the degeneracy of the corresponding eigenvalue.

For example if a Hamiltonian were invariant under the triangle group then there would be a set of nondegenerate eigenfunctions invariant under the operations of the group. There would be a set of nondegenerate eigenfunctions which remain unchanged under e, g_4, g_5 and change sign under g_1, g_2, or g_3. Finally there would be a set of doubly degenerate eigenfunctions transforming according to T^3.

Given two representation T^1 and T^2 of G on V_1 and V_2 we can form a new representation, the *direct product representation*:

$$(T^1 \otimes T^2)(g)(v_1 \otimes v_2) = T^1(g)v_1 \otimes T^2(g)v_2.$$

In terms of matrix elements, if we have $g \to T^1_{li}(g), g \to T^2_{kj}(g)$, then $[T^1 \otimes T^2(g)]_{lk,ij} = T^1_{li}(g)T^2_{kj}(g)$. The character of the direct product is $\chi_{T^1 \otimes T^2}(g) = \sum_{l=1}^n \sum_{k=1}^n T^1_{ll}(g) T^2_{kk}(g) = \chi_{T^1}(g)\chi_{T^2}(g)$. Since equivalent representations have the same character we immediately see that $T^1 \otimes T^2$ is equivalent to $T^2 \otimes T^1$.

If $\{T^\lambda\}$ is a complete set of inequivalent irreducible representations of G, then $T^1 \otimes T^2$ decomposes as

$$T^1 \otimes T^2 = \sum_{\lambda \in \hat{G}} a_\lambda T^\lambda.$$

This expansion is called the *Clebsch–Gordan* series.

Direct product representations occur in selection rule analysis in quantum mechanics. Viz. if the wave functions Ψ_λ are viewed as basis for the representation of the appropriate symmetry group, then the integral $\int \Psi_1 \Psi_2$ will be nonzero only if the integrand is invariant under all operations of the group, i.e. T^I occurs in the decomposition $T^1 \otimes T^2 = = \sum a_\lambda T^\lambda$. If the characters are real, then this occurs only if $T^1 = T^2$. Similarly one can analyze selection rules of the form $\int \Psi_\lambda f \Psi_{\lambda'}$.

Although we have discussed examples of group representations only for finite groups, the theory for finite groups immediately extends to compact (Lie) groups. We turn to this topic in Chapter 1. The study of group representations for noncompact groups is a little more complicated; but as we will see for the noncompact groups of interest in physics, the theory is readily accessible as we describe in Chapters 12 and 13.

0.4. Asymptotics in Statistical Mechanics

Quantum statistical mechanics assumes that the probability that a quantum mechanical system described by H is in state E_λ is given by $\exp(-E_\lambda\beta)/\sum_\lambda \exp(-E_\lambda\beta)$. Here β is a parameter to be determined momentarily as $1/kT$ where k is Boltzmann's constant and T is temperature. The thermodynamic internal energy is then given by $E=\sum_\lambda E_\lambda \exp(-E_\lambda\beta)/Z$, where $Z=\sum_\lambda \exp(-E_\lambda\beta)$. Clearly $E=kT^2\times$ $\times(\partial/\partial T)\log Z$. The specific heat is

$$C_v=\left(\frac{\partial E}{\partial T}\right)=\frac{\partial}{\partial T}\left[kT^2\left(\frac{\partial}{\partial T}\log Z\right)\right].$$

Z is called the *partition function* and the sum is taken over all the allowed energy level of the system. We could write Z equivalently as $Z=\sum g_\lambda\times$ $\exp(-E_\lambda\beta)$ where g_λ is the degeneracy of the λth level and the sum is over the distinct λ.

Consider the harmonic oscillator with classical Hamiltonian $H=(p^2/2m)+(kq^2/2)$. The associated Schrödinger operator $H=-(\hbar/2m)(\mathrm{d}^2/\mathrm{d}q^2)+(kq^2/2)$ has eigenvalues $E_n=\hbar\sqrt{(k/m)}(n+\frac{1}{2})=$ $=h\nu(n+\frac{1}{2})$, where $\nu=(1/2\pi)\sqrt{k/m}$ and $n=0, 1, 2,\ldots$. Thus the partition function for the harmonic oscillator is given by

$$Z_{HO}(\beta)=\sum_{n=0}^{\infty}\exp(-E_n\beta)=e^{-h\nu\beta}\left\{\frac{1}{1-e^{-h\nu\beta}}\right\}.$$

For a classical system governed by a Hamiltonian $H(p,q)=$ $=H(p_1,\ldots,p_n,q_1,\ldots,q_n)$ the *classical partition function* is defined by

$$Z=\frac{1}{h^n}\int\ldots\int e^{-\beta H(p,q)}\,\mathrm{d}p_1\ldots\mathrm{d}p_n\,\mathrm{d}q_1\ldots\mathrm{d}q_n.$$

One of the basic principles of quantum statistical mechanics is that the high temperature limit of Z should be the classical partition function – i.e.

$$Z(\beta)\sim Z_{\mathrm{cl}}(\beta)\quad\text{as}\quad\beta\to 0.$$

Explicitly in a one dimensional system with $H(p,q)=(p^2/2)+V(q)$, this means

$$\sum\exp(-E_n\beta)\sim\frac{1}{h}\int_{-\infty}^{\infty}\!\!\int\exp\left(-\beta\left(\frac{p^2}{2}+V(q)\right)\right)\mathrm{d}p\,\mathrm{d}q. \qquad (*)$$

If we let $B(\lambda)$ denote the area of the region $H(p,q)=p^2/2+V(q)\leqq\lambda$, we can rewrite $(*)$ as

$$Z(\beta)\sim\frac{1}{h}\int_0^\infty\exp(-\beta\lambda)\,\mathrm{d}B(\lambda). \qquad (**)$$

Assuming that the Hardy–Littlewood–Tauberian theorem applies to $H(p, q)$, (**) is equivalent to the statement that

$$N(\lambda) = \sum_{\lambda_n < \lambda} 1 \sim B(\lambda)/h \quad \text{as} \quad \lambda \to \infty.$$

For the harmonic oscillator we have $Z(\beta) = \frac{1}{2}(\sinh(h\nu\beta/2)$. Using the expansion

$$\frac{1}{\sinh(2z)} = \frac{-1}{2z} + \sum_{j=0}^{\infty} (-1)^j s_j z^{2j-1},$$

where

$$s_0 = 1,\ s_k = 2^{2k}\left(\frac{2^{2k-1} - 1}{(2k)!}\right)B_k,$$

where B_k are Bernoulli numbers $(B_1 = \frac{1}{6}, B_2 = \frac{1}{30}, \ldots)$ we see that as $\beta \to 0$

$$Z(\beta) \sim 1/h\nu\beta.$$

The reader can check that the right-hand side coincides with the integral

$$\frac{1}{h}\int_{-\infty}^{\infty}\int \exp\left(-\left(\frac{p^2}{2m} + \frac{q^2k}{2}\right)\beta\right)\mathrm{d}p\,\mathrm{d}q,$$

thus verifying the limiting rule of quantum statistical mechanics for the harmonic oscillator. The area of the ellipse $B(\lambda)$ is $2\pi\lambda\sqrt{(m/k)} = (\lambda/\nu)$. Thus $N(\lambda) \sim \lambda/h\nu$ as $\lambda \to \infty$ for the harmonic oscillator.

The perfect gas partition function is described by the energy levels of a particle in a box of dimensions (X, Y, Z), which are

$$E_{n_1,n_2,n_3} = \frac{h^2}{8m}\left(\frac{n_1^2}{X^2} + \frac{n_2^2}{Y^2} + \frac{n_3^2}{Z^2}\right).$$

Thus

$$Z = \sum_{n_1=1,n_2=1,n_3=1}^{\infty} \exp\left(-\frac{h^2}{8m}E_{n_1,n_2,n_3}\right).$$

Using the high temperature limit

$$Z_A = \sum_{n=1}^{\infty} \exp\left(-\frac{h^2}{8m}\frac{\hbar^2}{A}\right) \sim \left(\frac{2\pi m}{\beta h^2}\right)^{1/2} A$$

as $\beta \to 0$ (which will be discussed later) we have $Z \sim (2\pi m/\beta h^2)^{3/2}V$ where $V = XYZ$ is the volume of the box. One of the relationships of thermodynamics is that the force in the x direction is

$$F_x = \frac{N}{\beta}\frac{\partial \log Z}{\partial X} = N/\beta X.$$

Thus the pressure is the force per unit area $P = F_x/YZ = N/\beta V$. We know a perfect gas obeys $PV = NkT$. Thus we have identified $\beta = 1/kT$.

Once there are interactions the Hamiltonian becomes

$$H = \sum_{i=1}^{N} \frac{p_i^2}{2m} + \sum_{i<j} v_{ij},$$

where $v_{ij} = v(|r_i - r_j|)$. At low densities the equation of state of a perfect gas is approximated by

$$PV/NkT \approx 1 - \frac{N}{2V}\int_0^\infty \mathrm{d}r\, 4\pi r^2 f_{12}(r),$$

where $f_{12}(r) = \exp(-\beta v_{12}(r)) - 1$, where $v_{12}(r)$ is the intermolecular potential. The coefficient of N/V is called the *second virial coefficient*.

The next example from quantum statistical mechanics is the planar rigid rotator whose Schrödinger equation is

$$\frac{-\hbar^2}{2I}\frac{\partial^2}{\partial\varphi^2} = E_l\Psi,$$

where $E_l = (\hbar^2/2I)l^2$, $l \in Z$; E_l has multiplicity two for $E_l \neq 0$. The partition function is

$$Z_p = \sum_{l\in Z} \exp(-E_l\beta) = \sum_{l\in Z} e^{-\pi l^2 \tilde{\beta}}$$

which is just the Jacobi theta function $\vartheta_3(\tilde{\beta}, 0)$ where $\tilde{\beta} = \beta\hbar^2/\pi^2 I$. The high temperature asymptotics are $Z_p(\beta) = ((\hbar^2/2I\pi)\beta)^{-1/2} + (e^{-1/\beta})$ as $\beta \to 0$.

Consider now the rigid rotator with two masses m_1 and m_2 at a fixed intermolecular distance R with spins s_1 and s_2. This system has a moment of inertia $I = \mu R^2$ where $\mu = m_1 m_2/(m_1 + m_2)$ is the reduced mass. If the polar coordinates of one atoms are a, ϑ, φ where $a = (m_2 R)/(m_1 + m_2)$ and the other atom has b, ϑ, φ, where $b = (-m_1 R)/(m_1 + m_2)$, then the kinetic energy of the first particle is

$$\frac{m_1}{2}\left[\left(\frac{\mathrm{d}x_1}{\mathrm{d}t}\right)^2 + \left(\frac{\mathrm{d}x_2}{\mathrm{d}t}\right)^2 + \left(\frac{\mathrm{d}x_3}{\mathrm{d}t}\right)^2\right] =$$

$$= \frac{m_1 a^2}{2}\left[\left(\frac{\mathrm{d}\vartheta}{\mathrm{d}t}\right)^2 + \sin^2\vartheta\left(\frac{\mathrm{d}\varphi}{\mathrm{d}t}\right)^2\right].$$

Similarly the kinetic energy for the second particle is given by $m_1 \to m_2$ and $a \to b$. Thus the total kinetic energy is

$$T = \frac{I}{2}\left[\left(\frac{d\vartheta}{dt}\right)^2 + \sin^2\vartheta\left(\frac{d\varphi}{dt}\right)^2\right],$$

where $I = m_1 a^2 + m_2 b^2$. This is precisely the kinetic energy of a single particle of mass I confined to the surface of a sphere of unit radius. Taking the Laplacian in spherical coordinates we have the Schrödinger equation

$$\frac{1}{\sin\vartheta}\frac{\partial}{\partial\vartheta}\left(\sin\vartheta\frac{\partial\Psi}{\partial\vartheta}\right) + \frac{1}{\sin^2\vartheta}\frac{\partial^2\Psi}{\partial\varphi^2} + \frac{8\pi^2}{h^2}IE\Psi = 0.$$

As is well known the solutions of this eigenvalue problem are the normalized spherical harmonics $Y_{1\pm m}(\vartheta, \varphi)$ with eigenvalues $E_j = (h^2/8\pi^2 I)j(j+1)$, $j = 0, 1, 2, \ldots$; the angular momentum can have components along a particular axis with values $m\hbar, m = j, j-1, \ldots, -j$. Thus the rotational state of energy E_j is $(2j+1)$-degenerate. The rotational partition function is

$$Z_{\text{rot}} = \frac{1}{\sigma}\sum_{j=0}(2j+1)\exp(-yj(j+1)),$$

where $y = \hbar^2/2IkT$. The factor σ is a symmetry number being 2 for a symmetrical molecule AA and 1 for an unsymmetrical molecule AB. Mulholland and Fowler in 1928 developed the high temperature expansion for the rigid rotator and showed that

$$Z_{\text{rot}} = \frac{1}{\sigma y}\left(1 + \frac{y}{3} + \frac{y^2}{15} + \cdots\right) \quad \text{as} \quad \beta \to 0.$$

We leave it for the reader to check the limiting rule for the planar rigid rotator and the rigid rotator.

The next example is the asymmetric rotator which has three unequal principal moments of inertia. The kinetic energy is $K = \frac{1}{2}(I_1\omega_1^2 + + I_2\omega_2^2 + I_3\omega_3^2)$ where the ω's are the angular velocities of rotation around the principal axes, (x, y, z). The energy levels here are somewhat complicated; however, the reader can check that the classical partition function in this case is

$$Z^{cl}_{\text{Asymrot}} = \frac{\pi^{1/2}}{\sigma}\prod_{i=1}^{3}\left[\frac{8\pi^2 I_i}{h^2}\right]^{1/2}.$$

Stripp and Kirkwood in 1951 calculated the next higher order term in the high temperature expansion:

$$Z_{\text{Asymrot}} = \left[\frac{\pi}{a_x a_y a_z}\right]^{1/2} \left[1 + \left(\frac{1}{12} 2a_x + 2a_y + 2a_z - \right.\right.$$
$$\left.\left. - \frac{a_x a_y}{a_z} - \frac{a_y a_z}{a_x} - \frac{a_z a_x}{a_y}\right) + \cdots\right]$$

where $a_u = (\hbar^2/2I_u)$, $u = x, y, z$. If two moments of inertia are identical we have the result for a symmetric top (say $a_x = a_y = a$):

$$Z_{\text{symtop}} = \left[\frac{\pi^{1/2}}{a a_z^{1/2}}\right]\left[1 + \frac{1}{12}\left(4a - \frac{a^2}{a_z}\right) + \ldots\right].$$

The next higher order term for the symmetric top is

$$Z_{\text{symtop}} = \left(\frac{\pi^{1/2}}{a a_z^{1/2}}\right)\left[1 + \frac{1}{12}\left(4a - \frac{a^2}{a_z}\right) + \right.$$
$$\left. + \frac{1}{480}\left(32a^2 - 24\frac{a^3}{a_z} + 7\frac{a^4}{a_z^2}\right)\ldots\right].$$

This symmetric top analysis is due to Viney and Kassel in 1933. The next higher order term for the asymmetric top was specified by Kirkwood and Stripp.

The Schrödinger equation for the symmetric top is

$$\frac{1}{\sin\vartheta}\frac{\partial}{\partial\vartheta}\left(\sin\vartheta\frac{\partial\Psi}{\partial\vartheta}\right) + \frac{1}{\sin^2\vartheta}\frac{\partial^2\Psi}{\partial\varphi^2} + \left(\frac{\cos^2\vartheta}{\sin^2\vartheta} + \frac{a}{a_z}\right)\frac{\partial^2\Psi}{\partial\psi^2} -$$
$$- \frac{2\cos\vartheta}{\sin^2\vartheta}\frac{\partial^2\Psi}{\partial\psi\,\partial\varphi} + \frac{8\pi^2 aE\Psi}{h^2} = 0.$$

Using the substitution $\Psi = \Theta(\vartheta)e^{iM}\varphi e^{iK\psi}$, it can be shown that

$$E_{J,K} = \frac{\hbar^2}{2}\left[J\frac{(J+1)}{I} + K^2\left(\frac{1}{I_z} - \frac{1}{I}\right)\right],$$

where $J = j + \frac{1}{2}|K + M| + \frac{1}{2}|K - M|$

$$j = 0, 1, 2, \ldots,\quad J = 0, 1, 2, \ldots,$$
$$K = 0, \pm 1, \pm 2, \ldots, \pm J,$$
$$M = 0, \pm 1, \pm 2, \ldots, \pm J.$$

For details see Pauling and Wilson P8a. The partition function for the symmetric top can be expressed in the form

$$Z_{\text{symtop}} = \frac{1}{\sigma} \sum_{k=-\infty}^{\infty} e^{-(a^{z}-a)k^2} \sum_{j=|k|}^{\infty} (2j+1) e^{-aj(j+1)}.$$

In the case $I_1 = I_2 = I_3$ we have the rigid body with energy levels

$$E_{JKM} = \frac{\hbar^2}{2} \frac{J(J+1)}{I},$$

i.e. the eigenvalues of the Schrodinger operator $H = (-\hbar^2/2I)\, \Delta_{SO(3)}$ where $\Delta_{SO(3)}$ is the Laplacian on $SO(3)$ which we introduced in Section 0.2.2.

The partition function is then

$$Z_{RB}(\beta) = \sum (2j+1)^2 \exp\left(\frac{-\hbar^2}{2I} \beta j(j+1)\right)$$

and the high temperature expansion is

$$Z_{RB}(\beta) = \frac{\pi^{1/2}}{y^{3/2}} [1 + \tfrac{1}{4} y + \tfrac{1}{32} y^2 + \ldots],$$

using Viney's result or Kirkwood's result. We note that we can identify $SO(3)$ with the real projective space $RP(3)$.

Thus we see that the spectroscopists had developed the spectral geometry expansions for the compact symmetric spaces S^1, S^2, and $RP(3)$ prior to 1933. Furthermore, in the language of spectral geometry it seems important to know whether nature would require two underlying manifolds or configuration spaces to be equivalent if the spectra of the corresponding Schrödinger operators (i.e., Laplace–Beltrami operators) coincide; i.e. is the geometry of the underlying configuration space completely characterized by the spectrum? We turn to the general study of statistical mechanics and spectral geometry in Chapters 16 and 17.

As we have just noted, the elementary examples which we have developed involve compact symmetric spaces. Symmetric spaces play an important role in physics. For the benefit of the reader we review how these spaces arise. Let G be a compact connected Lie group. Let ϑ be an involution on G with full fixed set K. We say in this case that (G, K) is a *symmetric pair*. Let K_0 be the identity component of K. Then G/K is a symmetric space. The Lie algebra $\mathfrak{g}$ of G decomposes into a natural direct sum $\mathfrak{g}=\mathfrak{k}\oplus\mathfrak{p}$ where

$$\mathfrak{k}=\{X\in\mathfrak{g}\,|\,s(X)=X\}$$
$$\mathfrak{p}=\{X\in\mathfrak{g}\,|\,s(X)=-X\},$$

with s the differential action of ϑ on $\mathfrak{g}$, i.e. $\dot{\vartheta}$. Here $[\mathfrak{k},\mathfrak{k}]\subset\mathfrak{k}$, $[\mathfrak{p},\mathfrak{k}]\subset\mathfrak{p}$ and $[\mathfrak{p},\mathfrak{p}]\subset\mathfrak{k}$. A maximal subalgebra $\mathfrak{h}$ of $\mathfrak{p}$ is abelian and is called a *Cartan subalgebra.* The dimension of $\mathfrak{h}$ is called the *rank* of (G,K).

Define $\eta\colon G\to G$ by $\eta(g)=g.\vartheta(g^{-1})$. Then $\eta(gK)=\eta(g)$; so η is constant on left cosets of K. η induces the imbedding $G/K\to G\colon gK\to g.\vartheta(g^{-1})$, g in G. Thus we can view G/K as a submanifold M of G. (M is in fact a totally geodesic submanifold of G. Geodesics through e coincide with the 1-parameter subgroups of G, which lie in M.)

EXAMPLE 0.4.1. Let $G=U(n)$ and let $\vartheta A=\bar{A}$ (the complex conjugate of $A\in U(n)$). The full fixed set of ϑ is $O(n)$ with identity component $SO(n)$. $U(n)/O(n)$ is a symmetric space. The Lie algebra of $U(n)$ is the set of $n\times n$ skew-Hermitian matrices. $\mathfrak{k}$ is the set of all real $n\times n$ skew symmetric matrices. $\mathfrak{p}$ is the set of all $n\times n$ symmetric purely imaginary matrices. The diagonal matrices in $\mathfrak{p}$ form a maximal abelian subalgebra. $U(n)/O(n)$ can be identified with the manifold of all symmetric matrices in $U(n)$. Viz. since it is imbedded in $U(n)$ by $g.O(n)\to g.\vartheta(g^{-1})=g.\bar{g}^{-1}=gg^t$, then the coset space $U(n)/O(n)$ is mapped into symmetric matrices in $U(n)$.

EXAMPLE 0.4.2. Let $G=SO(p+q)$ and let

$$I(p,q)=\begin{pmatrix}-I(p) & 0\\ 0 & I(q)\end{pmatrix}.$$

Take $p\geqq q$. Take the involution $\vartheta(X)=I(p,q)XI(p,q)$. K is $SO(p,q)\cap(O(p)\times O(q))$ and $K_0=SO(p)\times SO(q)$. The real Grassmannian $G_{p+q,p}$ is the symmetric space $G_{p+q,p}=SO(p+q)/K$.

In this example

$$\mathfrak{k}=\left\{\begin{pmatrix}A & 0\\ 0 & B\end{pmatrix}\middle|\begin{matrix}A & p\times p \text{ skew-symmetric matrix}\\ B & q\times q \text{ skew-symmetric matrix}\end{matrix}\right\}$$

$$\mathfrak{p}=\left\{\begin{pmatrix}0 & X\\ -X^t & 0\end{pmatrix}\middle| X \quad p\times q \text{ real matrix}\right\}.$$

A maximal abelian subspace is spanned by matrices of the form $(E_{i,p+i}-E_{p+i,i})i=1,\ldots,q$ where $E_{i,j}$ is the $n\times n$ matrix with 1 at location i,j and zero elsewhere.

0.5. More spectral geometry

In the language of the mathematician the partition function is the trace of the density matrix

$$\rho(\beta, x, y) = \sum_{k=0}^{\infty} \exp(-\lambda_k \beta) \Psi_k(x) \otimes \overline{\Psi_k(y)}, \qquad (*)$$

where $\{\lambda_n, \Psi_n\}$ is the set of eigenfunctions and eigenvalues for the Hamiltonian H. In the example above, H is a self adjoint operator, nominally the Laplace–Beltrami operator Δ, on a compact Riemannian manifold M. Before pursuing this connection we must introduce a bit more notation. Associated to any linear differential operator

$$P(x, D) = \sum_{|\alpha| \leqq m} a_\alpha(x) D_x^\alpha,$$

where $x \in R^N$ and

$$D_x^\alpha = (-i)^{|\alpha|} \frac{\partial^{\alpha_1}}{\partial x_1^{\alpha_1}} \cdots \frac{\partial^{\alpha_N}}{\partial x_N^{\alpha_N}}$$

with $|\alpha| = \sum_{i=1}^N \alpha_i$, is the *symbol* of P defined by

$$\sigma(P) = p(x, \xi) = \sum_{|\alpha| \leqq m} a_\alpha(x) \xi^\alpha,$$

where $\xi \in R^N$. $\sigma(P)$ is a polynomial of order m. E.g. if $P = (1/i)a(x)(\partial/\partial x)$ then $\sigma(P) = a(x)\xi$; or if $Q - b(x)$, then $\sigma(Q) = b(x)$.

The Laplacian Δ on R^N,

$$\Delta = \sum_{i=1}^{N} \frac{\partial^2}{\partial x_i^2},$$

has the symbol $\sigma(\Delta) = -|\xi|^2$. The leading order symbol of a linear differential operator is $\sigma_p(P) = \sum_{|\alpha|=m} a_\alpha \xi^\alpha$. The Laplacian as we know is an elliptic differential operator. This can be directly characterized in terms of $\sigma_p(P)$; viz. P is elliptic if $\sigma_p(P) \neq 0$ for all $\xi \neq 0$. Clearly Δ satisfies this condition. For a Laplace–Beltrami operator Δ the leading order symbol is $\sigma_p(\Delta) = \sum_{i,j} g^{ij} \xi_i \xi_j$.

For any nonnegative selfadjoint elliptic partial differential operator H acting on a Hilbert space $\mathscr{H}$ the eigenspaces $E_\lambda = \{\Psi \in \mathscr{H} \mid H\Psi = \lambda \Psi\}$ are finite dimensional and $E_\lambda = 0$ except for a discrete set of nonnegative λ's. The Hilbert space $\mathscr{H}$ is spanned by the E_λ. The partition function

$Z(\beta)=\sum_{\lambda}e^{-\beta\lambda}\dim E_{\lambda}$ can be shown to converge for every $\beta\to 0$. If we set $\rho(\beta)=\exp(-H\beta)$, then it can be shown that $\rho(\beta)$ is a well defined family of bounded operators acting on $\mathscr{H}$ and $\rho(\beta)$ satisfies Bloch's equation

$$\frac{\partial}{\partial\beta}\rho(\beta)+H\rho(\beta)=0$$

with $\rho(0)=1$. Before examining the high temperature asymptotic expansion (i.e. $\beta\to 0$), we note that we could equally as well study the *zeta function* of H defined by

$$\zeta_H(s)=\sum_{\lambda}\lambda^{-s}\dim E_{\lambda},$$

(where we assume $H>0$ so $\lambda\neq 0$). The zeta function of H is related to the partition function by

$$\zeta_H(s)=\frac{1}{\Gamma(s)}\int_0^{\infty}\beta^s Z_H(\beta)\frac{\mathrm{d}\beta}{\beta}.$$

In particular we note that if $Z_p(\beta)$ is the planar partition function and if $\zeta_R(s)$ is the Riemann zeta function $\zeta_R(s)=\sum_{n=1}^{\infty}n^{-s}$, then

$$\zeta_R(2s)=\frac{\pi}{\Gamma(s)}\int_0^{\infty}\beta^{s-1}[Z_p(\beta)-\tfrac{1}{2}]\,\mathrm{d}\beta.$$

As a second example of a zeta function in statistical mechanics consider the particle of mass m in the box of width a whose eigenvalues are

$$\lambda_n=\frac{n^2h}{8ma^2},\quad n=1,2,3\ldots.$$

The zeta function for this system is

$$\zeta_{PB}(s)=\sum_{n=1}^{\infty}\lambda_n^{-s}=\left(\frac{h}{8ma^2}\right)^{-s}\sum_{n=1}^{\infty}n^{-2s}=\left(\frac{h}{8ma^2}\right)\zeta_R(2s).$$

It is known that $\zeta_R(s)$ is analytic on C except for a simple pole at $s=1$. $\zeta_R(s)$ has 'trivial' zeros $\zeta_R(-2n)=0, n\geqq 1$. The 'nontrivial' zeros are located inside $0\leqq\mathrm{Re}(s)\leqq 1$; denoting such a zero by r, the Riemann hypothesis is that $\mathrm{Re}(r)=\frac{1}{2}$ for all nontrivial zeros.

As a third example of a zeta function consider the harmonic oscillator with energy levels $\lambda_n=h\nu(n+\frac{1}{2})$, $n=0,1,2\ldots$. The zeta function for the harmonic oscillator is then

$$\zeta_{HO}(s) = (h\nu)^{-s} \sum_{n=0} (n + \tfrac{1}{2})^{-s}$$
$$= (h\nu)^{-s}(2^s - 1)\zeta_R(s).$$

We return to the study of partition functions for the case $H = \Delta_M$ for a manifold M of dimension d. As we have seen, the examples of quantum mechanical systems given above are described by Hamiltonians which are Laplace–Beltrami operators on Riemannian manifolds M. In this case it can be shown that the density matrix series (∗) converges uniformly on compact subsets of $(0, \infty) \times M \times M$ to the fundamental solution of the Bloch operator $(\partial/\partial\beta) - H$ and we have $\int_M \operatorname{Tr} \rho(\beta; x, x) = \sum \exp(\lambda\beta)$. Moreover, for any positive integer N we have an expansion for ρ of the form

$$\operatorname{Tr} \rho(\beta; x, x) = (4\pi\beta)^{-d/2}\{1 + \beta k_1(x) + \cdots + \beta^N k_N(x)\} + \mathcal{O}(\beta^{N-(d/2)+1}) \quad \text{as} \quad \beta \to 0,$$

where $k_j(x)$ are C^∞ functions on M. In particular we see that

$$\int_M \operatorname{Tr} \rho(\beta; x, x) = (4\pi\beta)^{-d/2} \operatorname{Vol}(M) + \mathcal{O}(\beta^{1-(d/2)}).$$

As M. Kac has made popular, this expresses the fact that the volume of M and the dimension of the manifold can be 'heard' from the spectrum $\{\lambda_n\}$. If we set $a_j = 1/(\operatorname{Vol}(M) \int_M k_j(x)$, then the high temperature expansion for the partition function arises from the study of spectral geometry:

$$Z = \sum \exp(\lambda_n \beta) = (4\pi\beta)^{-d/2} \operatorname{Vol}(M)\{1 + \beta a_1 + \cdots + \beta^N a_N\} + (\beta^{N-(d/2)+1})\beta \to 0.$$

The coefficients k_j in the expansion are the traces on the diagonal of forms $U_j(x, y)$ which are determined as follows. Take an open set U of a point m in M and introduce normal coordinates in U so that $g_{ij}(m) = \delta_{ij}$ and m has coordinates $(0, \ldots, 0)$. Let $F(r(x, y))$ be a function of y depending only on the geodesic distance r of y from x and let α be a smooth function on U. Then it is easily checked that

$$\Delta_y(F(r)\alpha) = \left(\frac{\mathrm{d}^2 F}{\mathrm{d}r^2}(r) + \frac{d-1}{r}\frac{\mathrm{d}F}{\mathrm{d}r} + \frac{1}{2g}\frac{\mathrm{d}g}{\mathrm{d}r}\frac{\mathrm{d}F}{\mathrm{d}r}\right)\alpha + {} + \frac{2}{r}\frac{\mathrm{d}F}{\mathrm{d}r}\nabla r \frac{\mathrm{d}}{\mathrm{d}r}(\alpha) + F(r)\Delta\alpha, \qquad \binom{*}{*}$$

where $\mathrm{d}/\mathrm{d}r$ denotes differentiation along the geodesic and $g(y) = \det(g_{ij}(y))$.

If we set

$$H_N(\beta, x, y) = \frac{\exp(-r^2/4\beta)}{(r\pi\beta)^{d/2}} \left(\sum_{i=0}^{N} \beta^i U_i(x, y) \right)$$

then by using $\left(\begin{smallmatrix} * \\ * \end{smallmatrix}\right)$ we have

$$\left(\frac{\partial}{\partial\beta} - \Delta_y\right) H_N(\beta, x, y) = \frac{\exp(-r^2/4\beta)}{(4\pi\beta)^{d/2}} \sum_{i=0}^{N} \left\{ \frac{r^2}{4\beta^2} + \frac{i - d/2}{\beta} \frac{-r^2}{4\beta^2} + \right.$$
$$+ \frac{1}{2\beta} + \frac{d-1}{2\beta} + \left. \frac{r}{4g\beta} \frac{\mathrm{d}g}{\mathrm{d}r} \right) \beta^i U_i(x, y) +$$
$$\left. + \beta^{i-1} \nabla r \frac{\mathrm{d}}{\mathrm{d}r} U_i(x, y) - \beta^i \Delta_y U_i(x, y) \right\}.$$

Equating the coefficient of

$$\frac{\exp(-r^2/4\beta)}{(4\pi\beta)^{d/2}} \beta^{i-1} \quad \text{in} \quad \left(\frac{\partial}{\partial\beta} - \Delta\right) H_N$$

to zero we have

$$\left(i + \frac{r}{4g} \frac{\mathrm{d}g}{\mathrm{d}r}\right) U_i + \nabla r \frac{\mathrm{d}}{\mathrm{d}r} U_i - \Delta_y U_{i-1} = 0$$

or since

$$\Delta_y r^2 = \frac{d}{2} + \frac{1}{4} \frac{r}{g} \frac{\mathrm{d}g}{\mathrm{d}r}$$

we have

$$\frac{\mathrm{d}}{\mathrm{d}r} U_i(x, y) + \left(i - \frac{d}{2} + \frac{1}{4} \Delta_y r^2(x, y) \right) U_i(x, y) =$$
$$= \Delta_y U_{i-1}(x, y), \ 0 \leqq i \leqq N.$$

This system of equations for U_i has a unique solution in a sufficiently small neighbourhood of the diagonal in $M \times M$ with initial condition $U_0(x, x) = I.H_k$ is called the parametrix of the Bloch equation. The study of spectral geometry then proceeds to show that $k_j(x) = \mathrm{Tr}(U_j(x, x))$ can be expressed as a polynomial in the partial derivatives at x of the components of the metric tensor and hence as a polynomial in the components of the curvature tensor and its covariant derivatives. Using invariant theory it can be shown that $k_1 = \alpha_1 \tau$, $k_2 = \alpha_2 \tau^2 + \alpha_3 |\rho|^2 + \alpha_4 |R|^2 + \alpha_5 \Delta\tau$ where τ

is the scalar curvature of M, $|\rho|$ is the norm of the Ricci tensor ρ, and $|R|$ is the norm of the curvature tensor. The α_i are constants which have been determined:

$$\alpha_1 = 1/6, \quad \alpha_2 = 1/72, \quad \alpha_3 = -1/180, \quad \alpha_4 = 1/180,$$
$$a_5 = 1/30.$$

Using the normalization of Berger *et al.* we present a table for the cases S^1, S^2, and S^3:

	S^1	S^2	S^3
$\|R\|^2$	0	4	12
$\|\rho\|^2$	0	2	12
$\|\tau\|^2$	0	4	36
τ	0	2	6
Vol	2π	4π	$2\pi^2$

Thus the high temperature asymptotics are:

$$\begin{aligned} Z_{S^1} &= (4\pi\beta)^{-1/2}(1 + \beta 0 + \beta^2 0 + \cdots) \\ Z_{S^2} &= (4\pi\beta)^{-1}(1 + \beta/3 + \beta^2/15 + \cdots) \quad \text{as} \quad \beta \to 0. \\ Z_{S^3} &= (4\pi\beta)^{-3/2}(1 + \beta + \beta^2/6 + \cdots) \end{aligned}$$

The first two examples are precisely the cases of a planar rigid rotator and a rigid rotator. We leave it to the reader to check the case RP(3).

Kac has a beautiful interpretation which leads one to expect that

$$Z(\beta) \sim \frac{\mathrm{Vol}(M)}{(4\pi\beta)^{d/2}} \quad \text{as} \quad \beta \to 0.$$

Viz. if we consider the Bloch equation as describing the Einstein equation for Brownian motion, then as $\beta \to 0$ the Brownian particle has had no time to 'feel' the boundary. Thus we should expect that

$$\rho \sim \frac{\exp(-r^2/4\beta)}{(4\pi\beta)^{d/2}} \quad \text{as} \quad \beta \to 0.$$

We will elaborate on this type of result in Chapter 16. Accepting it for now, we have if

$$\rho(\beta, x, y) = \sum \exp(-\lambda_j\beta)\Psi_j(x)\Psi_j^*(y)$$

then

$$\sum \exp(-\lambda_j\beta)\,|\Psi_j|^2 \sim \frac{1}{(4\pi\beta)^{d/2}} \quad \text{as} \quad \beta \to 0.$$

Integrating we have

$$Z(\beta) \sim \frac{\operatorname{Vol}(M)}{(4\pi\beta)^{d/2}} \quad \text{as} \quad \beta \to 0.$$

Applying the Hardy–Littlewood Tauberian theorem we have the classical result of Weyl regarding the asymptotic behavior of the number of eigenvalues less than λ, $N(\lambda)$:

$$N(\lambda) = \sum_{\lambda_j < \lambda} 1 \sim \frac{\operatorname{Vol}(M)}{(2\sqrt{\pi})^d} \frac{\lambda^{d/2}}{\Gamma\left(\frac{d}{2}+1\right)} \quad \text{as} \quad \lambda \to \infty.$$

Consider briefly the case that Ω is an open set in $M = R^n$ or M is a Riemannian manifold. We have just noted that $\lambda^{-n/2} N(\lambda) = C_n \operatorname{Vol}(\Omega)$ as $\lambda \to \infty$. We will now cite an interesting relationship between $N(\lambda)$ and the number $N_\alpha(V)$ of nonpositive eigenvalues of the Schrödinger equation $-\Delta_M + V(x)$ on $L^2(M)$ which are $\leqq \alpha \leqq 0$: viz. by the min-max principle it follows that

$$N(\lambda) \leqq N_\alpha((\alpha - \lambda)\mathfrak{X}_\Omega) \quad \text{for all} \quad \alpha \leqq 0,$$

where $\mathfrak{X}_\Omega$ is the characteristic function of Ω. Lieb has shown that

$$N_\alpha(V) \leqq L_n \int_M (V(x) - \alpha)_-^{n/2}\, dx \quad \text{for all} \quad \alpha$$

and V, when M is R^n or when M is a homogeneous manifold of curvature $\leqq 0$; and $\dim(M) \geqq 3$. Here $V_- = \frac{1}{2}(|V| - V)$. L_n is a constant. Estimates for L_n have been made by Lieb and Cwikel. In particular combining these results we have the following nonasymptotic bounds on $N(\lambda)$:

(1) $N(\lambda) \leqq D_n \lambda^{n/2} \operatorname{Vol}(\Omega)$ for all $\lambda \geqq 0$, for all $\Omega \subset M$ where $M = R^n$ or M is homogeneous manifold of curvature $\leqq 0$.

(2) $N(\lambda) \leqq (D_n \lambda^{n/2} + E_n) \operatorname{Vol}(\Omega)$ for all $\lambda \geqq 0$ and for all $\Omega \subset M$ where M is compact. Here D_n, E_n are independent of λ and Ω, depending only on M.

We return now to the general setup and assume that the Hamiltonian H acts on a Hilbert space $\mathscr{H}$ and there is a symmetry group of the underlying configuration space with a representation T on $\mathscr{H}$ which commutes with H. As we have noted in Section 0.3, the eigenspaces E_λ of H provide unitary representations of G, i.e. $T|E_\lambda$. These representations are not necessarily irreducible. Given an irreducible representation μ of G we are interested in the multiplicity $v_\mu(\lambda)$ of μ in E_λ and the asymptotic behavior of $N_\mu(t) = \sum_{\lambda \leqq t} v_\mu(\lambda)$ as $t \to \infty$. This is just a generalization of Weyl's question to which it reduces when $G = e$ and H is the Laplacian.

0.6. Statistical mechanics and representation theory

The reader should by now see that the partition function and quantum statistical mechanics are related to spectral geometry. He should also begin to see the connections with representation theory. Now we want to more intimately connect the symmetries of the dynamical system to the study of the partition function. Consider a system for which $H = \Delta$ and $M = G$ is a compact Lie group, as the case $G = S^1$ above; then the Bloch equation becomes

$$\frac{\partial \rho_G}{\partial \beta} = \Delta_G \rho_G.$$

An elementary use of Fourier analysis on compact Lie groups shows that since ρ_G is invariant by inner automorphisms of G, ρ_G admits a Fourier expansion

$$\rho_G(\beta, g) = \sum_{\lambda \in \hat{G}} a_\lambda(\beta) \mathfrak{x}_\lambda(g)$$

where $\mathfrak{x}_\lambda$ are the characters associated to the irreducible unitary representations λ of G of degree $d(\lambda)$. It is easily checked that $a_\lambda(\beta) = d(\lambda) \exp(-\lambda\beta)$ where $H\Psi = \lambda\Psi$. Thus the density matrix and correspondingly the partition function are determined by G. This result extends to quantum mechanical systems where $H = \Delta_M$ with M being a symmetric space, $M = G/K$. We will show in the chapters which follow that the spectra and thus the partition function again are determined by the appropriate representations of G. Moreover, we will develop the high temperature asymptotics, in detail, for symmetric spaces of rank one. The interesting twist in the history of science in this subject is that the complete story of the high temperature asymptotics of compact symmetric spaces of rank 1 arises from the above-mentioned work of the statistical mechanician Fowler and his student Mulholland for the case of a rigid rotator.

0.7. Transformation groups in physics

Definition 0.7.1. A *topological transformation group or* an *action* is a triple (G, M, f) consisting of a topological group G, a Hausdorff space M and a continuous map $f: G \times M \to M$ satisfying (1) $f(g_1, f(g_2, x)) = f(g_1 g_2, x)$; and (2) for g fixed $f(g, x)$ is a homeomorphism of M onto M. We usually denote $f(g, x)$ simply by gx. If M is a smooth manifold and if f is a smooth map, then (G, M, f) is a *smooth action* and M is called a *G-manifold.*

DEFINITION 0.7.2. Let (G, M, f) be a transformation group and x in M. The *isotropy* or *stability* group G_x is defined by $G_x = \{g \in G | gx = x\}$. The set $Gx = \{gx | g \in G\}$ is called an *orbit* of x.

Two orbits of M are either disjoint or coincide; so M is partitioned by its orbits. We let M/G denote the set of orbits or the orbit space. The topology of M/G is chosen so that the natural projection $p: M \to M/G$ is an open map. The reader can check that when G is compact, p is also a closed map.

DEFINITION 0.7.3. An action is called *effective* if $gx = x$ for all x in M implies $g = e$. An action is called *free* if $G_x = \{e\}$.

EXAMPLE 0.7.3a. Let $G = Z_2$ act on S^n by $g(x_0, \ldots, x_n) = (x_0 g, \ldots, x_n g)$, $(x_0, \ldots, x_n) \in R^{n+1} \backslash 0$, $\sum x_j^2 = 1$. This action is free with orbit space $S^n/Z_2 = RP(n)$.

DEFINITION 0.7.4. A *transitive* action of G on M means that for every x, y in M there is at least one g in G such that $gx = y$; i.e. $Gx = M$ for every x in M. M is then called a *homogeneous space*.

We leave it to the reader to show that every topological group G can act transitively on the coset space G/H where H is a closed subgroup of G; thus G/H is a homogeneous space. And if (G, M, f) is a transitive action with G a compact Lie group, then $p: G \to M$ given by $p(g) = gx$ induces a diffeomorphism $p': G/G_x \to M$.

EXAMPLE 0.7.5 An orthonormal k-frame v^k in C^n is an ordered set of k independent orthonormal vectors. Let $V_{n,k}(C)$ denote the set of all orthonormal k frames in C^n. The multiplication of matrices $G \times G \to G$ where $G = U(n)$, induces a transitive action of $U(n)$ on $V_{n,k}(C)$, which we denote by $U(n)V_{n,k}(C)$. Viz. any fixed orthonormal k frame v_0^k can be mapped into any other v^k by an element of G. If we let $v_0^k = = (e_1, \ldots, e_k) \in V_{n,k}(C)$ where

$$e_i = \begin{bmatrix} 0 \\ \vdots \\ 1 \\ \vdots \\ 0 \end{bmatrix} i^{\text{th}} \in C^n,$$

then $G_{v_0^k} = I_k \times U(n-k)$, which we identify with $U(n-k)$. Thus we have $V_{n,k}(C) = U(n)/U(n-k)$. Note that $V_{n,n-1}(C) = U(n)$.

Similarly we have $V_{n,k}(R) = O(n)/O(n-k)$. In particular $V_{n,1}(R) = S^{n-1}$. Since we have the exact sequences $O \to SO(n) \to O(n) \to S^0 \to O$ (i.e. $O(n) = SO(n) \times S^0$) we have $V_{n,k}(R) = SO(n)/SO(n-k)$. Similarly since $O \to SU(n) \to U(n) \to S^1 \to O$ is exact, we have $V_{n,k}(C) = SU(n)/SU(n-k)$.

Homogeneous spaces arise naturally in the study of physics. Unfortunately this connection is often not exploited. We have already seen examples of S^1, S^2, S^3, and $RP(3)$ in statistical mechanics. An example which we will treat later in this volume is the n-dimensional Kepler problem or hydrogen atom. Here the Hamiltonian is $H = (p^2/2m) - (k/q)$ and the phase space is the cotangent bundle $T^*(R^n \setminus \{0\})$. The surfaces of constant negative energy Σ_H, where $H = -a^2$, are connected but noncompact. After removing the singularity at the origin and compactifying the resulting energy manifold, we have a new energy surface $\tilde{\Sigma}_H$ which is the unit tangent bundle of the n-sphere S^n. Such a space is known to be diffeomorphic to the Stiefel manifold $V_{n+1,2} = SO(n+1)/SO(n-1)$. Furthermore the regularized energy surface is naturally the principal circle bundle over a Grassman manifold of oriented planes; we denote this relationship by the mappings $SO(2) \to V_{n+1,2}(R) \to G'_{n+1,2}(R)$. In general $G'_{n,k}(R) = SO(n)/SO(k)SO(n-k)$. Finally we note that $G'_{n+1,2}(R)$ is diffeomorphic to a space $Q_{n-1}(C)$, the complex quadric – i.e. $Q_{n-1}(C)$ is defined by the homogeneous equation $\sum_{i=0}^{n} z_i^2 = 0$ for homogeneous coordinates $\{z_i\}$ of a point in complex projective space $CP(n)$. The spaces $Q_{n-1}(C)$ are special symplectic manifolds which are called Hodge manifolds. We will find that Hodge manifolds play an important role in the study of geometric quantization. These structures will be studied in Chapters 10 and 11. They also occur in the study of instantons (v. e.g. Perelomov P12a); however, we will not examine that connection in this volume.

Returning now to the general topic of homogeneous spaces we note that two homogeneous space G/H and G/H' are homeomorphic if $H' = gHg^{-1}$; thus, we see that to determine all homogeneous spaces for G we need only determine all possible conjugacy classes of closed subgroups of G.

EXAMPLE 0.7.6. For $G = SU(2)$ the proper closed subgroups are $U(1)$, normalizer of $U(1)$, $NU(1)$, cyclic subgroups C_n, $n = 1, 2, \ldots$ of order n, subgroups $\tilde{D}_{2n}$ whose factor groups $\tilde{D}_{2n}/Z_2$ are the dihedral groups of order $2n$, $n = 1, 2, \ldots$; subgroup $\tilde{T}$ whose factor group $\tilde{T}/Z_2$ is the tetrahedral group of order 12; the subgroup $\tilde{O}$ whose factor group $\tilde{O}/Z_2$ is isomorphic to the octohedral group of order 24; and the subgroup $\tilde{Y}$ whose factor group $\tilde{Y}/Z_2$ is the icosohedral group of order 60. Thus, the homogeneous spaces for $SU(2)$ are $SU(2)=S^3$, $SU(2)/C_n \simeq L(n,1)$ (lens spaces), $SU(2)/\tilde{D}_{2n}$, $SU(2)/\tilde{T}$, $SU(2)/\tilde{O}$, $SU(2)/\tilde{Y}$ all of dimension 3; $SU(2)/U(1)=S^2$ and $SU(2)/NU(1)=RP(2)$ of dimension 2; and the zero dimension space $SU(2)/SU(2)$.

The interaction of orbit spaces and representation theory has been studied in physics and mathematics. For this material the reader is referred to Janich JS1 and Mickelsson–Niederle M33a.

DEFINITION 0.7.7. Let M be a G-space and let H be a closed subgroup of G. The set $(H)=\{gHg^{-1}|g\in G\}$ is called the *orbit type* of H.

Since $G_{gx}=gG_xg^{-1}$ for any x in M and g in G, the isotropy subgroups of points on the same orbit have the same orbit type.

DEFINITION 0.7.8. The set $\{(G(x))|x\in M\}$ is called the *orbit structure* of M. The orbit structure is said to be *finite* if there are only finitely many different orbit types. We set $M_{(H)}=\{x\in M|(G_x)=(H)\}$. If (H_1) and (H_2) are two orbit types we say $(H_1)\leqq(H_2)$ if there are representatives H_1, H_2 in (H_i) $i=1,2$ respectively such that $H_1\supset H_2$.

THEOREM 0.7.9. If M is a connected G-manifold and G is a compact Lie group then there is an absolute maximal orbit type (H) among the orbits of M. $M_{(H)}$ is called the *principal orbit type* of M and $M_{(H)}$ is an open dense subset of M.

DEFINITION 0.7.10. Let M be a G-manifold where G is a compact Lie group. Let $N_x=TM_x/TG(x)_x$ be the normal space to the orbit G_x at point x. Then for every g in G the differential of the map $g: M\to M$ induces an automorphism of N_x – i.e. we have representation $\sigma_x: G_x\to GL(N_x)$. $[G_x, \sigma_x]$ is called the *slice type* of M.

The reader can check that the slice type is constant along orbits in M. The set of all slice types of a G-manifold M can be given a natural partial order; and the set with this partial order is called the *slice diagram* of M. If the orbit space M/G is connected, the slice diagram, which we denote by $\Delta(G, M)$, has a unique largest element, $[H, \sigma]$ where σ is the trivial representation. If M is compact, $\Delta(G, M)$ has a finite number of slice types. For more information regarding orbit and slice structures, see Janich JS1.

0.8. Fiber bundles

The concepts and theory of fiber bundles have been reviewed several times recently for physicists. We will not repeat that treatment here; we will only introduce some basic notation. For a principal G-bundle over M, denoted by $P(M, G)$, with projection $\pi: P \to M$, G acts as a right transformation group on P and acts simply transitively on each fiber, $G_x = \pi^{-1}(x)$ of P over x in M; and there is a local section $s: U \to P$ with $s \circ \pi = 1$ on a neighborhood U of any point x in M. We write the action of G as multiplication $P \times G \to \to P, (p, g) \to pg$. Right translation R_g gives a diffeomorphism $R_g: G_x \to G_x$ on each fiber G_x. Moreover there exists an open covering $\{U_i\}$ of M and a system of local sections $\{s_i\}$ $s_i: U_i \to P$ with $\pi s_i = 1$. Thus there is a system of maps $\{g_{ij}\}$ $g_{ij}: U_i \cap U_j \to G$ where $s_j(x) = s_i(x) g_{ij}(x)$ for x in $U_i \cap U_j$. The g_{ij} satisfy $g_{ii}(x) = e$ for x in U_i and $g_{ij}(x) g_{jk}(x) = g_{ik}(x)$ for x in $U_i \cap U_j \cap U_k$. If $\mathbf{G}$ denotes the sheaf of germs of local G-valued functions on M, then the 1-cocycle $\{g_{ij}\}$ determines a cohomology class ξ in $H^1(M, \mathbf{G})$. For further details *v.* Hirzebruch H22.

If F is a left G-manifold, i.e. there is an action $G \times F \to F, (g, y) \to gy$, then the *associated bundle* $B = P \times_G F$ with fiber F is defined as the quotient space $P \times F$ by the equivalence reaction $(pg, y) \sim (p, gy)$, $p \in P$, $g \in G$, $y \in F$. The natural projection $P \times F \to B$, $(p, y) \to py$ is characterized by $(pg)y = = p(gy)$. The projection $\pi': B \to M$ is defined by $\pi'(py) = \pi(p)$. Thus we can regard a point p in P as a diffeomorphism $p: F \to F_x$, $y \to py$ where $F_x = \pi^{-1}(x)$ is a fiber of B over $x = \pi(p) \in M$.

A fiber bundle $B(M, E, G)$ is called a *vector bundle* if the fiber E is a vector space and G acts on E as a linear transformation group. Thus given a principal bundle $P(M, G)$ and a representation $\rho: G \to GL(E)$ of G on a vector space E, then $B = P \times_{\rho(G)} E$ becomes a vector bundle.

0.9. Orbit spaces in Lie algebras

A major topic to be studied in this volume is the differential geometry of orbit spaces on Lie algebras. We turn to this topic in Chapter 7. We consider a few simple examples to help familiarize the reader with the study of orbit spaces.

Consider the Lie group $G = SU(n)$ of $n \times n$ complex matrices g which satisfy $gg^* = I$ where g^* is the complex conjugate and I is the identity matrix. The tangent space TG_e can be identified with the space of $n \times n$

skew Hermitian matrices, $\mathfrak{g}$. The exponential map $\exp A = \sum_{n=0}^{\infty} A^n/n!$ of a complex matrix A thus defines a map $\exp: \mathfrak{g} \to G$. Associated to each matrix A in $\mathfrak{g}$ there is a one-parameter subgroup, $t \to \exp(tA)$, of G. It is easy to check that for any compact Lie group the geodesics $\gamma(t)$ in G with $\gamma(0) = e$ are precisely the one parameter subgroups of G. Thus any geodesic γ in G with $\gamma(0) = e$ can be written uniquely as $\gamma(t) = \exp(tA)$ for some A in $\mathfrak{g}$.

$\mathfrak{g}$ admits a natural inner product (which defines a Riemannian metric on G) viz. $\langle A, B \rangle = \operatorname{Re} \operatorname{Tr}(AB^*)$ for A, B in $\mathfrak{g}$. There is a natural action of G on $\mathfrak{g}$ given by the map $Ad(g)A = gAg^{-1}$ for A in $\mathfrak{g}$ and g in G. The inner product $\langle A, B \rangle$ is invariant by $Ad(g)$. To see this, set $A' = Ad(g)A$, $B' = Ad(g)B$; then since $A'B'^* = gAg^{-1}(gBg^{-1})^* = gAB^*g^{-1}$ we have $\operatorname{Tr}(A'B'^*) = \operatorname{Tr}(AB^*)$ or $\langle A', B' \rangle = \langle A, B \rangle$. We note that the length of the geodesic $\gamma(t) = \exp(tA)$ form $t = 0$ to $t = 1$ is then $|A| = \sqrt{\operatorname{Tr}(AA^*)}$

We are interested in the orbits $M = G.A$ where $g.A$ denotes the adjoint action on $A \in \mathfrak{g}$. As an example of such an orbit space, let $G = U(n)$ and let $I_{n_1,\ldots,n_k}$ be a diagonal matrix with the same first n_1 entries, the same n_2 second entries, etc. Let $A = i \times I_{n_1,\ldots,n_k} \in \mathfrak{g}$. The stability subgroup at A, i.e. $K = \{g \in G \mid g.A = A\}$ is just $U(n_1) \times \ldots U(n_k)$ where $n_1 + \cdots + n_k = n$. Thus the orbit space $M = G.A$ is diffeomorphic to G/K (viz. take the map $g \to g.A$ from M to G/K). This homogeneous space is denoted $W(n_1, \ldots, n_k) = U(n)/U(n_1) \times \cdots \times U(n_k)$ and is called the *complex flag manifold*. If $k = 2$ we get the Grassmann manifolds and if $n_1 = 1$ and $n_2 = n - 1$ we get the complex projective spaces.

Let $\mathfrak{g} = \mathfrak{so}(n)$ denote the space of all $n \times n$ skew symmetric matrices and let $A \in \mathfrak{g}$ be the element

$$A = \begin{pmatrix} 0 & I_n \\ -I_n & 0 \end{pmatrix}.$$

Then the orbit space of $SO(2n)$ on A is $SO(2n)/U(n_1) \times \cdots \times U(n_k)$.

If we let

$$A = \begin{pmatrix} 0 & I_n & 0 \\ -I_n & 0 & \vdots \\ & 0 \quad \ldots & 0 \end{pmatrix} \in \mathfrak{so}(2n+1)$$

then the $SO(2n+1)$ orbit through A is $SO(2n+1)/U(n_1) \times \cdots \times U(n_k) x1$. Finally if we

$$A = \begin{pmatrix} iI_n & 0 \\ 0 & -iI_n \end{pmatrix} \in \mathfrak{sp}(n)$$

then the $Sp(n)$ orbit space through A is $Sp(n)/U(n_1) \times \cdots U(n_k)$.

A maximal abelian subgroup of G is called a *maximal torus*. If it is denoted by T and its Lie algebra is denoted by $\mathfrak{t}$, then a point A in $\mathfrak{t}$ is called a *general point* if $\mathfrak{g}_A = \{X \in \mathfrak{g} | \; [X, A] = 0\} = \mathfrak{t}$. Orbits of general points are interesting for one can show that these orbits are of maximal dimension. E.g. in the case $\mathfrak{g} = U(n)$ and $n_1 = n_2 = \cdots = n_k = 1$ and A is a general point then $M = G.A = U(n)/T$ where $T = e^{i\vartheta_1} \times \cdots \times e^{i\vartheta_n}$.

For orbit spaces $M_A = G.A$ the tangent space to M_A at A is given by $T_A = \{Z | Z = [Y, A] \; Y \text{ in } \mathfrak{g}\} = ad(A)\mathfrak{g}$. And the normal space at A is given by $N_A = \{Y \in \mathfrak{g} | [Y, A] = 0\}$, i.e. the *centralizer* of A in $\mathfrak{g}$. For a proof of the first statement the reader can check that $(d/dt)(Ade^{tY} A) = [Y, A]$ for Y in $\mathfrak{g}$; the second fact follows since $\langle [X, Y], Z \rangle = -\langle Y, [X, Z] \rangle$.

In several cases below we will study the adjoint action of subgroups of G on Lie subalgebras of $\mathfrak{g}$. Consider the case that (G, K) is a symmetric space – i.e. K is a full fixed set of an involution σ on G. Then the Lie algebra $\mathfrak{g}$ of G splits as $\mathfrak{g} = \mathfrak{k} + \mathfrak{p}$. The Cartan subalgebras are then maximal abelian subalgebras $\mathfrak{h}$ of $\mathfrak{p}$ whose dimension is the rank of (G, K). In this case it can be shown that K acts on $\mathfrak{p}$ by conjugation and we study the orbit spaces $K.A$ for A in $\mathfrak{p}$.

As an example consider the involution σ in $U(n)$ given by $\sigma X = X^*$. In this case $K = O(n)$ and $\mathfrak{p}$ is the set of all $n \times n$ purely imaginary symmetric matrices. Let $A = iI_n \in \mathfrak{p}$. Then $K/K_A = G(n_1, \ldots, n_k) = O(n)/O(n_1) \times \ldots \times O(n_k)$ (i.e. the real flag manifolds).

As a second example consider the Stiefel manifolds. Take $G = SO(n)$ where $n = 2p + q$ The involution is $X \to I(p + q, q) X I(p + q, p)$ on G where

$$I(p+q, p) = \begin{pmatrix} 1 & & & & & \\ & \ddots & & & & \\ & & 1 & & & \\ & & & -1 & & \\ & & & & \ddots & \\ & & & & & -1 \end{pmatrix} \begin{matrix} p+q \\ \\ \\ \\ \\ p \end{matrix}$$

(columns: $p+q$, p).

K is then $\{O(p + q) \times O(p)\} \cap SO(2p + q)$.

We let

$$k = \begin{pmatrix} X & 0 \\ 0 & Y \end{pmatrix} \begin{matrix} p+q \\ p \end{matrix}$$

$$\begin{matrix} p+q & p \end{matrix}$$

represent an element in K. The Lie algebra of G can be decomposed as $\mathfrak{g} = \mathfrak{k} + \mathfrak{p}$ where

$$\mathfrak{p} = \left\{ \begin{array}{c} \begin{pmatrix} 0 & * \\ * & 0 \end{pmatrix} \begin{array}{l} p+q \\ p \end{array} \\ \begin{array}{cc} p+q & p \end{array} \end{array} \right\}.$$

If we take

$$P = \begin{pmatrix} 0 & I_p \\ -I_{p+q} & 0 \end{pmatrix} \in \mathfrak{p},$$

then the orbit of P under the adjoint action of K is K/K_P where

$$K_P = \left\{ \begin{array}{c} \begin{pmatrix} X & 0 & 0 \\ 0 & y & 0 \\ 0 & 0 & X \end{pmatrix} \in SO(n) \\ \begin{array}{ccc} p & q & p \end{array} \end{array} \right\}.$$

We can check that $V_{p+q,p} = K/K_P$. Viz. K acts on g in $SO(p+q)$ by

$$\begin{pmatrix} X & 0 \\ 0 & y \end{pmatrix} \cdot g = Xg \begin{pmatrix} y & 0 \\ 0 & I_q \end{pmatrix}^{-1}$$

This action is transitive and induces a transitive action on $SO(q+p)/SO(q)$. K leaves $SO(q)$ invariant if

$$X \begin{pmatrix} I_p & 0 \\ 0 & SO(q) \end{pmatrix} = \begin{pmatrix} I_p & 0 \\ 0 & SO(q) \end{pmatrix} \begin{pmatrix} Y & 0 \\ 0 & I_q \end{pmatrix}.$$

Thus X is of the form

$$\begin{pmatrix} Y & 0 \\ 0 & v \end{pmatrix} \quad \text{and so} \quad k = \begin{pmatrix} Y & & \\ & v & \\ & & Y \end{pmatrix} \in K_P.$$

0.10. Scattering Theory and Statistical Mechanics

The basic examples in this survey chapter up to now deal with the theory of partial differential operators on compact spaces. Many aspects of physics, however, require the study of operators on noncompact spaces. One elementary example of this subject is the scattering theory of the Schrödinger equation on the real line. This scattering theory formed the background for the original development of Selberg's trace formula on noncompact spaces. Later Fadeev, Lax, Phillips, and others applied scattering theory to the study of the Selberg trace formula. We will study this topic in Chapter 18. To provide the reader with some background in scattering theory we review some elementary concepts.

Consider the energy of a particle in one dimension, $H=(1/2m)p^2+$ $+V(x)$. The Schrödinger equation is

$$\frac{h^2}{2m}\frac{\mathrm{d}^2\Psi}{\mathrm{d}x^2}+[E-V(x)]\Psi=0.$$

Setting $q(x)=(2mV/h^2)(x)$ and $(hk)^2=2mE$ we have $(H-k^2)\Psi=0$ where $H=H_0+q(x)$ and $H_0=-(\mathrm{d}^2/\mathrm{d}x^2)$. If $q(x)$ is smooth with compact support, say contained in an interval $[A, B]$ then for $x\to+\infty$ Ψ will behave asymptotically like $\exp(ikx)$ where $k\in C\backslash\{0\}$. Similarly as $x\to-\infty$, $\Psi(x)$ behaves like $a(k)\exp(ikx)+b(k)\exp(-ikx)$ with $a(k)$ and $b(k)$ holomorphic on $C\backslash\{0\}$. We must also have $\mathrm{d}\Psi/\mathrm{d}x$ behaving like $ik(ae^{ikx}-be^{-ikx})$. Multiplying $(H-k^2)\Psi=0$ by Ψ^* and $(H-k^2)\Psi^*=0$ by Ψ, we have

$$\frac{\mathrm{d}}{\mathrm{d}x}\left(\Psi^*\frac{\mathrm{d}\Psi}{\mathrm{d}x}-\Psi\frac{\mathrm{d}\Psi^*}{\mathrm{d}x}\right)=0$$

i.e.

$$\det\begin{pmatrix}\Psi^* & \Psi\\ \frac{\mathrm{d}\Psi^*}{\mathrm{d}x} & \frac{\mathrm{d}\Psi}{\mathrm{d}x}\end{pmatrix}=\text{const.}$$

The Wronskian has limiting value $2ik$ as $x\to\infty$ and $2ik(aa^*-bb^*)$ as $x\to-\infty$; thus we have $|a|^2-|b|^2=1$, which implies $a\neq 0$. Similarly we can consider the solution $\Psi(x,k)$ of $(H-k^2)\Psi=0$ with asymptotic behavior e^{-ikx} as $x\to-\infty$ and $a_-e^{-ikx}+b_-e^{ikx}$ as $x\to\infty$. The constancy of the Wronskian of two solutions of the second order differential equation implies that (1) $a(k)=a_-(k)$; (2) the scattering matrix

$$S(k)=\frac{1}{a(k)}\begin{pmatrix}1 & b_-(k)\\ b(k) & 1\end{pmatrix}$$

is unitary for $k\in R\backslash\{0\}$, and (3) for $k\in R\backslash\{0\}$ $S(-k)=\overline{S(k)}$. Finally the only zeros of $a(k)$ with $\operatorname{Im} k\geqq 0$ are pure imaginary numbers of the form $i\mu_l$, $\mu_l>0$, where $(H+\mu_l^2)\Psi=0$ has a nontrivial square integrable solution. We set $c_l=\|\Psi_+(.,i\mu_l)\|_{L^2}^{-1}$.

The Schrödinger operator H is essentially self adjoint and H has a finite number of eigenvalues $-\mu_1^2, \ldots, -\mu_n^2 < 0$ of multiplicity one and a continuous spectrum $[0, +\infty[$.

We saw that a was nonzero, thus replacing $\Psi(x)$ by Ψ/a we have a solution $\Psi(x)$ whose asymptotic behavior is

$$\Psi(x) \sim e^{ikx} + r_+(k)e^{-ikx} \quad x \to -\infty,$$

$$\Psi(x) \sim t(k)e^{ikx} \quad \text{as} \quad x \to +\infty,$$

where $t(k) = 1/a$ and $r_+(k) = b/a$. In this case the Wronskian condition implies $|t|^2 + |r_+|^2 = 1$: i.e. for a particle coming from the left the probability that it gets over the potential is $|t^2|$ and $|r_+|^2$ is the probability that it is reflected back.

The inverse scattering problem is that given $r_-(k) = b_-/a$ and (μ_l, c_l), $l = 1, \ldots, n$, can one reconstruct $q(x)$. For suitable potentials this problem is solvable.

Scattering theory and quantum statistical mechanics can be related as follows. As opposed to the cases above, since H no longer has purely discrete spectra, $\rho(\beta) = \exp(-\beta H)$ no longer is a trace class operator. However, it can be shown that the operator $\exp(-\beta H) - \exp(-\beta H_0)$ is of trace class (i.e. has a well-defined trace): viz.

$$Z(\beta) = \mathrm{Tr}(e^{-\beta H} - e^{-\beta H_0}) = \sum_{i=1}^{n} e^{\beta\mu_i^2} + \frac{1 - \det(S(0))}{4} + \\ + \frac{1}{2\pi i} \int_0^\infty e^{-\beta\alpha^2}\, \mathrm{Tr}(S^*(\alpha)S^{-1}(\alpha))\, d\alpha.$$

In quantum statistical mechanics the small quantum corrections at high temperature were treated in this case by the Wigner–Kirkwood expansion. The Wigner–Kirkwood expansion involves expanding $\mathrm{Tr}(e^{-\beta H})$ in powers of $\hbar(d^2/dx^2)$ – i.e. the kinetic energy is consider small compared with the potential energy. The quantum corrections are then obtained by evaluating limits $\hbar \to 0$. Our philosophy in this volume is not to consider $\hbar \to 0$ limits when one is interested in high temperature limits $\beta \to 0$. In particular the high temperature limit of the partition function $Z(\beta)$ can be obtained

$$Z(\beta) \sim (4\pi\beta)^{-1/2} \sum_{j=1}^{\infty} a_j \beta^j, \tag{*}$$

where a_j are integrals of polynomials in q. In particular one can show that

$$a_1 = - \int_{-\infty}^{\infty} q(x)\,dx,$$

$$a_2 = \tfrac{1}{2} \int_{-\infty}^{\infty} q^2(x)\,dx,$$

$$a_3 = -\tfrac{1}{6} \int_{-\infty}^{\infty} (q(x) + \tfrac{1}{2}(q'(x)^2))\,dx, \qquad\qquad (*)$$

etc.

We compare the result of the second virial coefficient as obtained by the Wigner–Kirkwood method. Here

$$B_2 = \lim_{\hbar \to 0} -\left(\frac{4\pi\lambda}{2!}\right)^{1/2} \mathrm{Tr}(e^{-\beta H} - e^{-\beta H_0})$$

$$= -\frac{1}{2!}\int dr(e^{-\beta q(r)} - 1)$$

$$= -\frac{1}{2!}\int dr(-\beta q(r)) + \text{higher order terms in } \beta.$$

Here $H_0 = (\hbar^2/2\mu)\nabla^2$, μ is the reduced mass of the two interacting particles, $q(r)$ is the interaction potential between the two particles.

The functionals $a_j(q)$ in $(*)$ are related to another topic of current interest in physics – the theory of the Korteweg–de Vries equation. Consider the antisymmetric bilinear form on the Schwartz space $\mathscr{S}(R)$ given by

$$\Omega(f_1, f_2) = \frac{1}{2} \int_{x \leq y} (f_1(x)f_2(y) - f_1(y)f_2(x))\,dx\,dy.$$

This gives a symplectic like structure to the space $\mathscr{S}(R)$. As we will develop below, on any symplectic space we can consider the associated vector field X_F to a function $F : \mathscr{S}(R) \to C$. In the present case X_F is given by $X_F(q)(x) = = (d/dx)G_F(q)(x)$ where for $F(q) = \int_{-\infty}^{\infty} P(q(x), \ldots q^{(\alpha)}(x))\,dx$ with P a polynomial in $\alpha + 1$ variables with constant term zero, then

$$G_F = \sum_{i=0}^{\infty} (-1)^i \left(\frac{d}{dx}\right)^i \frac{\partial P}{\partial q^{(i)}}.$$

X_F then satisfies the relation $\Omega(X_F, q) = \int_{-\infty}^{\infty} G_F q\,dx$. There is also a natural Poisson bracket $\{,\}$ on this generalized symplectic space. The equation of evolution given by $dq/dt = X_F(q)$ becomes the object of study. E.g. if $F(q) = a_2(q) = \tfrac{1}{2}\int q^2$, then the evolution equation becomes $\partial q/\partial t = \partial q/\partial x$;

while if $F(q) = -3a_3(q) = \frac{1}{2}\int[q^3 + \frac{1}{2}(q')^2]$ then the evolution equation is $\partial q/\partial t = 3q(\partial q/\partial x) - (\partial^3 q/\partial x^3)$ i.e. the $K\,\mathrm{d}V$ equation. The amazing relationship between the study of this equation and the associated scattering theory or quantum statistical mechanics is that the functionals $a_j(q)$ form an involutive system of first integrals for the KdV equation – i.e. for $i,\ j = 1, 2, 3, \ldots, \{a_i,\ a_j\} = 0$. One can then show that $r_-(k)(t) = r_-(k)(0)\exp(4ik^3 t)$, $\mu_l(t) = \mu_l(0)$, and $c_l(t) = c_l(0)\exp(2\mu_l^3 t)$. These equations thus provide the integrability of the $K\,\mathrm{d}V$ equation.

0.11. Quantum Field Theory

We briefly introduce the reader to the type of problems in quantum field theory that we treat later in this volume in Chapter 19. Consider a scalar field with Lagrangian

$$L = -\tfrac{1}{2}g^{\mu\nu}\partial_\mu\varphi\partial_\nu\varphi - \tfrac{1}{2}\xi R\varphi^2 - \tfrac{1}{2}m^2\varphi^2,$$

where R is the scalar curvature of spacetime, ξ is a real number and m is the mass. Here we use units with $\hbar = c = 1$ and we try to adopt the physicists', notation in this section. The field equation associated to L is

$$-g^{\mu\nu}\nabla_\mu\nabla_\nu\varphi + (m^2 + \xi R)\varphi = 0.$$

where ∇_μ is the covariant derivative. The Green's function $G(x, x')$ is the solution of the equation

$$(-\nabla^\mu\nabla_\mu + \xi R + m^2)G(x, x') = [-g(x)]^{-1/2}\,\delta(x - x').$$

If we view $G(x, x')$ as the matrix element in a Hilbert space

$$G(x, x') = \langle x|G|x'\rangle$$

then formally we want to find H with $HG = 1$. This is solved by setting $G = H^{-1} = \int_0^\infty i\,\mathrm{d}s\,e^{-isH}$ or

$$G(x\,x') = \int_0^\infty i\,\mathrm{d}s\langle x|e^{-isH}|x'\rangle.$$

If we set $|x, s\rangle = e^{isH}|x\rangle$, then $\Psi(x, s) = \langle x, s|\Psi\rangle$ satisfies the 'Schrödinger' equation

$$\frac{i\partial\Psi}{\partial s}(x, s) = H(x)\Psi,$$

where $H(x) = -\nabla^\mu\nabla_\mu + \xi R + m^2$. This formally describes a fictitious particle with mass $\frac{1}{2}$ moving in a 4-dimensional space under the potential $\xi R + m^2$. The parameter s plays the role of time. Since it does not change when the 'space' coordinates x^μ are transformed it is called proper time.

The integrand of $(*)$ is written $\langle x, s | x', 0 \rangle$ and the problem is to solve

$$\frac{i\partial}{\partial s}\langle x, s | x', 0\rangle = [-\nabla_\alpha \nabla^\alpha + \xi R + m^2]\langle x, s | x', 0\rangle \qquad (**)$$

with

$$\lim_{s\to 0} \langle x, s | x', 0\rangle = (-g)^{1/2}\,\delta(x - x')$$

In Minkowski space where

$$g_{\mu\nu} = \eta_{\mu\nu} = \begin{pmatrix} -1 & & & 0 \\ \vdots & 1 & & \vdots \\ & & 1 & \\ 0 & & & 1 \end{pmatrix}$$

then the solution to $(**)$ is easily seen to be

$$\begin{aligned}\langle x, s | x', 0\rangle &= \frac{i}{(4\pi i s)^2} e^{-im^2 s} \exp\left(\frac{i}{4s}\eta_{\alpha\beta}(x^\alpha - x^{\alpha'})(x^\beta - x^{\beta'})\right) \\ &= \frac{i}{(4\pi i s)^2} e^{-im^2 s} \exp\left(\frac{i\tau^2}{4s}\right),\end{aligned}$$

where

$$\tau = \int_0^s \mathrm{d}s' \left(g_{\alpha\beta}\frac{\mathrm{d}x^\sigma}{\mathrm{d}s'}\frac{\mathrm{d}x^\beta}{\mathrm{d}s'}\right)^{1/2}$$

is the proper arc length along the geodesic from x to x'.

Using normal coordinates in a neighborhood of x (i.e. $y^\mu = \tau\xi^\mu$ where $\xi^\mu = (\mathrm{d}x^\mu/\mathrm{d}\tau)_{x'}$ is the tangent to the geodesic at x') we have a locally inertial frame at x'. We can assume the metric at x' (i.e. at $y = 0$) in the normal coordinates is $g_{\mu\nu}(0) = \eta_{\mu\nu}$. We now assume that $\langle x, s | x', 0\rangle$ is of the form

$$\langle x, s | x', 0\rangle = \phi U,$$

where $\phi = i(4\pi i s)^{-2}\exp(-im^2 s)\exp(i\tau^2/4s)$ is independent of ξ^μ. In normal coordinates one finds that $F = (-g(x))^{1/4}\, U(x, x'; is)$ then must satisfy

$$\frac{i\partial F}{\partial s} = -(-g)^{1/4}\nabla^\mu \nabla_\mu[(-g)^{1/4}F] + \xi R F - \frac{i\tau}{s}\frac{\partial F}{\partial \tau}. \qquad (**)$$

The approach to solving (∗∗) near $s=0$ involves expanding F as

$$F(x, x', is) = 1 + isf_1(x, x') + (is)^2 f_2(x, x') + \cdots$$

and equating coefficients of equal powers of s. This gives

$$f_1(x, x') = L(1) - \xi R(x) - \tau\frac{\partial f_1}{\partial\tau}(x, x')$$

$$f_2(x, x') = \frac{1}{2}\left[L(f_1) - \xi R(x) f_1(x, x') - \tau\frac{\partial f_2}{\partial\tau}(x, x')\right],$$

where

$$L(f) = (-g)^{1/4}\partial_\mu\{\sqrt{-g}(x)g^{\mu\nu}(x)\partial_\nu[(-g(x))^{-1/4}f(x, x')\}$$

with $\partial_\mu = \partial/\partial y^\mu$. We need the coincidence limits of f as $x \to x'$ – i.e. as $y \to 0$. One approach is to use the Taylor series for g in normal coordinates. In this case one finds that $(-g)^{-1/4} = 1 + \frac{1}{12}R_{\alpha\beta}y^\alpha y^\beta + o(y^3)$ and

$$\sqrt{-g}\, g^{\mu\nu} = \eta^{\mu\nu} + o(y^2),$$

thus

$$\lim_{y\to 0} L(1) = \tfrac{1}{6}\eta^{\mu\nu}R_{\mu\nu} = \tfrac{1}{6}R.$$

Since $\tau = 0$ at $y = 0$ we have $f(x, x') = (\frac{1}{6} - \xi)R(x')$.

Zeta functions arose naturally in our discussion of quantum statistical mechanics. Hawking and Dowker have introduced the concept of generalized ζ-functions in field theory. Viz. if $H|\varphi_n\rangle = \lambda_n|\varphi_n\rangle$ and if we use the completeness relation $\sum|\varphi_n\rangle\langle\varphi_n| = 1$, we see that formally

$$G^\nu = H^{-\nu} = \sum_n \frac{|\varphi_n\rangle\langle\varphi_n|}{\lambda_n^\nu}.$$

Thus we have a generalized zeta function

$$\zeta(\nu) = \operatorname{Tr} G^\nu = \sum\langle\varphi_n|G^\nu|\varphi_n\rangle = \sum\lambda_n^{-\nu}.$$

Formally we see that

$$\zeta(\nu) = \operatorname{Tr} G^\nu = \Gamma(\nu)^{-1}\int \mathrm{d}^4x\sqrt{-g}\int_0^\infty \mathrm{i}\mathrm{d}s(is)^{\nu-1}\langle x|e^{-isH}|x\rangle.$$

Since

$$\langle x|e^{-isH}|x\rangle = e^{-im^2 s}\frac{i}{(4\pi is)^2}[1 + isf_1(x, x) + \\ + (is)^2 f_2(x, x) + \cdots],$$

we have

$$\zeta(0)=i(4\pi)^{-2}\int \mathrm{d}^4x\sqrt{-g}\,[f_2(x,x)-m^2f_1(x,x)+\tfrac{1}{2}m^4].$$

In field theory the energy momentum tensor $T^{\mu\nu}(x)$ is related formally to the action

$$S=\int \mathrm{d}^4xL=-\tfrac{1}{2}\int \mathrm{d}^4x\sqrt{-g}\,\varphi(x)H(x)\varphi(x)$$

by

$$T^{\mu\nu}(x)=\frac{2}{\sqrt{-g(x)}}\frac{\delta S}{\delta g_{\mu\nu}(x)}.$$

The role of the partition function in quantum field theory is played by $Z=\int \mathrm{d}[\varphi]e^{iS}=e^{iW}$. Assuming that $H|\varphi_n\rangle=\lambda_n|\varphi_n\rangle$ is a complete set of eigen-functions where

$$\int \mathrm{d}^4x\sqrt{-g}\,\varphi_n(x)\varphi_m(x)=\delta_{mn},$$

we have

$$S=-\tfrac{1}{2}\int \varphi H\varphi\sqrt{-g}\,\mathrm{d}^4x=-\tfrac{1}{2}\sum_n c_n^2\lambda_n.$$

Thus we can view Z as an integral over the c_n, i.e.

$$\int \mathrm{d}[\varphi]e^{iS}=\mu\prod_n\int \mathrm{d}c_n\exp(-\tfrac{1}{2}ic_n^2\lambda_n)=$$

$$=\left\{\prod_n\left[\frac{2\pi\mu^2}{i\lambda_n}\right]\right\}^{1/2}=\det(iH/2\pi\mu^2)^{-1/2}$$

This implies that

$$W=i\log\det\left(\frac{iH}{2\pi\mu^2}\right)=\frac{i}{2}\zeta'(0)-i\zeta(0)\log(-2\pi i\mu^2).$$

Using the functional variation of S we have

$$\int \mathrm{d}^4x\sqrt{-g}\,\langle T^\mu_\mu\rangle=-i\zeta(0)=(4\pi)^{-2}\int \mathrm{d}^4x\sqrt{-g}\,f_2(x,x).$$

Problems

Exercise 0.1. Show for a finite group G that the number of inequivalent irreducible representations of G is equal to the number of conjugacy classes of G; show that the sum of the squares of the dimensions of the inequivalent irreducible representations of G is equal to the order of G.

EXERCISE 0.2. Show that the space of all minimal geodesics from e to $-e$ in $SU(2m)$ is homeomorphic to the Grassman manifold $G_m(C^{2m})$ of all m dimensional subspaces of C^{2m}.

EXERCISE 0.3. Show that the Green's function for

$$\left(-\frac{\hbar^2}{2m}\frac{d^2}{dx^2}+\frac{m^2\omega^2x^2}{2}\right)\Psi=E\Psi$$

is

$$G(x,x';E)=\sqrt{\frac{m}{\omega\pi\hbar^3}}\Gamma(-v)D_v(-y_<)D_v(y_>)$$

where $y=\sqrt{2m\omega/\hbar}\,x$, $E/\hbar\omega=v+\frac{1}{2}$, $D_v(y)$ is the parabolic cylinder function. Here $y_<$ and $y_>$ are the lesser and greater of y and y'. Show that the poles of G are given by the poles of $\Gamma(-v)$ which in turn determine the energy levels $E=\hbar\omega(v+\frac{1}{2})$, $v=0,1,2,\ldots$. What are the residues at these poles?

EXERCISE 0.4. Show that the Green's function of the Kepler problem (i.e. the radial Schrödinger equation for the hydrogen atom)

$$\left(-\frac{\hbar^2}{2m}\frac{1}{r^2}\frac{d}{dr}\left(r^2\frac{d}{dr}\right)+\frac{A}{r^2}-\frac{B}{r}\right)\Psi=E\Psi$$

with $A=(\hbar^2/2m)l(l+1)$, 1 nonnegative integer, is of the form

$$G(r,r';E)=\frac{m}{\hbar^2k}\frac{\Gamma(s-v+1)}{\Gamma(2s+2)}(rr')^{-1}M_{v,s+\frac{1}{2}}(2kr_<)W_{v,s+\frac{1}{2}}(2kr_>).$$

Here $M_{v,s+\frac{1}{2}}$, $W_{v,s+\frac{1}{2}}$ are Whittaker functions with $k=-2mE/\hbar$, $v=mB/\hbar^2k$, $s(s+1)=2mA/\hbar^2$. Show that the poles of $\Gamma(x-v+1)$ determine the bound state spectra

$$E=\frac{-2mB}{\hbar^2}\left(2n+1+\sqrt{\frac{8mA}{\hbar^2}+1}\right)^{-2}.$$

EXERCISE 0.6. Use the Euler–Maclaurin summation formula

$$\sum_{v=0}^{n}f(v)=\int_0^n f(x)\,dx+\tfrac{1}{2}\{f(n)-f(0)\}+\sum_{k=1}^{\lambda}(-1)^{k-1}\frac{B_k}{(2k)!}\times$$
$$\times\{f^{(2k-1)}(n)-f^{(2k-1)}(0)\}+R_\lambda,$$

where

$$B_k = 4k \int_0^\infty \frac{t^{2k-1}\mathrm{d}t}{e^{2\pi t} - 1}$$

is the Bernoulli number, R_λ is the remainder, to show that

$$F(\sigma,\bar{\sigma}) = \sum_{j=0}^{\infty} (2j+1)e^{-\sigma j(j+1)} + 2\sum_{j=1}^{\infty}\sum_{n=1}^{j} (2j+1) \times$$
$$\times\, e^{-\sigma j(j+1) - \bar{\sigma} n^2}$$
$$\sim \frac{\sqrt{\pi}}{\sigma\sqrt{a}} + \frac{\sqrt{\pi}}{a^{3/2}}\left(\frac{\sigma}{4} + \frac{\bar{\sigma}}{3}\right) + \frac{\sqrt{\pi}}{a^{5/2}}\left(\frac{\sigma^3}{32} + \frac{\sigma^2\bar{\sigma}}{15} + \frac{\sigma\bar{\sigma}^2}{12}\right) + \cdots,$$

where $a = \sigma + \bar{\sigma}$.

EXERCISE 0.7. Let

$$F(\sigma) = e^{\sigma/4} \sum_{j=0} (2j+1)^2 e^{-\sigma(j+1/2)^2}.$$

Noting that

$$F(\sigma) = -2e^{\sigma/4}\frac{\partial}{\partial\sigma}\vartheta_2(0, e^{-\sigma})$$

where

$$\vartheta_2(z,q) = 2\sum_{n=0}^{\infty} q^{(n+1/2)^2}\cos(2n+1)z, \qquad q = e^{\pi i\tau},$$

use the identity

$$\vartheta_2(0, e^{-\sigma}) = \sum_{-\infty}^{\infty} (-1)^n e^{-\pi^2 n^2/\sigma}$$

to show that

$$F(\sigma) = \frac{\sqrt{\pi}}{\sigma^{3/2}}e^{\sigma/4} = \frac{\sqrt{\pi}}{\sigma^{3/2}} + \frac{\sqrt{\pi}}{4\sqrt{\sigma}} + \frac{\sqrt{\pi}\sqrt{\sigma}}{32} + \frac{\sqrt{\pi}\,\sigma^{3/2}}{384} + \cdots.$$

EXERCISE (Quartic Oscillator) 0.8. Consider the quartic oscillator with Hamiltonian $H = c^{4/3}((-\mathrm{d}^2/\mathrm{d}x^2) + x^4)$ and spectrum $\{\lambda_n\}$. show that

$$Z(\beta) = \mathrm{Tr}(\exp(-\beta H)) \sim \sum_{n=0}^{\infty} a_n \beta^{i_n},$$

where

$$i_n = \tfrac{3}{4}(2n-1), \qquad a_0 = \frac{3}{8\pi}\Gamma(\tfrac{3}{4}), \qquad a_1 = \frac{-1}{6}\Gamma(\tfrac{1}{4}), \ldots.$$

Show that the zeta function for the quartic oscillator $\zeta(s)$ is meromorphic in C with simple poles at $(-i_n)$ with residues $a_n/\Gamma(-i_n)$. Show that $\zeta(-n)=0$ for $n\in N$. Let P be the space relection (parity) operator and consider the alternating spectral function $Z_p(\beta) = \mathrm{Tr}(P\exp(-\beta H)) = \sum(-1)^n \times \times \exp(-\beta\lambda_n)$. Show that $Z_p(\beta) \sim \sum d_n \beta^{3n}/(3n)!$, where $d_0 = \frac{1}{2}$, $d_1 = = -(9c^4/4), \ldots$ Show that $\zeta(1) = 1(+(\cos\pi/3)^{-1}\zeta_p(1)$.

EXERCISE 0.9. Consider group $SU(2)$ with 1-parameter subgroups (i.e. curves $t \to g(t)$ with $g(t+s) = g(t)g(s)$):

$$g_1(t) = \begin{pmatrix} \cos\frac{t}{2} & i\sin\frac{t}{2} \\ i\sin\frac{t}{2} & \cos\frac{t}{2} \end{pmatrix}$$

$$g_2(t) = \begin{pmatrix} \cos\frac{t}{2} & -\sin\frac{t}{2} \\ \sin\frac{t}{2} & \cos\frac{t}{2} \end{pmatrix}$$

$$g_3(t) = \begin{pmatrix} e^{it/2} & 0 \\ 0 & e^{-it/2} \end{pmatrix}.$$

Show $a_i = \mathrm{d}g_i(t)/\mathrm{d}t$ satisfy

$$[a_1, a_2] = a_3, [a_2, a_3] = a_1, [a_3, a_1] = a_2.$$

Let $f(u)$ be a function on $SU(2)$ and let $R(g(t))f(u) = f(ug(t))$. Then $\hat{A}_i$ is defined by

$$\frac{\mathrm{d}f}{\mathrm{d}t}(ug(t)) = \frac{\partial f}{\partial\varphi}\varphi'(0) + \frac{\partial f}{\partial\vartheta}\vartheta'(0) + \frac{\partial f}{\partial\psi}\psi'(0),$$

where $\varphi(t)$, $\vartheta(t)$, and $\psi(t)$ are Euler angles of $ug(t)$. (E. g. if $u = u(\varphi, \vartheta, \psi)$ has Euler angles φ, ϑ, ψ; then $ug_3(t)$ has Euler angles $\varphi, \vartheta, \psi + t$.)

$$\hat{A}_{g(t)} = \varphi'(0)\frac{\partial}{\partial\varphi} + \vartheta'(0)\frac{\partial}{\partial\vartheta} + \psi'(0)\frac{\partial}{\partial\psi}.$$

Using the comment above, we have

$$\hat{A}_3 = \hat{A}_{g_3}(t) = \frac{\partial}{\partial \psi}.$$

Show

$$\hat{A}_1 = \cos\psi \frac{\partial}{\partial \vartheta} + \frac{\sin\psi}{\cos\vartheta}\frac{\partial}{\partial \varphi} - \operatorname{ctg}\vartheta \sin\psi \frac{\partial}{\partial \psi};$$

$$\hat{A}_2 = -\sin\psi \frac{\partial}{\partial \vartheta} + \frac{\cos\psi}{\sin\vartheta}\frac{\partial}{\partial \varphi} - \operatorname{ctg}\vartheta \cos\psi \frac{\partial}{\partial \psi}.$$

Show

$$\Delta = \hat{A}_1^2 + \hat{A}_2^2 + \hat{A}_3^2$$
$$= \frac{1}{\sin\vartheta}\left(\frac{\partial}{\partial \vartheta}\sin\frac{\vartheta\partial}{\partial\vartheta}\right) + \frac{1}{\sin^2\vartheta}\left(\frac{\partial^2}{\partial\varphi^2} - 2\cos\vartheta\frac{\partial^2}{\partial\varphi\partial\psi} + \frac{\partial^2}{\partial\psi^2}\right).$$

Δ is called the Laplace operator on group $SU(2)$.

EXERCISE 0.10. Let $z_0, \ldots z_n$ be the homogeneous coordinate system of $CP(n)$. The variety $Q(n-1)$ in $CP(n)$ given by

$$\sum_{i=0}^{n} z_i^2 = 0$$

is called the $(n-1)$-dimensional complex quadric. Show $Q(n-1)$ is isometric to $SO(n+1)/SO(n-1)$. Show that the Stiefel manifold $S = SO(n+1)/SO(2) \times SO(n-1)$ is a principal circle bundle over $Q(n-1)$. (We will study later in this volume the fact that $Q(n-1)$ is a compact Kähler manifold which is a Hodge manifold. For supplementary reading regarding these facts and this example the reader is directed to S. S. Chern, *Complex Manifolds without Potential Theory* (van Nostrand, New York, 1965)).

EXERCISE 0.11 Let $H = (1/2m)(p^2 + m^2\omega^2 q^2)$ be the Hamiltonian of the harmonic oscillator with eigenfunctions

$$\varphi_n(q) = (m\omega/\pi\hbar)^{1/4}(n!)^{-1/2}\, D_n(q\sqrt{2m\omega/h}).$$

Note that Mehler's identity states that

$$\sum_{n=0}^{\infty} \varphi_n(q)\varphi_n^*(Q)e^{-i(n+\frac{1}{2})(t-T)} = \left(\frac{m\omega}{2i\pi\hbar\sin\vartheta}\right)^{1/2} \times$$
$$\times \exp\left[\frac{im}{2\hbar\sin\vartheta}\{(q^2+Q^2)\cos\vartheta - 2qQ\}\right] = S(q, Q, t-T),$$

where $\vartheta = \omega(t - T)$, which physicists write as $\langle q|Q\rangle$. Show that S satisfies $S(q, q_1, t - t_1) = \int S(q, Q, t - T)S(Q, q_1, T - t_1)\,\mathrm{d}Q$. Show that $S(q, Q, t - T)$ satisfies the Schrödinger equation determined by H: (i.e. $H(q, (\hbar/i)(\partial/\partial q))\langle q|Q\rangle + (\hbar/i)(\partial/\partial t)\langle q|Q\rangle = 0$). (Cf. E. T. Whittaker, *On Hamilton's Principal Function in Quantum Mechanics*, Proc. Roy. Soc. Edinburgh A LXI (1940/41), pp. 1–19.)

map $U:G\to$ {unitary

$_2) = U(g_1)U(g_2)$ for all

function $U(g)f:G\to H$

d to be *invariant* if $U(g)f$

restricting U to a closed

ation and is denoted $U|_K$.

s $(U_i, H_i)i = 1, 2$, of G the

n $U = U_1 \oplus U_2$ acting on

ch has no invariant sub-

$(U_i, H_i)i = 1, 2$ are called

$\to H_2$ such that $VU_1(g) =$

is called an *intertwining*

of U_1 and U_2 is denoted by

direct sum of irreducible

lucible.

itary representation and H

degree of U.

of equivalence classes of

DEFINITION 1.1.9. We shall denote this set by $\hat{G}$.

The theme of geometric quantization is to characterize $\hat{G}$, or distinguished subsets of $\hat{G}$, by objects from differential geometric notions from classical mechanics.

The set $\hat{G}$ is often very simple. For example the group of rotations $SO(2)$ of 2×2 real matrices of the form

$$g = \begin{pmatrix} \cos\varphi & -\sin\varphi \\ \sin\varphi & \cos\varphi \end{pmatrix}$$

has irreducible unitary representations $U(g) = \exp(in\,\varphi)$, n in Z.

EXAMPLE 1.1.10. $\widehat{SO}(2) = Z$.

EXAMPLE 1.1.11. The rotation group $G = SO(3)$ is the group of all 3×3 real matrices g such that $g^t g = I$ and $\det(g) = 1$. The covering group of $SO(3)$ is $SU(2)$, the group of matrices

$$\begin{pmatrix} a & b \\ -\bar{b} & \bar{a} \end{pmatrix}$$

where $|a|^2 + |b|^2 = 1$. The irreducible unitary representations of $SU(2)$ are denoted by $D^{(u)}$ and are realized on the $(2u+1)$-dimensional space of polynomials of order $2u$; $f(z) = \sum_{j=0}^{2u} c_j z^j$. Here g in $SU(2)$ acts on f in $D^{(u)}$ by

$$U(g)f(z) = (bz + \bar{a})^{2u} f\left(\frac{az - \bar{b}}{bz + \bar{a}}\right).$$

As is well known $D^{(u)}$ for $2u = 0, 1, 2, \ldots$ parameterizes the set $\hat{G}$ of all irreducible unitary representations of $G = SU(2)$, whereas for $u = 0, 1, 2, \ldots D^{(u)}$ parameterizes $\hat{G}$ for $G = SO(3)$.

1.2. INDUCED REPRESENTATIONS

A standard technique for the construction of unitary representations is the induction method.

DEFINITION 1.2.1. Let K be a closed subgroup of a nice group G. Let $(L, H(L))$ be a representation of K. Let F be the set of functions $f : G \to H(L)$ such that

(i) $f(kg) = L(k)f(g)$ for all k in K, g in G.

(ii) $g \to (f(g), f(g))_{H(L)}$ is a Borel function for which

$$\int_{G/K} (f(g), f(g))_{H(L)}\, d\mu < \infty.$$

We let F also denote the associated Hilbert space under the inner product

$$(f_1, f_2) = \int_{G/K} (f_1(g), f_2(g))_{H(L)} \, d\mu.$$

Then

$$(U^L(g_1)f)(g_2) = f(g_1 g_2)$$

for f in F defines a unitary representation. (U^L, F) is called the *induced representation.*

We have assumed that an invariant measure was provided. For a discussion of the quasi-invariant case see reference V1.

The simplest example of an induced representation is the case $K = \{e\}$ and L is the identity representation I.

DEFINITION 1.2.2. The induced representation so defined, U^I, is called the *left regular representation.*

Two properties of induced representations are noted next.

THEOREM 1.2.3. $U^{L \oplus M} = U^L \oplus U^M$ whenever L and M are representations of the same closed subgroup K of G.

Thus U^L cannot be irreducible unless L is irreducible. However, even if L is irreducible U^L may not be. Viz. the regular representation U^I is never irreducible (unless G is trivial). The complexity of the regular representation is simply expressed as a corollary to the Frobenius reciprocity theorem which we present next.

DEFINITION 1.2.4. Given an irreducible unitary representation T of G and a direct sum reduction of a completely reducible unitary representation U of G, then the number of times that T occurs in this reduction is called the *multiplicity* of T in U and it is denoted $\dim \operatorname{Hom}_G(T, U)$.

The Frobenius reciprocity theorem relates induction and restriction as follows:

THEOREM (Frobenius Reciprocity) 1.2.5. Let L be an irreducible unitary representation of a closed subgroup K in G and let U be an irreducible unitary representation of G. Then

$$\dim \operatorname{Hom}_G(U, U^L) = \dim \operatorname{Hom}_K(L, U|_K).$$

Taking G to be a compact group and $K = \{e\}$ in this theorem we see that

COROLLARY 1.2.6. $\dim \operatorname{Hom}_G(U, U^I) = \dim H(U)$.

This corollary is also part of the Peter–Weyl theory to be presented below.

EXAMPLE 1.2.7. The regular representation arises naturally in quantum mechanics. Viz. let the quantum system be specified by a Hamiltonian H acting on a Hilbert space of wave functions $\psi \in L^2(G)$. The regular representation $(U, L^2(G))$ as noted above is in general reducible. Now assume that G is a symmetry group of H – i.e.

$$U(g)\mathsf{H} = \mathsf{H}U(g) \qquad \text{for all} \qquad g \text{ in } G.$$

In this case the eigenspace

$$E(\lambda) = \{\psi \in L^2(G) | \mathsf{H}\psi = \lambda\psi\}$$

is invariant under U. $E(\lambda)$ may not be irreducible. We say that G is a *complete symmetry group* if $E(\lambda)$ are irreducible. In that case $E(\lambda)$ are parametrized by λ in $\hat{G}$ and the degeneracy of the eigenspace, i.e. $\dim E(\lambda)$, is equal to the degree of λ.

DEFINITION 1.2.8. The *character* of a representation U of G is the function $\mathfrak{x}_U(g) = \mathrm{Tr}(U(g)) : G \to C$.

EXAMPLE 1.2.9. For the representation $D^{(u)}$ of $SO(3)$

$$\mathfrak{x}_u(g) = \frac{\sin((u + \frac{1}{2})\varphi)}{\sin(\varphi/2)},$$

where g is a rotation through the angle φ about a fixed axis.

We note that $\mathfrak{x}_U$ is a *central function* on G – i.e. $\mathfrak{x}_U(g_1 g g_1^{-1}) = \mathfrak{x}_U(g)$; and $\mathfrak{x}_U$ depends only on the equivalence class of U.

1.3. SCHUR AND PETER–WEYL THEOREMS

In this section we turn to the Schur theorems and the Peter–Weyl theory for compact groups.

Before proceeding we recall that every locally compact topological group has a left Haar measure μ unique up to a multiplicative constant; thus $\mu(gB) = \mu(B)$ for any g in G and Borel subset B of G. If the Haar measure is also right invariant (i.e. $\mu(Bg) = \mu(B)$), we say the group is *unimodular*. It is simply checked that a compact topological group is unimodular. Usually we write the biinvariant measure on a compact topological group as $\mathrm{d}g$ and normalize it so that $\int_G \mathrm{d}g = 1$.

Let $\{U_\lambda(g)\}$, λ in $\hat{G}$, be a complete set of pairwise inequivalent irreducible unitary representations of a compact group G. If $U_\lambda(g)$ acts on space $H(\lambda)$ with orthonormal basis $e_1, \ldots, e_N$ and inner product $\langle , \rangle$, we set

$$u_{ij}^\lambda(g) = \langle e_i, U_\lambda(g) e_j \rangle.$$

THEOREM (Schur Orthogonality) 1.3.1.

$$\int_G u^{\alpha}_{ij}(g) u^{\beta}_{mn}(g)\, dg = 0 \qquad \text{if} \qquad (\alpha, i, j) \neq (\beta, m, n)$$

and

$$\int_G |u^{\lambda}_{ij}|^2\, dg = 1/d(\lambda),$$

where $d(\lambda) = \dim H(\lambda) = \text{degree}(U_\lambda)$.

THEOREM 1.3.2. χ_U determines U up to unitary equivalence; and there is a 1–1 correspondence between $\hat{G}$ and the set of irreducible characters.

As a generalization of Fourier coefficients we have

DEFINITION 1.3.3. For an irreducible unitary representation U in the class λ in $\hat{G}$ and f in $L^2(G)$ we set

$$U_\lambda(f) = \int_G f(g) U_\lambda(g)\, dg.$$

THEOREM (Peter–Weyl) 1.3.4. Let G be a compact group. Then:

(i) Every irreducible unitary representation of G is finite dimensional.

(ii) (Plancherel formula) for f in $L^2(G)$

$$\|f\|_2^2 = \int_G |f(g)|^2\, dg = \sum_{\lambda \in \hat{G}} d(\lambda) \operatorname{Tr}(U_\lambda(f) U_\lambda(f)^*)$$
$$= \int_{\hat{G}} \operatorname{Tr}(U_\lambda(f) U_\lambda(f)^*)\, d\mu(\lambda).$$

(iii) for f in $L^1(G) \cap L^2(G)$

$$f(g) = \int_{\hat{G}} \operatorname{Tr}(U_\lambda(f) U_\lambda(g)^*)\, d\mu(\lambda).$$

In particular we have the decomposition of the delta function on G:

$$f(e) = \int_{\hat{G}} \operatorname{Tr} U_\lambda(f)\, d\mu(\lambda)$$
$$= \sum_{\lambda \in \hat{G}} d(\lambda) \operatorname{Tr} U_\lambda(f).$$

1.4. LIE GROUPS AND PARALLELIZATION

The compact groups of interest are also C^∞ manifolds such that the group product and inversion are C^∞ maps – i.e. they are Lie groups. The Lie algebra $\mathfrak{g}$ associated to G is isomorphic with the space of left-invariant vector fields on G.

EXAMPLE 1.4.1. The Lie algebra of $G = SO(2)$ is just

$$so(2) = \left\{ \begin{pmatrix} 0 & -a \\ a & 0 \end{pmatrix} a \text{ in } R \right\}.$$

The Killing form is

$$B(X_a, Y_b) = -\tfrac{1}{2}\mathrm{Tr}(X_a Y_b) = ab,$$

where

$$X_a = \begin{pmatrix} 0 & -a \\ a & 0 \end{pmatrix}, \qquad Y_b = \begin{pmatrix} 0 & -b \\ b & 0 \end{pmatrix}.$$

The exponential map exp: $\mathfrak{g} \to G$ is just

$$\begin{pmatrix} 0 & -a \\ a & 0 \end{pmatrix} \to \begin{pmatrix} \cos a & -\sin a \\ \sin a & \cos a \end{pmatrix}.$$

Finally

$$T_g SO(2) = \left\{ V_\varphi = \begin{pmatrix} 0 & -a \\ a & 0 \end{pmatrix} g \,\middle|\, a \quad \text{in } R \text{ and} \right.$$
$$\left. g = \begin{pmatrix} \cos\varphi & -\sin\varphi \\ \sin\varphi & \cos\varphi \end{pmatrix} \right\}.$$

Clearly $T_e(SO(2)) \simeq so(2)$.

This example generalizes to all Lie groups, viz. that they are parallelizable – where the parallelization is given by a basis of left invariant vector fields. In other words, the tangent bundle TG is trivial for parallelizable manifolds like G. The example is also one of the 3 parallelizable spheres: S^1, S^3, and S^7. The associated real projective spaces $RP(1) \simeq S^1$, $RP(3)$, and $RP(7)$ are also parallelizable.

EXAMPLE 1.4.2. The Lie algebra $\mathfrak{g}$ of $G = SO(3)$ is the set of all 3×3 real matrices A such that $A^t = -A$. A basis for $\mathfrak{g}$ is given by

$$L_1 = \begin{pmatrix} 0 & -1 & 0 \\ 1 & 0 & 0 \\ 0 & 0 & 0 \end{pmatrix}, \qquad L_2 = \begin{pmatrix} 0 & 0 & 1 \\ 0 & 0 & 0 \\ -1 & 0 & 0 \end{pmatrix},$$

$$L_3 = \begin{pmatrix} 0 & 0 & 0 \\ 0 & 0 & -1 \\ 0 & 1 & 0 \end{pmatrix},$$

where

$$[L_1, L_2] = L_3, \qquad [L_3, L_1] = L_2 \qquad \text{and} \qquad [L_2, L_3] = L_1.$$

The Lie algebra $su(2)$ of the covering group $SU(2)$ of $SO(3)$ is the set of 2×2 complex skew-Hermitian matrices A of trace zero. $su(2)$ has a basis given by

$$J_1 = \tfrac{1}{2}\begin{pmatrix} i & 0 \\ 0 & -i \end{pmatrix}, \qquad J_2 = \tfrac{1}{2}\begin{pmatrix} 0 & 1 \\ 1 & 0 \end{pmatrix} \quad \text{and} \quad J_3 = \tfrac{1}{2}\begin{pmatrix} 0 & i \\ -i & 0 \end{pmatrix}$$

which satisfy the commutation relations

$$[J_1, J_2] = J_3, \qquad [J_1, J_3] = -J_2 \qquad \text{and}$$
$$[J_2, J_3] = -J_1.$$

1.5. Spectral Theory and Representation Theory

In the latter part of this book, several examples of representation theory in physics are considered. Nearly all of these examples involve the mixture of representation theory and spectral theory.

Recall from the introduction that a G-vector bundle is a pair of manifolds M, E, a smooth map $\pi: E \to M$, an effective G-vector space V such that for an open covering $\{U_i\}$ of M, there are diffeomorphisms h_i: $\pi^{-1}(U_i) \to U_i \times V$ which map the fiber $\pi^{-1}(m)$ onto $m \times V$ and smooth maps g_{ij}: $U_i \cap U_j \to G$ such that $h_i h_j^{-1}(m \times v) = m \times g_{ij}(m)v$ for all m on $U_i \cap U_j$ and v in V. We assume the reader is familiar with the notion of sheaves (v. Hirzebruch H22). If we let $\mathbf{G}$ denote the sheaf for which the space of sections $\Gamma(U, \mathbf{G})$ is the group of smooth functions from U to G; then the elements g_{ij} determine a cocycle $\{g_{ij}\}$ in $Z^1(\{U_i\}, \mathbf{G})$; and as is well known the isomorphism classes of fiber bundles over M with structure group G and fiber V are in a natural 1–1 correspondence with the cohomology set $H^1(M, \mathbf{G})$. A smooth map s: $M \to E$ such that $\pi \circ s = 1$ is called a *section.* We let $\Gamma(E)$ or $C^\infty(E)$ denote the set all smooth sections.

We are interested in vector bundles with a Hermitian structure given on each fiber E i.e. we are given a Hermitian form h_m on each fiber such that for cross sections s_1, s_2 we have a smooth map $m \to h_m(s_1(m), s_2(m))$. In this case we say that E is a Hermitian *vector bundle*. For a Hermitian vector bundle E the space of continuous cross sections with compact support, $C_0(E)$, can be given a pre-Hilbert space structure by setting

$$\langle s_1, s_2 \rangle = \int_M h_m(s_1(m), s_2(m))\, dm,$$

where dm is the volume element on M. We let $L^2(E)$ denote the Hilbert space completion. Thus $U(g)f(m) = g(f(g^{-1}(m)))$ for f in $L^2(E)$, g in G and m in M defines a unitary representation of G in $L^2(E)$.

DEFINITION 1.5.1. Given a self adjoint operator H in $L^2(E)$ with domain $D(H)$ we say that H *commutes* with G if $U(g)D(H) \subset D(H)$ and $U(g)H(\Psi) = HU(g)\Psi$ for g in G and Ψ in $D(H)$.

Consider the spectral resolution $t \to H(t)$ of $H: H = \int_{-\infty}^{\infty} t\, dH(t)$. If we identify $H(t)$ with its image in $L^2(E)$, then $H(t)$ is G-invariant.

DEFINITION 1.5.2. For $\tau \in \hat{G}$ we define the multiplicity of τ in $H(t)$ by $N(t) = \dim \operatorname{Hom}_G(V, H(t))$ where V is a representative G-module for τ.

Normally H is generated by an elliptic operator P: viz. H is the self-adjoint extension of P in $L^2(E)$. Thus the spectrum of H consists of positive eigenvalues λ with finite multiplicities, say $n(\lambda)$. Thus in that case

$$N(t) = \sum_{\lambda \leqq t} n(\lambda).$$

In general we have

THEOREM 1.5.3. $\lim_{t\to\infty} N(t) = \dim \operatorname{Hom}_G(V, L^2(E)) =$ multiplicity of τ in $L^2(E)$.

Proof. Since we have the canonical isomorphism $\operatorname{Hom}(V, L^2(E)) = L^2 \otimes V^*$ we have the orthogonal decomposition $L^2(E) \otimes V^* = H(t) \otimes V^* \oplus \sum_{j=1}^{\infty} (H(t+j-1)) \otimes V^*$. Using the orthogonal projection on the G-invariant elements $\mathsf{P}: L^2(E) \to L^2(E)^G$ given by $\mathsf{P}f = [1/\operatorname{Vol}(G)] \int_G U(g) f\, dg$, we see that P maps each element of the sum into itself. Taking dimensions we have $\dim \operatorname{Hom}_G(V, L^2(E)) = N(t) + \sum_{j=1}^{\infty} \dim \operatorname{Hom}_G(V, H(t+j) - H(t+j-1))$ and the result follows.

In general we consider a compact Riemannian manifold M and a differential operator $P: C^\infty(E) \to C^\infty(E)$. If P is strongly elliptic and positive of order $2k$ (for the definitions v.i.) then P has eigenvalues $0 \leqq \lambda_1 < \lambda_2 \ldots$ tending to finity and each eigenspace $P_\lambda = \{s \in C^\infty(E) \mid Ps = \lambda s\}$ has finite dimensions. We have seen in the introduction that in statistical mechanics or quantum mechanics we are interested in the asymptotic behavior of

$$N(t) = \sum_{\lambda \leqq t} \dim P_\lambda.$$

Weyl first considered this problem in the case that M is a bounded open set in Euclidean space, E is the trivial bundle and P is the Laplacian acting on functions. Weyl showed that

$$N(t) \sim \text{const.}\, t^{\dim M/2k} \quad \text{as} \quad t \to \infty,$$

where the constant is given by an integral over the unit cotangent bundle $T^*(M)$ and it involves the symbol of P.

Representation theory enters the problem when we consider E to be a G-vector bundle where G acts by isometries on M, unitarily on the fibers of E and P commutes with the action of G on $C^\infty(E)$. In this case P_λ are unitary G-modules. As discussed in Section 0.5 we are interested in the multiplicity $v_\rho(\lambda)$ of a fixed irreducible unitary representation ρ of G in P_λ. Thus instead of $N(t)$ we study $N_\rho(t) = \sum_{\lambda \leqq t} v_\rho(\lambda)$. Since $v_\rho(\lambda) = \dim P_\lambda$ if $G = e$, this is an appropriate generalization.

Extending the notion of symbol of differential operators $P: C^\infty(E) \to C^\infty(E)$, the principal symbol $\sigma(P)$ of P is defined as an element of $C^\infty(\text{Hom}(\pi^* E, \pi^* E)$ where π: $T^*M \to M$ is the natural projection. That is for P in M, v in T^*M, $e \in \pi^* E$ we choose $\varphi \in C^\infty(M)$ such that $\varphi(p) = 0$, $d\varphi_p = v$, and $f \in C^\infty(E)$ with $f(p) = e$; then

$$\sigma(P)(v)(e) = P\left(\left(\frac{-i}{k}\right)^k \varphi^k f\right)(p).$$

P is then *elliptic* if $\sigma(P)(v)$ is an isomorphism for v in $T^*(M)$. P is called *strongly elliptic* if k is even and $\sigma(P)(v)$ is positive definite for v in $T^*(M)$.

Let $T_G^*(M) = \{\xi \in T^*M \mid \xi(X) = 0$ for all X in $TM_{\pi(\xi)}$, the tangent space to $G\pi(\xi)$, $\pi: T^*M \to M\}$. We say that P is *transversally elliptic* if $\sigma(P)(\xi)$ is an isomorphism for ξ in $T_1^*M \cap T_G^*M$.

Consider now a compact Riemannian manifold and a Hermitian G-vector bundle over M where G acts on M by isometries and unitarily on fibers. Let P be a transversally strongly elliptic differential operator of order $2k$ acting on the space of smooth section. Assume that P commutes with the action of G; and we assume further that P is essentially self-adjoint on $L^2(E)$, which is automatically true if P is strongly elliptic and a symmetric operator on $C^\infty(E)$.

From Theorem 1.5.3 we see that τ occurs in $L^2(E)$ iff it occurs in some spectral subspace $H(t)$ of H. However, if G acts transitively the asymptotic behavior of N is completely independent of H. If G is not transitive we must consider the union of principal orbits M_0 in M. In this case it can be shown that

THEOREM 1.5.4. (Brunig–Heintze). If M is a compact Riemannian manifolds and P is strongly elliptic and *symmetric on the space of smooth sections* of E and if $v_\tau(\lambda)$ = multiplicity of τ in the eigenspace of P_λ with eigenvalues λ for τ in $\hat{G}$ and if we set $N_\tau(t) = \sum_{\lambda \leqq t} v_\tau(\lambda)$, then N_τ is finitely-valued and as $t \to \infty$

$$N_\tau(t) \sim \frac{t^{m/2k}}{m(2\pi)^m} \int_{M_0} \frac{1}{\mathrm{Vol}(Gx)} \int_{T_1{}^*M_0 \cap T_G{}^*M} \mathrm{Tr}_{(E_x + V^*)} G_{\sigma(P)} \times$$
$$\times (\xi)^{-m/2k} \otimes id_{V^*}) \, \mathrm{d}w(\xi) \, \mathrm{d}\mu(x),$$

where $m = \dim(M_0/G)$, $\sigma(P)(\xi)$ is the symbol of the kth order operator P, and $\mathrm{d}w$ denotes the volume element induced by Lebesque measure.

We set $E' = \bigcup_{x \in M_0} E_x^{G_x}$ = union of all elements in $E|M_0$ that are invariant under the isotropy group of their base point. Then under fairly general conditions on P, N (v. ibid.) we have

THEOREM 1.5.5. If P is a generalized Laplacian – i.e. $\sigma(P) = |\xi|^2 \, id_{E_{\pi(\xi)}}$ then $N(t) = \dim H^G(t)$ satisfies

$$N(t) \sim t^{m/2k} \, \mathrm{Vol}(B(m)) \dim E' \, \mathrm{Vol}(M_0/G)/(2\pi)^m,$$

where $B(m)$ is the m-dimensional unit ball.

This result generalizes a result of Weyl for bounded open sets in R^n and the Minakshisundaram–Pleijel formula for Laplacians on functions on Riemannian manifolds. Viz.

COROLLARY 1.5.6. Let $G = \{e\}$, M compact; then $E' = E, m = n, T_G^*M = T^*M$ and

$$N(t) \sim \frac{t^{n/2k}}{n(2\pi)^n} \int_M \int_{(T_1^* M_x)} \mathrm{Tr}_E(\sigma(P)(\xi))^{-n/2k} \dim w(\xi) \, d\mu(\chi).$$

We now turn to the more general question of the relationship of spectral theory and the decomposition of unitary representations. Recall first the result due to Gårding:

THEOREM 1.5.7. If (U, H) is a representation of a connected locally compact separable Lie group G and H_∞ is the space spanned by $U(f)h$ for h in H and f in $A_0(G)$ (i.e. the space of smooth functions on G with compact support) then H_∞ is dense in H; setting $\mathrm{d}U(X)U(f)h = U(Xf)h$ then $X \to \mathrm{d}U(X)$ is a representation of the universal enveloping algebra $\mathscr{U}$ of $\mathfrak{g}$

with common dense domain H_∞; the representation $\mathrm{d}U$ is symmetric – i.e. symmetric elements $X = X^+$ in $\mathscr{U}$ are represented by symmetric operators $\mathrm{d}U(X)$ in H.

The operators $\mathrm{d}U(X)$ are quantum mechanical observables iff they are essentially self adjoint – i.e. their closures are self adjoint. There are several results relating quantum observables and representation theory. E.g. Segal has shown that if X is in the center $\mathscr{Z}$ of $\mathscr{U}$ that $\mathrm{d}U(X)$ is essentially self adjoint if $X = X^+$.

Generalizing this result the Maurins have developed a general decomposition theory of unitary representations.

THEOREM 1.5.8 (Maurin–Maurin). Let (U, H) be a unitary representation of a connected locally compact group G. Then (U, H) admits a direct integral decomposition $(\int_\Lambda U(\lambda), \int_\Lambda H(\lambda))$ where almost all $H(\lambda)$ are common eigenspaces of the operators $\mathrm{d}U(X)$ for $X = X^+$ in $\mathscr{Z}: \langle\psi, \mathrm{d}U(X)e(\lambda)\rangle = = X(\lambda)\langle\psi, e(\lambda)\rangle, e(\lambda) \in H(\lambda), \psi \in \Psi$, where $\Psi \subset H \subset \Psi'$, is a Gelfand triplet.

Recall that a Gel'fand triplet is a triple of spaces $\Phi \subset H \subset \Phi'$ where H is a Hilbert space, Φ is a nuclear space dense in H with a continuous imbedding $\Phi \hookrightarrow H$, and Φ' is the strong dual of Φ. The construction of the appropriate Gel'fand triplet is presented in Maurin M21, Theorem 18, Chapter 6. The stated theorem follows directly from Maurin's Theorem 19, ibid. The development of quantum mechanics a la Dirac using Gel'fand triplets has been explained several times by physicists; v. e.g. Bohm Blla.

For the case $M = G/K$ with K a closed subgroup of G and an invariant measure on M, then for the unitary representation U of G on $L^2(M)$ we say

DEFINITION 1.5.9. The *Laplace operators* on $M = G/K$ are operators of the form $\mathrm{d}U(X)$ for $X = X^+ \in \mathscr{Z}$.

Thus the Laplace operators are defined on the Garding space H_∞ of $U, L^2(M)$). They are essentially self adjoint, their closures commute with U and for any X_1, X_2 in $\mathscr{Z}$, $\mathrm{d}U(X_1)$ and $\mathrm{d}U(X_2)$ are strongly commuting (v.M21).

The following is a generalization of Cartan's theorem which is covered in Exercise 1.2:

THEOREM 1.5.10. (Maurin–Maurin). $L^2(M)$ admits a direct integral decomposition $L^2(M) = \int_\Lambda H(\lambda)\, \mathrm{d}\mu(\lambda)$ where the elements of $H(\lambda)$ are common eigendistributions of the Laplace operators on M. Here $e(\lambda)$ are the generalized spherical functions in Theorem 1.5.8.

Let G be a locally compact group and let Γ be a closed subgroup of G. Let (j, V) be a unitary representation of Γ in $(V, \langle | \rangle)$. Let $p: G \to G/\Gamma$ be the canonical imbedding $p(g) = g\Gamma$. On G/Γ there is a quasi invariant measure μ giving rise to a continuous strictly positive function ρ which satisfies:

(i) $\rho(g\gamma) = \rho(g)\rho(\gamma)$.

(ii) $\int_G \rho(g) f(g) dg = \int_{G/\Gamma} d\mu(g\Gamma) \int_\Gamma f(g\gamma) dv(\gamma)$ for f in $A_0(G)$, $v =$ Haar Measure on Γ. Let H^j be the Hilbert space given by measurable functions $f: G \to V$ which satisfy:

(i) $f(g\gamma) = \rho(\gamma)^{1/2} j(\gamma) f(g)$.

(ii) $\int_{G/\Gamma} \rho(g)^{-1} \| f(g) \|_V^2 d\mu(g\Gamma) < \infty$ under the norm

$$(f_1 | f_2)_{H^j} = \int_{G/\Gamma} \rho(g)^{-1} \langle f_1(g) | f_2(g) \rangle d\mu(g\Gamma).$$

Here the representation U^j is defined by

$$U^j f(x) = f(g^{-1}x) \text{ for } g, x \text{ in } G, f \text{ in } H^j.$$

DEFINITION 1.5.11. Let $\eta: H_\infty \to V$ be a linear map which is continuous and satisfies:

(i) $\eta(U(\gamma^{-1})\varphi) = \rho(\gamma)^{1/2} j(\gamma) \eta(\varphi)$, $\varphi \in H_\infty, \gamma \in \Gamma$;

(ii) $f_{\eta,\varphi}(g) = \eta(U(g^{-1})\varphi) \in H^j$, $\| \eta(U^{-1}(.)U(\psi)h \|_{H^j} \leqq c(K) \| \psi \|_\infty \| h \|$ for ψ in $D^0(K) = \{ f \in A_0(G), \operatorname{supp} f \subset K, K \text{ compact} \}$ then η is called a (U,j)-*automorphic form.* Denote the space of (U,j)-*automorphic forms* by $A(U,j)$.

THEOREM 1.5.12 (Maurin–Maurin). $(U^j, H^j) = \oplus (U^j | H_k, H_k)$ where $U^j | H_k$ are irreducible and have finite multiplicities. The space of continuous intertwining operators $\operatorname{Hom}_G(U, U^j)$ is isomorphic to $A(U,j)$ for U in $\hat{G}$. And $\dim A(U,j) =$ multiplicity of U in U^j.

DEFINITION 1.5.13. Assume there is a compact subgroup $K \subset G$; then (G, K) is said to satisfy the *Gelfand condition* if the convolution algebra $A_0(K \backslash G/K)$ is commutative. If the Gelfand condition holds the common eigenfunctions of $A_f \eta = \int_G f(g) U^j(g) \eta \, dg = \lambda_f \eta$ for f in $A_0(K \backslash G/K)$ are called *Tamagawa automorphic forms.*

If η is a Tamagawa automorphic form then $\omega(t) = (U^j(t^{-1})\eta | \eta)_{H^j}(x)$ is a positive definite zonal spherical function. And as is well known every positive definite zonal spherical function defines a unitary representation of class 1. Call it (U_η, H_η). One can then show that the Tamagawa automorphic forms pick out subrepresentations of U^j of class 1; each such representation contains only one Tamagawa automorphic form up to a factor; and forms of different eigenspaces of A_f lead to inequivalent representations of class 1. Viz.:

THEOREM 1.5.14 (Maurin–Maurin). Let η in H^j be a Tamagawa automorphic form. Denote by H^j_η the U^j-invariant subspace of H^j having η as a cyclic vector and let U^j_η be the restriction $U^j|H^j_\eta$. Then (U^j_η, H^j_η) is an irreducible representation, U^j_η is equivalent to U_η, hence U^j_η is of class 1; the mapping $T_\eta: H_\eta \to H^j_\eta$ is an isometric intertwing operator $T_\eta \in \mathrm{Hom}_G(U_\eta, U^j_\eta)$ where $T_\eta\psi = \int \psi(t)U^j_\eta(t)\eta\,\mathrm{d}t$; every Tamagawa automorphic form η defines a (U_η, j)-automorphic form where $\eta(\varphi) = \int_G \varphi(g)\eta(g)\,\mathrm{d}g$, φ in $A_0(G)$; and if an element η in H^j defines a zonal spherical function $\omega(g) = (U^j(g^{-1})\eta|\eta)_{H^j}$ then η is a Tamagawa automorphic form.

Now we turn to the generalization of a theorem of Roelcke. Consider the unitary representation (U^j, H^j) and the projection $\bar{P} = \int_K U^j(k)\,\mathrm{d}k$. It can be checked that $\bar{P}$ commutes with $\mathrm{d}U^j(X)$, X in $\mathscr{Z}$. The common generalized eigenvectors of $\mathrm{d}U^j(X)|\bar{P}H^j$ are called *automorphic forms.* Since the operators $\mathrm{d}U^j(X)$ commute with right translations, the operators $\mathrm{d}U^j(X)|\bar{P}H^j$ can be considered as an invariant differential operators on $K\backslash G$. Let $\bar{P}\Psi \subset \bar{P}H^j \subset \bar{P}'\Psi'$ be the Gelfand triplet given by $\bar{P}$ action on $\Psi \subset H^j \subset \Psi'$.

THEOREM 1.5.15 (Roelcke–Maurin). The Hilbert space $H = \bar{P}H^j$ can be decomposed into a direct integral $\int_\Lambda H(\lambda)$ where almost all $H(\lambda) \subset \bar{P}'\Psi'$ are Hilbert spaces of automorphic forms. Thus the unitary presentation (U^j, H^j) admits a decomposition $(U^j, H^j) = \int_\Lambda (U^j(\lambda), H^j(\lambda))\,\mathrm{d}\lambda$ where almost all $H(\lambda) \subset \Psi'$ are common eigenspaces of the commuting algebra associated to $\mathrm{d}U(X)$, $X \in \mathscr{Z}$. And the projections $\bar{P}'\Psi'$ of left K-invariant elements of Ψ' are spaces of automorphic forms.

Returning to the case of G/K and $K\backslash G/K$ we note that the zonal spherical functions are the eigen functions of the 'radial' parts of the Laplace operators on G/K. Recall that the rank of $G/K = \dim(K\backslash G/K)$, which is equal to one for S^2. For Euclidean spaces or Riemannian globally symmetric spaces M of rank one, the only differential operators which are invariant under the group of all isometries are polynomials in the Laplace–Beltrami operators Δ_M. E.g. for the sphere $S^2 = SO(3)/SO(2)$ with metric $\mathrm{d}s^2 = \mathrm{d}\vartheta^2 + (\sin\vartheta)^2\mathrm{d}\psi^2$, the eigen functions of $\Delta_{S^2} = \partial^2/\partial\vartheta^2 + \cot\vartheta\,\partial/\partial\vartheta + {} + 1/\sin^2\vartheta)\partial^2/\partial\psi^2$ are the spherical functions and the zonal spherical functions are the eigen functions of the radial part $\Delta = \mathrm{d}^2/\mathrm{d}\vartheta^2 + \cot\vartheta\,\mathrm{d}/\mathrm{d}\vartheta$, viz. the Legendre polynomials. Thus we see that the Fourier series expansions given by the Cartan theorems in Exercises 1.2 or 1.3 or their generalizations given in the Maurin theorems are really spectral expansions for the appropriate Laplace operators.

PROBLEMS

EXERCISE 1.1. Show that if $M = G/K$, then the induced representation U^L as defined in 1.2.1 is equivalent to

$$T(g)f(m) = J_L(M, g)f(mg)$$

for f in $L^2(M, H(L))$, where J_L satisfies

$$J_L(m, g_1)J_L(mg_1, g_2) = J_L(m, g_1g_2)$$
$$J_L(m, e) = 1 \text{ for all } m \text{ in } M.$$

Show that $k \to I(k) = J_L(m_0, k)$ is a representation of $K = \{g \text{ in } G | gm_0 = m_0\}$.

EXERCISE 1.2. Let U be an irreducible representation of group G on V_λ. Let K be a subgroup of G. A representation U is called of *class one* with respect to K is there a nonzero vector v in V_λ which is invariant with respect to U and $U|K$ is unitary. If (U, V_λ) is a class one representation with respect to K and there is only one invariant vector v, K is called *massive*. We let $\hat{G}_1$ denote the set of equivalence classes of irreducible unitary representations of class one. Let $Z(V_\lambda)$denote the subspace of V_λ of vectors v such that $U(k)\ v = v$ for all k in K. Assume $\dim Z(V_\lambda) = r$ and let $e_1, \ldots, e_r$ be the invariant basis vectors. The matrix elements $t_\lambda^{ij}(g)$, $i, j = 1, \ldots, r$ are the *zonal spherical functions.*

Clearly if $v(g)$ is a zonal spherical function then $v(kgk') = v(g)$ for all k, k' in K.

Let G be a compact Lie group and let K be a massive subgroup. Show that functions on $M = G/K$ have an expansion of the form

$$f(g) = \sum_{\lambda \in \hat{G}_1} \sum_{k=1}^{d(\lambda)} c_{k1}^\lambda t_\lambda^{k1}(g)$$

where $t_\lambda^{k1}(g) = (U_\lambda(g)e_k, e_1)$, e_1 is the invariant element of the class one representation and

$$c_{k1}^\lambda = \mathrm{d}(\lambda) \int f(g)\overline{t_\lambda^{k1}}(g)\, \mathrm{d}g.$$

EXERCISE 1.3. Let G and K be as in the last exercise. Show that any function on $K\backslash G/K$ can be decomposed into a zonal spherical series of the form $f(g)=\sum_{\lambda\in\hat{G}_1} c(\lambda)t_\lambda^{11}(g)$ where $c(\lambda)=\mathrm{d}(\lambda)\int f(g)\overline{t_\lambda^{11}}(g)\,\mathrm{d}g$.

EXERCISE 1.4. Show that if G is a compact group and f is a central function on G, then f admits a decomposition of the form $f(g)=\sum_{\lambda\in G} c(\lambda)\chi(g)$ where $c(\lambda)=\int f(g)\overline{\chi(g)}\,\mathrm{d}g$.

EXERCISE 1.5. Let A be a G-module and let $C^n(G,A)$ denote the set of maps $G\times\cdots\times G\to A$. Define the coboundary map $d^n: C^n\to C^{n+1}$ by

$$\begin{aligned} d^n f(g_1,\ldots,g_{n+1}) &= g_1 f(g_2,\ldots,g_{n+1})+\\ &+\sum_{i=1}^{n}(-1)^i f(g_1,\ldots,g_i g_{i+1},\ldots g_{n+1})+\\ &+(-1)^{n+1} f(g_1,\ldots,g_n). \end{aligned}$$

Let $H^q(G,A)=\ker(d^q)/\mathrm{Im}(d^{q-1})$ denote the cohomology group. Clearly $H^0(G,A)=A^G$. A 1-cocycle is a map $f:G\to A$ which satisfies $f(gg'')=$ $=gf(g')+f(g)$. Such an f is called a crossed *homorphism.* A crossed homomorphism is a coboundary if there is an a in A such that $f(g)=ga-a$ for all g in G. Show that if G acts trivially on A then $H^1(G,A)=\mathrm{Hom}(G,A)$. A2-cocycle is a map $f:G\times G\to A$ such that

$$gf(g',g'')-f(gg',g'')+f(g,g'g'')-f(g,g')=0.$$

Such an f is called a *system of factors.*

Let H be a subgroup of G and let A be an H-module. Define the induced module I_H^G as the space of maps $f: G\to A$ such that $f(gh)=h^{-1}f(g)$ for all h in H, g in G. G acts on I_H^G by the formula $(gf)(x)=f(g^{-1}x)$. Using the canonical map m: $I_H^G\to A$ given by $mf=g(1)$ we have a homomorphism $I_H^G(A)\to A$. Show that if B is a G-module then m induces an isomorphism $\mathrm{Hom}_G(B,I_H^G(A))\simeq\mathrm{Hom}_H(B,A)$. Show that the related homomorphism

$$H^q(G,I_H^G(A))\simeq H^q(H,A) \tag{*}$$

is an isomorphism. Show that $H^q(G,I_H^G(A))=0$ for $q\geqq 1$.

We leave it to the reader to extend these cohomology groups to locally compact separable groups where $C^n(G,A)$ are Borel functions from G^n to A. In particular the reader should relate the isomorphism (*) in the case $q=1$ and the Mackey imprimitivity theorem.

Chapter 2

Euclidean Group

2.1. The Euclidean group and semidirect products

Let A and H be two groups such that for each h in H there is a map t in $\mathrm{Hom}(H, \mathrm{Aut}(A))$. If we write $t_h(a) = h[a]$ we see that A is an H-space – i.e. $h_1 h_2[a] = h_1[a]h_2[a]$. The product space $G = H \times A$ is made into a group by defining the multiplication law

$$(h, a)(h', a') = (hh', at_h(a')).$$

The identity of G is $(e = e_H, e_A)$ and the inverse is $(h,a)^{-1} = (h^{-1}, h^{-1}[a^{-1}])$.

Definition 2.1.1. G is called the *semidirect product* of H and A with respect to t and is denoted by $G = H \circledS A$.

Elementary topology shows that if A and H are locally compact groups and if $H \times A \to A$ is continuous, then under the product topology G is a locally compact topological group.

Example 2.1.2. The Euclidean group $E(n)$ is the semidirect product of $A = R^n$ and $H = SO(n)$ with $t_h(a) = hah^{-1}$. Note that in this case A is a closed abelian normal subgroup of G and H is a closed subgroup of G with $A \cap H = \{e\}$.

The set $\hat{A}$ of irreducible unitary representations of A in this example is just the character group of A. H acts on A by $H \times \hat{A} \to \hat{A} : (h, \mathfrak{x}) \to \mathfrak{x}(h^{-1}[a])$.

The following construction leads to a theorem of Mackey. For each H-orbit $\mathcal{O}$ in $\hat{A}$ select a point $\mathfrak{x}$ on $\mathcal{O}$ and an irreducible unitary representation V of the stability subgroup $H_{\mathfrak{x}} = \{h \in H | h(\mathfrak{x}) = \mathfrak{x}\}$. Defining $(\mathfrak{x}V)(ah) = \mathfrak{x}(a)V(h)$ on $AH_{\mathfrak{x}}$ we see that $\mathfrak{x}V$ is an irreducible representation of $A \circledS H_{\mathfrak{x}}$. We form the induced representation $U^{\mathfrak{x}V}$.

Theorem (Mackey) 2.1.3. (i) The induced representation $U^{\mathfrak{x}V}$ is irreducible.

(ii) $U^{\mathfrak{x}V}$ is equivalent to $U^{\mathfrak{x}'V'}$ if the orbits of $\mathfrak{x}$ and $\mathfrak{x}'$ coincide.

(iii) If the H-orbit of $\hat{A}$ is smooth (v. Mackey MS1), then each irreducible unitary representation of G is equivalent to some $U^{\mathfrak{x}V}$.

This theorem provides the first example of a characterization of $\hat{G}$ by orbit spaces, which we shall find plays a key role in geometric quantization. Viz. we have the identification $\hat{G} = \hat{A}/ad(H)$.

EXAMPLE 2.1.4. Let $G = R^2 \circledS SO(2) = E(2)$. Here $\hat{R}^2 = \{\mathfrak{x}_a(x) = \exp(iax)\}$. $SO(2)$ acts on $\mathfrak{x}_a$ by taking $\mathfrak{x}_a \to \mathfrak{x}_b$ where $b = ga$. Then $\mathfrak{x}_a$ and $\mathfrak{x}_b$ lie on the same orbit iff $|a|^2 = |b|^2$. The orbits are circles of radius r with $|a| = r$. For each such circle orbit with nonzero radius $H_{\mathfrak{x}_a} = 1$ and $U^{\mathfrak{x}_a}$ is irreducible. $U^{\mathfrak{x}_a}$ are inequivalent for distinct radii of the orbits.

The point orbit of radius zero has $H_{\mathfrak{x}_0} = H$. Thus the induced representations from $A \circledS H_{\mathfrak{x}_0}$ to $A \circledS H$ are just the representations $n, h \to L(h)$ in $\widehat{SO}(2)$.

Since the set $(0, r)$ in R^2 with $r \geqq 0$ is a Borel set, we have an exhaustive list of the irreducible unitary representations of G.

We now present the same example and its covering group from the point of view of Kirillov and then develop the geometry of this example relating it to quantum theory.

EXAMPLE 2.1.5. Let G be the simply connected covering group of $E(2)$. G can be realized as the set $A \times H$ where $A = R$, $H = C$ with the product rule $(t, z)(t', z') = (t + t', \exp(it)z' + z)$. G has the matrix realization given by

$$(t, z) \to \begin{pmatrix} \exp(it) & z \\ 0 & 1 \end{pmatrix}.$$

The centre of G is the set $\{(2\pi n, 0)\}$. In this realization $E(2)$ is given by $g = (t, z)$ where $t \in R/2\pi Z$. Using the matrix realization of $E(2)$ given by $g = (t, z = x + iy)$ acting on $w = (u, v, 1)$ in R^3 where

$$gw = \begin{pmatrix} \cos t & -\sin t & x \\ \sin t & \cos t & y \\ 0 & 0 & 1 \end{pmatrix} \begin{pmatrix} u \\ v \\ 1 \end{pmatrix}$$

we find that the basis of the Lie algebra of $E(2)$ is given by

$$e_1 = (\mathrm{d}g/\mathrm{d}x)|t = x = y = 0, \qquad e_2 = (\mathrm{d}g/\mathrm{d}y)|t = x = y = 0$$
$$e_3 = (\mathrm{d}g/\mathrm{d}t)|t = x = y = 0.$$

These elements satisfy the commutation relations

$$[e_3, e_1] = e_2 \qquad [e_3, e_2] = -e_1$$
$$[e_1, e_2] = 0$$

and is clearly a solvable Lie algebra. The Lie algebra is identified to $R \times C$

by mapping $(t, \mathfrak{z}) \to te_3 + \mathrm{Re}(\mathfrak{z})e_1 + \mathrm{Im}(\mathfrak{z})e_2$ with the product $(t_1, \mathfrak{z}_1)$ $(t_2, \mathfrak{z}_2) = (0, i(t_1\mathfrak{z}_2 - t_2\mathfrak{z}_1))$. We set $\mathfrak{a} = \mathrm{R}e_1 + \mathrm{R}e_2$ and $\mathfrak{b} = \mathrm{R}e_3$. G is the semidirect product of $A = \exp(\mathfrak{a})$ and $H = \exp(\mathfrak{b})$. The dual basis on $\mathfrak{g}^*$ is denoted e_1^*, e_2^*, e_3^*.

Identifying the subspace $\mathfrak{a}$ by $t = 0$ we see that the dual $\mathfrak{g}^*$ of $\mathfrak{g}$ is just C–viz. $f((0, \mathfrak{z})) = \mathrm{Re}(f\mathfrak{z})$. The group G acts on $\mathfrak{a}^*$ by $gf((0, \mathfrak{z})) = f(g^{-1}(0, \mathfrak{z})g)$. This is the first example that we have seen in this book of the coadjoint action of G on $\mathfrak{a}^* \subset \mathfrak{g}^*$:

$$gf(Y) = f((Adg)^{-1}Y).$$

This action will be studied in further detail in Chapter 7. Clearly $gf(Y) = \exp(-it)\, f(Y)$ for $g = (t, z)$. The orbits $\mathcal{O}_f = Gf$ are just circles $|z| = r, r \geqq 0$, which we have seen in the last example.

We can extend the above identification of $\mathfrak{g}^*$ by mapping f in $\mathfrak{g}^*$ to $(s, \mathfrak{z})$ in $R \times C$ by the pairing.

$$f(t, \mathfrak{z}') = \langle (s, \mathfrak{z}), (t, \mathfrak{z}') \rangle = st + \mathrm{Re}(\mathfrak{z}\mathfrak{z}') \qquad \text{where} \qquad (t, \mathfrak{z}) \in \mathfrak{g}.$$

The orbits of the coadjoint action of G on $\mathfrak{g}^*$ are then seen to be cylinders C_r of radius $r > 0$ and the line C_0.

Let $f \in C_r$ – e.g. $f = re_1^*$. Then $\mathfrak{g}(f) = \mathrm{R}e_1$ and $G(f) = 2\pi Z \times \exp(\mathrm{R}e_1)$. Consider the Lie algebras $\mathfrak{h}$ of maximum dimension such that

$$\mathfrak{g}(f) \subset \mathfrak{h} \subset \mathfrak{g}$$

with $f|[\mathfrak{h}, \mathfrak{h}] = 0$. Such an algebra $\mathfrak{h}$ will have dimension $\frac{1}{2}(\dim \mathfrak{g} + \dim \mathfrak{g}(f))$. An algebra with these properties is called a *maximal subordinate subalgebra.* For the case at hand there is only one $\mathfrak{h}$ and it is clearly generated by $e_1 + e_2$. If we let $(0, \mathfrak{z})$, $\mathfrak{z}$ in C, represent a point of $\mathfrak{h}$, then a character of $\exp(\mathfrak{h})$ is defined by f by setting

$$\mathfrak{x}_r(\exp(0, \mathfrak{z})) = \exp(i\, \mathrm{Re}(r\mathfrak{z})).$$

Inducing by $\mathfrak{x}_r$ to $E(2)$ is now given by Mackey's technique which we outlined above. It is left to the reader to check what happens upon induction to G, the covering group of $E(2)$.

This is the first example of Kirillov's method of orbits. For further reference we will call the following a definition, although it is really a philosophy.

DEFINITION 2.1.6. Let f belong to $\mathfrak{g}^*$ and consider the maximal subordinate subalgebras $\mathfrak{h}, \mathfrak{g}(f) \subset \mathfrak{h} \subset \mathfrak{g}$. Let $\mathfrak{x}_f$ be the character of $H = \exp(\mathfrak{h})$

such that $d\mathfrak{x}_f = if$. Form the induced representation $U^{\mathfrak{x}_f,\mathfrak{h}}$. Then the method of orbits states that:

(i) $U^{\mathfrak{x}_f,\mathfrak{h}}$ should be independent of $\mathfrak{h}$–call it U^f.

(ii) U^f belongs to $\hat{G}$.

(iii) Every element of $\hat{G}$ is obtained by this method.

(iv) If $f_1, f_2 \in \mathfrak{g}^*$, then U^{f_1} is equivalent to U^{f_2} iff f_1, f_2 are on the same orbit in $\mathfrak{g}^*$ under the coadjoint action.

(v) thus we have a bijective correspondence between $\mathfrak{g}^*/G \simeq \hat{G}$.

From the above example we see that the Euclidean group satisfies Kirillov's method of orbits. Part of Kirillov's technique of proving that the representation depends only on the orbit involves Kirillov's character formula which we present next for the case at hand.

THEOREM (Kirillov) 2.1.7. Let $\mathcal{O}$ be the orbit in $\mathfrak{g}^*$ given by the cylinder of base $|\mathfrak{z}| = r$. Under the identification of $\mathfrak{g}^*$ with $R \times C$ let $f = (s, \mathfrak{z}) \in \mathfrak{g}^*$ and let $\varphi \in \mathcal{D}(\mathfrak{g})$. Set the fourier transform

$$F\varphi(f) = \int_{\mathfrak{g}} \exp(i\langle f, X\rangle)\varphi(X)\,dX.$$

$d\beta = (1/2\pi)dx\,ds$ determines the measure $\beta(f)$ on $\mathcal{O}$. Then for $r > 0$ we have

$$\begin{aligned}\operatorname{Tr} U^{\mathfrak{x}_r V}(\varphi) &= \int \varphi(X) U^{\mathfrak{x}_r V}(\exp X)\,dX \\ &= \int_{\mathcal{O}} F\varphi(f)\,d\beta(f) + \\ &\quad + \sum_{n\in Z} F\varphi((u+n, 0)) - \int_{-\infty}^{\infty} F\varphi((s, 0))\,ds,\end{aligned}$$

where $\mathfrak{x}_r\, V(2\pi k, z) = \exp(2\pi iku)\exp(i\operatorname{Re}(rz))$.

In particular if φ has its support where $t \neq 2\pi k$, $k \neq 0$, then the trace formula becomes

$$\operatorname{Tr} U^{\mathfrak{x}_r V}(\varphi) = \int_{\mathcal{O}} F\varphi(f)\,d\beta(f)$$

and it depends only on the orbit. We will see later that this is generic for the groups of interest.

COROLLARY 2.1.8.

$$\varphi(0) = \frac{1}{4\pi^2} \int_0^{\infty} \int_{|V|=1} \operatorname{Tr} U^{\mathfrak{x}_r V}(\varphi) r\,dr\,dV.$$

The group G is intimately related to many aspects of quantum mechanics as we now demonstrate.

2.2. Fock space, an introduction

Let $\mathscr{F}$ denote the Hilbert space of entire analytic functions on C with the inner product

$$(f, f') = \int f(z) f'(z)\, \mathrm{d}\mu(z),$$

where $\mathrm{d}\mu(z) = (1/\pi)\exp(-\bar{z}z)\,\mathrm{d}x\,\mathrm{d}y$.

Definition 2.2.1. $\mathscr{F}$ is called the *Fock space*.

Define the map $W: L^2(R) \to \mathscr{F}$ by

$$f(z) = W\psi(z) = \int W(z,q)\psi(q)\,\mathrm{d}q,$$

where $W(z,q) = \pi^{-1/4}\exp(-\tfrac{1}{2}(z^2 + q^2) + \sqrt{2}\, zq)$.

Theorem 2.2.2. *W is a unitary isomorphism of $L^2(R)$ and $\mathscr{F}$.*

The representation of $g = (t, c)$ in G, the simply connected covering group of $E(2)$, *on $\mathscr{F}$* is given as follows: the pure translation component $z \to z + c$ acts by

$$(V(c)f)(z) = \exp(\bar{c}(z - c/2)) f(z - c)$$

and the unitary part $z \to uz$, $u = \exp(it)$, acts by

$$(V(u)f)(z) = f(u^{-1}z);$$

then we set $V(g) = V(c)V(u)$.

Theorem 2.2.3. *$(V, \mathscr{F})$ is a projective representation of G on $\mathscr{F}$; i.e. V is strongly continuous, $\| Vf \| = \| f \|$; and*

$$V(g')V(g) = \mu(g', g)V(g'g),$$

where $\mu(g', g) = \exp(i\,\mathrm{Im}(c'(u'c))$, the projective multiplier, satisfies:

(i) $\mu(e, g) = \mu(g, e) = 1$.

(ii) $|\mu(g, g')| = 1$.

(iii) $\mu(gh, k)\mu(g, h) = \mu(g, hk)\mu(h, k)$.

(iv) μ is a Borel function on $G \times G$.

Writing $V(t)$ for $V(u)$ with $u = \exp(it)$ then $V(t)V(t') = V(t + t')$ and the representation $U(t) = W^{-1}V(t)W$ on $L^2(R)$ satisfies:

(1) $U(\pi) = P$, the parity operator, $(P\psi)(q) = \psi(-q)$ of quantum mechanics.

(2) $F = U(\pi/2)$ is the Fourier transform. In particular F is unitary, $F^2 = P$, and $F^4 = 1$.

If we write $c = (1/\sqrt{2})(x + iy)$ for x, y in R, then

$$T(x, y) = W^{-1}V(c)W$$

is a very interesting representation.

THEOREM 2.2.4. $(T(x, y)\psi)(q) = \exp(-iy(q - x/2))\psi(q - x)$.

Proof. The proof is an elementary check.

COROLLARY 2.2.5. $T(x', y')T(x, y) = \exp(i(yx' - xy')/2)T(x + x', y + y')$.

This representation in the corollary we shall see is just the integrated form of the Weyl commutation relations.

To motivate the geometry to be introduced in the next chapters we consider this example in more geometric detail. Consider the manifold $M = C \times R$ given by $M = \{(\mathfrak{z}, z) \in C^2 | \operatorname{Im} \mathfrak{z} = z^* z/4\}$ and consider the 1-form $\vartheta = \mathrm{d}\sigma + \frac{1}{2}(p\,\mathrm{d}q - q\,\mathrm{d}p) = \mathrm{d}\sigma + (1/4i)(\bar{z}\,\mathrm{d}z - z\,\mathrm{d}\bar{z})$. Then $\mathrm{d}\vartheta = \Omega = (1/2i)\mathrm{d}\bar{z} \wedge \mathrm{d}z$ is a closed $(1, 1)$ form. It is real and nonsingular. Thus it is a Kähler form. It can be written as $\Omega = i\partial\bar{\partial}(\frac{1}{2}|z|^2)$ where $\partial f = (\partial f/\partial z)\mathrm{d}z$ and $\partial^2 = \bar{\partial}^2 = \partial\bar{\partial} + \bar{\partial}\partial = 0$.

There is a complex coordinate $\mathfrak{z}$ such that $\vartheta = \mathrm{d}\sigma + \frac{1}{2}(p\,\mathrm{d}q - q\,\mathrm{d}p) = \mathrm{d}\mathfrak{z} - i\partial(\frac{1}{2}|z|^2)$, viz. $\mathfrak{z} = \sigma + (i/4)|z|^2$. Thus we consider the one form (a contact form) $\vartheta = \mathrm{d}\mathfrak{z} = i\,\partial f(z, z)$ where $\operatorname{Im}(\mathfrak{z}) = \frac{1}{2}f = \frac{1}{4}|z|^2$. $g = (t, c)$ in G acts on M by $g(z) = \exp(it)z + c$ and $g(\mathfrak{z}) = \mathfrak{z} + i(\bar{c}(\exp(-it)z + (c/2))$. Thus G preserves ϑ – i.e. $g_*\vartheta = \vartheta$, as well as Ω. This implies that

$$f(g(z), \overline{g(z)}) = f(z, \bar{z}) + h_g(z) + \overline{h_g(z)},$$

where $h_g(z)$ are holomorphic functions. We note that $\mathrm{d}(g(\mathfrak{z}) - \mathfrak{z}) = i\,\mathrm{d}h_g(z)$.

We define the representation of G on the space $\mathscr{F}$ of holomorphic functions f on C by

$$(V(g)f)(z) = \exp(h_g(z))f(g^{-1}(z)).$$

With the inner product

$$(f_1, f_2) = \frac{1}{\pi}\int_C \exp(-i(\mathfrak{z}^* - \mathfrak{z}))f_1(z)^* f_2(z)\Omega^n$$
$$= \frac{1}{\pi}\int_C \exp(-f(z, \bar{z}))f_1(z)f_2(z)\Omega^n,$$

then this is precisely the unitary representation defined above on the Fock space. The group G acts transitively on the Kähler manifold (C, Ω). A general result of Kobayashi shows that this representation is irreducible.

PROBLEMS

EXERCISE 2.1. Show that if $V(t) = \exp(-it\tilde{H})$ and $U(t) = \exp(-itH)$ then $\tilde{H} = WHW^{-1} = z(\partial/\partial z)$ and $H = \frac{1}{2}(p^2 + q^2 - 1)$ is the Hamiltonian of the harmonic oscillator. Here $Hf = \lim_{t\to 0} it^{-1}(V(t) - 1)f$ for f in $\mathscr{F}$.

EXERCISE 2.2. Show that the infinitesimal generator of the representation of the 1-parameter subgroup $g(t) = tc$ is $V(tc) = \exp(-itL)$ where $L = \bar{c}z - c(\partial/\partial z)$.

EXERCISE 2.3. Use the representation relation

$$V(u)V(c)V(u^{-1}) = V(uc)$$

to show that

$$FqF^{-1} = -p$$
$$FpF^{-1} = q,$$

where $p = -i(\partial/\partial q)$.

EXERCISE 2.3. Let $z = (x^1 + ix^2)/2$ be complex coordinate in R^2. set

$$\partial = \partial/\partial z,\ \bar{\partial} = \partial/\partial \bar{z}, \qquad M_B = z\partial - \bar{z}\bar{\partial},$$

and

$$M_F = z\partial - \bar{z}\bar{\partial} + \tfrac{1}{2}\begin{pmatrix} 1 & 0 \\ 0 & -1 \end{pmatrix}.$$

Show $[\partial, \bar{\partial}] = 0$, $[M_*, \partial] = -\partial$, $[M_*, \bar{\partial}] = \bar{\partial}$;$_* = B$ or F; thus $(\partial, \bar{\partial}, M_*)$ span the Lie algebra of the Euclidean motion group $E(2)$.

Let $\tilde{v}_1(z, \bar{z}) = e^{i1(\vartheta + \pi)}K_1(mr)$ where K_1 is the modified Bessel function of the second kind, $1 \in C$, $z = \frac{1}{2}re^{i\vartheta}$. Show $\tilde{v}_0$ is the fundamental solution of the Euclidean Klein–Gordon equation $(m^2 - \partial\bar{\partial})v = 0$ i.e. $(m^2 - \partial\bar{\partial})\tilde{v}_0(z, \bar{z}) = 2\pi\delta(x^1)\delta(x^2)$.

Chapter 3

Geometry of Symplectic Manifolds

3.1. Elementary Review of Lagrangian and Hamiltonian Mechanics: Notation

Space does not permit a leisurely introduction to the concepts of differential geometry. The reader is assumed to be familiar with the basic ideas. The notation which we have adopted is as follows. For a smooth or C^∞ manifold M we let $A = A(M)$ denote the ring of C^∞ functions on M; $V = V(M)$ denotes the A-module of all C^∞ vector fields on M; A^p denotes the A-module of p-forms on M.

A *vector field* X in V is then a map $X: A \to A$ which is R-linear and is a derivation – i.e. $X(fg) = X(f)g + fX(f)$ for f, g in A. The *Lie bracket* X, Y is defined by

$$[X, Y]f = X(Y(f)) - Y(X(f)) \text{ for } X, Y \text{ in } V.$$

Definition 3.1.1. A *p-form* φ is an A-multilinear map $\varphi: V x \cdots \times V \to A$. The *exterior product* of a p-form and a q-form is the map

$$A^p \times A^q \to A^{p+q}: (\varphi, \psi) \to \varphi \wedge \psi$$

which is A-bilinear, associative, and satisfies

(i) $1 \wedge \varphi = \varphi$.
(ii) if $\varphi_i \in A^1, (\varphi_1 \wedge \ldots \wedge \varphi_p)(X_1, \ldots, X_p) = \det(\varphi_i(X_j))$.
(iii) $\varphi \wedge \psi = (-1)^{pq} \psi \wedge \varphi$ for $\varphi \in A^p$, $\psi \in A^q$.

We often use the shorthand notation

$$\varphi^n = \underbrace{\varphi \wedge \ldots \wedge \varphi}_{n}.$$

A^p has the map

$$d: A^p \to A^{p+1}$$

called *exterior derivative* which is R-linear and satisfies

(i) $d(\varphi \wedge \psi) = d\varphi \wedge \psi + (-1)^p \varphi \wedge d\psi$ for φ in A^p and ψ in A^q. With this property d is said to be an *antiderivation.*

(ii) $d \cdot d = 0$.

(iii) $df(X) = Xf$ for f in A and X in V.

DEFINITION 3.1.2. The *inner product* of vector field X and k-form α is defined by

$$(i(X)\alpha)(X_1,\ldots,X_{k-1}) = \alpha(X, X_1,\ldots,X_{k-1}).$$

Thus $i\colon A^k \to A^{k-1}$ is an A-linear antiderivation which satisfies:

(i) $i(X)i(X) = 0$.

(ii) $i(X)1 = 0$.

(iii) $i(X)\varphi = \varphi(X)$ if $\varphi \in A^1$.

DEFINITION 3.1.3. The *Lie derivative* of α with respect to X is defined by

$$\mathscr{L}(X)\alpha(X_1,\ldots,X_k) = X\alpha(X_1,\ldots,X_k) - \\ - \sum_{i=1}^{k} (X_1,\ldots,[X,X_i],\ldots,X_k).$$

The following identities are easily checked

THEOREM 3.1.4.

(i) $\mathscr{L}$ is R-linear.

(ii) $\mathscr{L}(X)(\varphi \wedge \psi) = \mathscr{L}(X)\varphi \wedge \psi + \varphi \wedge \mathscr{L}(X)\psi$ for φ and ψ in A^p.

(iii) $\mathscr{L}(X)f = X(f)$.

(iv) $\mathscr{L}(X)\,\mathrm{d}f = d(Xf)$.

(v) $i([X, Y]) = \mathscr{L}(X)i(Y) - i(Y)\mathscr{L}(X)$.

(vi) $\mathscr{L}([X, Y]) = [\mathscr{L}(X), \mathscr{L}(Y)]$.

For reference and notation we recall deRham's theorem. For the exterior derivative $d\colon A^k(M) \to A^{k+1}(M)$, let

$$H^k_{\mathrm{dR}}(M, R) = \ker(d)/\mathrm{d}A^{k-1} \qquad \text{for} \qquad k = 0, 1, 2, \ldots.$$

Here $A^{-1} = 0$. Let $H^k(M, R)$ denote the Cech cohomology groups.

THEOREM (deRham) 3.1.5. $H^k_{\mathrm{dR}}(M, R)$ is canonically isomorphic to $H^k(M, R)$.

DEFINITION 3.1.6. We let $d.R$ denote the isomorphism $H^k_{\mathrm{dR}} \to H^k$. The common dimension is called the *Betti number*, $b_k(M) = \dim H^k(M, R)$.

Let $\quad T^k_x(M) = T_x(M) \oplus \cdots \oplus T_x(M)$ and $T^k(M) = \bigcup_{x \in M} T^k_x(M)$.

For a vector space E an E-valued k-form φ on M is a smooth map

$\varphi: T^k(M) \to E, (X_1, \ldots, X_k) \to \varphi(X_1, \ldots, X_k)$ which is multilinear and alternate with respect to $X_1, \ldots X_k$. We denote the space of E-valued k-forms by $A^k(M, E)$.

For a vector space E we let $E^{(k)}$ denote the exterior k-vector space over E. Thus we set $T^{(k)}(M) = \bigcup_{x \in M} T_x^{(k)}(M)$.

DEFINITION 3.1.8. Given a principle bundle $P(M, G)$ and a representation $\rho: G \to GL(E)$, an E-valued k-form φ: $T^k(P) \to E$ on P is said to be a *contravariant k-form on $P(M, G)$ of type (ρ, E)* if $\varphi R_g = \rho(g^{-1})\varphi$ for g in G.

3.2. CONNECTIONS ON PRINCIPAL BUNDLES

Let $G \to M \xrightarrow{\pi} B$ be a G-bundle over B with projection. Let $r(g)$ denote the right translation $r(g)m = mg$. We can view a point m in M as a map $m: G \to G_m: g \to mg$. Here G_m is the fiber of M over $x = \pi(m)$ in B. Let $\mathfrak{g}$ denote the Lie algebra of G. As noted previously $\mathfrak{g} \simeq T_e(G)$.

DEFINITION 3.2.1. Vector field X in $T(M)$ is called *vertical* if $\pi X = 0$.

Thus a vertical vector field X in $T_m(M)$ is given by $X = mY$ where $Y \in \mathfrak{g}$. This provides an injection

$$\lambda: M \times \mathfrak{g} \to A(M)$$

given by $\lambda(m, Y) \to mY$. But since $mg(g^{-1}Yg) = m(Y)g$ for all g in G we have an exact sequence, called the *Atiyah exact sequence*

$$0 \to L(B) \xrightarrow{\lambda} Q(B) \xrightarrow{\pi} T(B) \to 0$$

where $L(B) = B \times_{\mathrm{Ad}(G)} \mathfrak{g}$ and $Q(B) = T(M)/G$.

DEFINITION 3.2.2. A *connection* of $G \to M \to B$ is a 1-form in $A^1(M, \mathfrak{g})$ which satisfies

(i) $\omega \cdot r(g) = ad(g^{-1})\omega$.

(ii) $\omega(mY) = Y$ for m in M and Y in $\mathfrak{g}$.

The *curvature form* of ω is the 2-form Ω in $A^2(M, \mathfrak{g})$ given by

$$\begin{aligned} \Omega(X, Y) &= (\mathrm{d}\omega + \tfrac{1}{2}[\omega, \omega])(X, Y) \\ &= \mathrm{d}\omega(X, Y) + [\omega(X), \omega(Y)] \end{aligned}$$

for X, Y in $T_m(M)$.

THEOREM 3.2.3. The curvature form Ω satisfies
(i) $\Omega \cdot r(g) = ad(g^{-1})\Omega$.
(ii) $\Omega(X, Y) = 0$ if X is vertical.

THEOREM 3.2.4. A connection ω gives a splitting of the Atiyah exact sequence, i.e. $\omega: Q \to L$ such that $\omega \cdot \lambda = 1$.

Forming

$$0 \to \mathrm{Hom}(T, L) \to \mathrm{Hom}(Q, L) \to \mathrm{End}(L) \to 0$$

we have a cohomology sequence

$$\xrightarrow{\pi^*} H^0(B, \mathbf{Hom}(Q, L)) \xrightarrow{\lambda^*} H^0(B, \mathbf{End}(L)) \xrightarrow{\delta^*} H^1(B, \mathbf{Hom}(T, L)).$$

$$\omega \longrightarrow 1 \longrightarrow \delta^*(1)$$

Thus ω is a global section ω in $H^0(B, \mathbf{Hom}(Q, L))$ such that $\lambda^* \omega = 1$.

THEOREM 3.2.5. There exists a connection on M iff $\delta^*(1) = 0$.

COROLLARY 3.2.6. On a smooth principal bundle there always exists a C^∞-connection. On a complex analytic vector bundle the obstruction to the existence of a complex analytic connection is the (1, 1)-component of Ω.

If we set $V(M) = P \times_{\rho(G)} E$ and let $\tilde{A}^k(M, V)$ denote the module of all contravariant k-forms on $P(M, G)$ of type (ρ, E) and if

$$0 \to L(M) \to Q(M) \to T(M) \to 0$$

is the Atiyah sequence of $P(M, G)$; then an E-valued k-form $\vartheta: T^k(P) \to E$ on P is a linear map of vector bundles

$$T^{(k)}(P) \to P \times E, u \to (p, \vartheta(u))$$

where

$$u = \sum X_1 \wedge \ldots \wedge X_k \in T_p^{(k)}(P).$$

If ϑ is a contravariant form on $P(M, G)$ of type (ρ, E) then dividing $T^{(k)}(P)$ and $P \times E$ by G we can regard ϑ as a linear map of vector bundles over M, ϑ: $Q^{(k)}(M) \to V(M)$. Thus we have a natural bijection:

THEOREM 3.2.7. $\tilde{A}^k(M, V) \simeq \Gamma(M, \mathrm{Hom}(Q^{(k)}, V))$.

COROLLARY 3.2.8. If ω is a connection on $P(M, G)$ then the linear map of vector bundles can be regarded as a contravariant 1-form $\omega: T(P) \to \mathfrak{g}$

on $P(M, G)$ of type (ad, $\mathfrak{g}$), where the adjoint representation of G is ad:$G \to GL(\mathfrak{g})$ and is given by ad(x):$\mathfrak{g} \to \mathfrak{g}$, $A \to xAx^{-1}$, x in G.

3.3. Riemannian connections

Let (M, g) be a Riemannian manifold with metric g.

Definition 3.3.1. A *Riemannain covariant derivative* is a map $\nabla: V(M) \to \to \mathrm{End}(V(M))$ which satisfies:

(i) $\nabla_{X+Y}(Z) = \nabla_X Z + \nabla_Y Z$.
(ii) $\nabla_{fX} = f\nabla_X$.
(iii) $\nabla_X(Y+Z) = \nabla_X Y + \nabla_X Z$.
(iv) $\nabla_X(fY) = f\nabla_X Y + X(f)Y$.
(v) $\nabla_X Y - \nabla_Y X = [X, Y]$.
(vi) $X(g(Y,Z)) = g(\nabla_X Y, Z) + g(Y, \nabla_X Z)$ for X, Y, Z in $V(M)$ and f in $A(M)$.

If M has local coordinates $(q^1, \ldots, q^n)$ with vector fields $\partial_i = \partial/\partial q^i i = = 1, \ldots, n$, then we set $\nabla_i = \nabla_{\partial_i}$ and we define the Christoffel symbols Γ^k_{ij} by

$$\nabla_i \partial_j = \sum_k \Gamma^k_{ij} \partial_k.$$

Definition 3.3.2. Let $T = \frac{1}{2}g(\dot{q}(t), \dot{q}(t))$ be the kinetic energy and $V(q)$ be the potential energy; then the *Lagrange equations* are given by

$$\frac{\mathrm{d}}{\mathrm{d}t}\left(\frac{\partial(T-V)}{\partial \dot{q}^i}\right) - \frac{\partial(T-V)}{\partial q^i} = 0, \qquad i = 1, \ldots, n.$$

The curve $t \to q(t) = (q^1(t), \ldots, q^n(t))$ is called the *geodesic*.

Using $T = \frac{1}{2}\sum g_{ij}(q)\dot{q}^i \dot{q}^j$ we see that the geodesic equations are $\ddot{q}^l + \Gamma^l_{kj}\dot{q}^k \dot{q}^l = -g^{lk}\partial_k V$, $l = i, \ldots, n$; or in local coordinates of the tangent bundle $(q^1, \ldots, q^n, \dot{v}^1, \ldots, \dot{v}^n)$

$$\begin{aligned} \dot{q}^l &= v^l \\ \dot{v}^l &= -\Gamma^l_{kj} v^k v^l - g^{lk}\partial_k V. \end{aligned}$$

The Hamiltonian form of the geodesic equations is given by passing to the cotangent bundle via

$$(\dot{q}^k, v^k) \to (\dot{q}^k, p_k = \sum g_{kl} v^l).$$

Then the geodesic equations become

$$\begin{aligned} \dot{q}^k &= \sum g^{kl} p_l \\ \dot{p}_l &= -\tfrac{1}{2}\sum \partial_l g^{ab} p_a p_b - \partial_l V, \end{aligned}$$

or

$$\dot{q} = \partial H/\partial p$$
$$\dot{p} = -\partial H/\partial q$$

for the Hamiltonian $H = \frac{1}{2}\sum g^{ab}(q)p_a p_b + V(q)$.

EXAMPLE 3.3.3. The harmonic oscillator has Lagrangian form

$$\ddot{q} + v^2 q = 0$$

and Hamiltonian form

$$\dot{q} = p$$
$$\dot{p} = -v^2 q,$$

where $H = \frac{1}{2}(p^2 + v^2 q^2)$. (Here the symplectic form is $\Omega = \mathrm{d}p \wedge \mathrm{d}q$. For the definition see Section 3.4.)

EXAMPLE 3.3.4. The geodesic flow on the Poincaré upper half plane $\mathscr{P}$ is given as follows. $\mathscr{P}$ is an example of a Clifford–Klein surface with constant curvature $K < 0$. Selecting $K = -1/r^2$ the metric is $\mathrm{d}s^2 = (\mathrm{d}q^1)^2 + \exp(2q^1/r)(\mathrm{d}q^2)^2$. The substitution $x = q^2$ and $y = r\exp(-q^1/r)$ gives the metric

$$\mathrm{d}s^2 = r^2\left(\frac{\mathrm{d}x^2 + \mathrm{d}y^2}{y^2}\right), \qquad y > 0.$$

For convenience we set $r = 1$. Thus

$$(g_{ab}) = \begin{pmatrix} 1/y^2 & 0 \\ 0 & 1/y^2 \end{pmatrix}.$$

The Christoffel symbols are easily checked to be

$$\Gamma^1_{11} = \Gamma^1_{22} = \Gamma^2_{12} = \Gamma^2_{21} = 0, \Gamma^2_{11} = 1/y$$

and

$$\Gamma^1_{12} = \Gamma^1_{21} = \Gamma^2_{22} = -1/y.$$

The geodesic equations in coordinates (x, y) are thus

$$\ddot{x} - \frac{2}{y}\dot{x}\dot{y} = 0$$

$$\ddot{y} + \frac{1}{y}(\dot{x}^2 - \dot{y}^2) = 0.$$

The solutions to these differential equations are

$$x(t) = a\tanh(t) + b$$
$$y(t) = a/\cosh(t)$$

or $x(t) = b$, $y(t) = \exp(t)$ – i.e. semicircles with center $(b, 0)$ and radii a or lines parallel to the y axis.

The Hamiltonian of this system is $H = (1/2y^2)(p_1^2 + p_2^2)$. To study the associated quantum mechanical system there must be a set of rules to replace the Hamiltonian on an arbitrary Riemannian manifold by a Schrödinger equation. The early work in quantum theory proposed the replacement:

$$-\sum g^{ab} p_a p_b \to \Delta,$$

where Δ is the Laplace–Beltrami operator

$$\Delta = \frac{1}{\sqrt{g}} \sum_{a,b} \partial_a [g^{ab} \sqrt{g}\, \partial_b].$$

For the geodesic flow on the Poincaré upper half plane the Schrödinger equation would then be

$$\frac{y^2}{2}\left(\frac{\partial^2}{\partial x^2} + \frac{\partial^2}{\partial y^2}\right)\psi_\lambda = E_\lambda \psi_\lambda.$$

Other derivations have suggested

$$\left(-\frac{\Delta}{2} + \frac{R}{6}\right)\psi_\lambda = E_\lambda \psi_\lambda,$$

where R is the scalar curvature; or replacing $\hbar$ appropriately the time dependent Schrödinger equation would be

$$i\hbar \frac{\partial \psi}{\partial t} = \frac{-\hbar^2 \Delta}{2}\psi + \frac{+\hbar^2 R}{6}\psi.$$

Example 3.3.5. Consider the Hamiltonian $H = \frac{1}{2}(p^2 + \exp(-2q))$. Then the Hamiltonian equations are

$$\dot{p} = -\exp(-2q)$$
$$\dot{q} = p,$$

which admit a solution $q(t) = \ln(\cosh t)$. This example is the most primitive Toda system. These systems will be studied in Chapter 7.

3.4. Geometry of Symplectic Manifolds

Definition 3.4.0. Let M^{2n} be a smooth manifold. If there is a tensor field J of type (1, 1) on M such that $J^2 = -I$, then M is said to be an *almost complex manifold*.

We will often use the following result:

Theorem 3.4.1. Let X be a vector field on M and let α be a k-form. If φ_t is the local 1-parameter group of diffeomorphisms generated by X, then

$$\frac{\mathrm{d}}{\mathrm{d}t}(\varphi_t^*\alpha) = \varphi_t^*(\mathscr{L}_X(\alpha)).$$

Definition 3.4.2. A manifold M is called an *almost symplectic* (AS) manifold if there is a 2-form Ω on M such that $\Omega^n \neq 0$.

Thus an AS manifold has a volume form $\tilde{v} = \Omega^n/n!$ and is orientable.

Definition 3.4.3. An *AS* manifold (M, Ω) is called *symplectic* if Ω is closed, $\mathrm{d}\Omega = 0$.

Example 3.4.4. The basic example is R^{2n} with $\Omega = \mathrm{d}\omega = \sum \mathrm{d}p_k \wedge \mathrm{d}q^k$. Clearly $\mathrm{d}\Omega = 0$ and $\Omega^n \neq 0$, since $\Omega^n = c\,\mathrm{d}q^1 \wedge \ldots \wedge \mathrm{d}q^n \wedge \mathrm{d}p_1 \wedge \ldots \wedge \mathrm{d}p_n$ is proportional to the standard volume form

Every smooth manifold with an almost complex structure admits a positive definite Riemannian metric such that $g(JX, JY) = g(X, Y)$ and the almost complex manifold with this metric is said to have *almost Hermitian structure*. If (M, g, Ω, J) is a symplectic manifold with almost Hermitian structure such that $g(X, Y) = \Omega(X, JY)$, then M is said to be an *almost Kähler manifold*.

Consider now the vector field $X = (\partial H/\partial p, -\partial H/\partial q)$ in local coordinates (q, p). Since $\mathscr{L}(X)p = -\partial H/\partial q$ and $\mathscr{L}(X)q = \partial H/\partial p$ we see that

$$\begin{aligned}\mathscr{L}(X)\Omega &= \mathrm{d}\left(-\frac{\partial H}{\partial q}\right) \wedge \mathrm{d}q + \mathrm{d}p \wedge \mathrm{d}\left(\frac{\partial H}{\partial p}\right) = \\ &= \left(\frac{-\partial^2 H}{\partial q \partial p}\mathrm{d}p + \frac{\partial^2 H}{\partial q^2}\mathrm{d}q\right) \wedge \mathrm{d}q + \\ &\quad + \mathrm{d}p \wedge \left(\frac{\partial^2 H}{\partial p^2}\mathrm{d}p + \frac{\partial^2 H}{\partial p \partial q}\mathrm{d}q\right) = 0.\end{aligned}$$

Thus we have

THEOREM (Liouville) 3.4.5. The volume element $\tilde{v}$ is invariant under the Hamiltonian vector field X, i.e. $\mathscr{L}(X)\tilde{v}=0$.

DEFINITION 3.4.6. If (M,Ω) is a symplectic manifold, X in $V(M)$ is called *symplectic vector field* if $\mathscr{L}(X)\Omega=0$. Let $\operatorname{symp}(M,\Omega)$ denote the set of symplectic vector fields on (M,Ω).

Since $\mathscr{L}([X,Y])=[\mathscr{L}(X),\mathscr{L}(Y)]$ we see that

THEOREM 3.4.7. $\operatorname{symp}(M,\Omega)$ is a Lie algebra.

DEFINITION 3.4.8. A diffeomorphism $f\colon M\to M$ of a symplectic manifold (M,Ω) onto itself is called a *symplectomorphism* if $f^*\Omega=\Omega$.

Clearly if $X\in\operatorname{symp}(M,\Omega)$ generates a 1-parameter group φ_t, then

$$\frac{\mathrm{d}}{\mathrm{d}t}\varphi_t^*\Omega=\varphi^*(\mathscr{L}(X)\Omega)=0$$

Thus $\varphi_t^*\Omega=\varphi_0^*\Omega=\Omega$.

We return now to the general case of almost symplectic manifolds.

THEOREM 3.4.9. If (M,Ω) is an AS manifold then there exist uniquely defined maps

$$p\colon A^1(M)\to V(M)\colon\varphi\to p(\varphi)$$
$$P\colon A(M)\to V(M)\colon f\to P(f)$$

given by

$$p(\varphi)g\Omega^n=n\varphi\wedge\mathrm{d}g\wedge\Omega^{n-1}$$
$$P(f)g\Omega^n=n\,\mathrm{d}f\wedge\mathrm{d}g\wedge\Omega^{n-1}.$$

Proof. This is trivial since $\Omega^n\neq 0$ and $p(\varphi)$ and $P(f)$ satisfy the axioms of vector fields.

DEFINITION 3.4.10. p, P are called *Poisson brackets* of the AS manifold (M,Ω).

THEOREM 3.4.11. Let p,P be Poisson brackets of AS manifold (M,Ω). Then

(i) $P(f)=p(\mathrm{d}f)$.

(ii) $P\colon A^1\to V$ is A-lincar.

(iii) $P: A \to V$ is R-linear.

(iv) $P(f)g = -P(g)f$.

THEOREM 3.4.12. If (M, Ω) is an AS manifold then p, P are characterized uniquely as follows:

(i) $X = p(\varphi)$ for φ in A^1 iff $i(X)\Omega = -\varphi$.

(ii) $X = P(f)$ iff $i(X)\Omega = -\mathrm{d}f$.

The proofs of these two theorems are elementary and are left as exercises.

COROLLARY 3.4.13. $p: A^1 \to V$ is a bijection with inverse $\mu(X) = p^{-1}(X) = -i(X)\Omega$ for X in $V(M)$.

We extend μ to an isomorphism of the module of p-contravariant tensors A_p onto the module of p-covariant tensors A^p.

THEOREM 3.4.14. If φ_1, φ_2 are p-forms we have

$$i(\mu^{-1}(\varphi_1))\varphi_2 = (-1)^p i(\mu^{-1}(\varphi_2))\varphi_1.$$

DEFINITION 3.4.15. The *symplectic adjoint* operator $\tilde{*}: A^p \to A^{2n-p}$ is given by

$$\tilde{*}\varphi = i(\mu^{-1}(\varphi))\tilde{v}.$$

Here $2n = \dim(M)$.

THEOREM 3.4.16. (i) $\tilde{*}^2 = Id$.

(ii) $\tilde{*}(\varphi \wedge \psi) = i(\mu^{-1}(\psi))(\tilde{*}\varphi)$.

Proof. (i) is clear and (ii) follows by

$$\begin{aligned} i(\mu^{-1}(\varphi \wedge \psi))\tilde{v} &= i(\mu^{-1}(\varphi) \wedge \mu^{-1}(\psi))\tilde{v} \\ &= i(\mu^{-1}(\psi))i(\mu^{-1}(\varphi))\tilde{v}. \end{aligned}$$

DEFINITION 3.4.17. Define L: $A^p \to A^{p+2}$ by $L(\varphi) = \Omega \wedge \varphi$. And set $\tilde{\Lambda} = \tilde{*}^{-1} L \tilde{*} = \tilde{*} L \tilde{*}$.

THEOREM 3.4.18. $\tilde{\Lambda}\varphi = i(\mu^{-1}(\Omega))\varphi$.

Locally the 2-form $\Omega = \sum \vartheta^\alpha \wedge \varphi^{\alpha+n}$ and $\mathrm{d}s^2 = 2\sum \vartheta^\alpha \vartheta^{\alpha+n}$ defines a metric. This leads to

THEOREM 3.4.19. If M is an AS manifold then there is a metric ds^2 which is defined by Ω so that M is an almost complex manifold; let J denote the almost complex structure. Then (M, Ω, J) is an almost Hermitian structure.

DEFINITION 3.4.20. J defines the map $C: A^p \to A^p$ given by

$$C\varphi(X_1, \ldots, X_p) = \varphi(JX_1, \ldots, JX_p).$$

THEOREM 3.4.21.

(i) $C\Omega = \Omega$.

(ii) $C(\varphi_1 \wedge \varphi_2) = C\varphi_1 \wedge C\varphi_2$.

(iii) $C(\varphi_1 + \varphi_2) = C(\varphi_1) + C(\varphi_2)$.

Given an almost Hermitian manifold, the metric ds^2 defines a metric adjoint $*: A^p \to A^{2n-p}$ (v. H22) which has the properties

(i) $*^2 = (-1)^p Id$.

(ii) $*(a_1\varphi_1 + a_2\varphi_2) = \bar{a}_1 * \varphi + \bar{a}_2 * \varphi_2$.

(iii) $*\varphi_1 \wedge \varphi_2 = \varphi_2 \wedge *\varphi_1$.

(iv) $[C, *] = 0$.

(v) If we set $\Lambda = *L*$, then $\Lambda = i(\mu^{-1}(\Omega))$ and $[C, \Lambda] = 0$.

The almost Hermitian and almost symplectic properties can be related as follows:

THEOREM 3.4.22.

(i) $* = (-1)^{[n(n-1)]/2} C^{-1} *$.

(ii) $\tilde{*}^2 = \tilde{*}^2 C^{-2}$.

(iii) Since $\Lambda = \tilde{*} L \tilde{*} = *CLC^{-1}*$ and $[C, L] = 0$ we have $\tilde{\Lambda} = *L* = \Lambda$. Thus Λ depends only on the AS structure.

The scalar product of two p-forms φ_1, φ_2 is defined by

$$(\varphi_1, \varphi_2) = *(\varphi_1 \wedge *\varphi_2).$$

One checks easily that

$$(\varphi_1, \varphi_2) = \overline{(\varphi_2, \varphi_1)}$$
$$(a\varphi_1 + b\varphi_2, \varphi_3) = \bar{a}(\varphi_1, \varphi_3) + \bar{b}(\varphi_2, \varphi_3).$$
$$(\varphi, \varphi) \neq 0 \text{ for } \varphi \neq 0.$$

With respect to (,) we have $(L\varphi, \psi) = (\varphi, \Lambda\psi)$ where φ is a p-form and ψ is a $(p+2)$-form. The scalar product of deRham is then formed by

$$(\varphi_1, \varphi_2)_{\mathrm{dR}} = \int_M *(\varphi_1, \varphi_2).$$

Assume now that (M, Ω) is a symplectic manifold. Then the *symplectic codifferential* $\tilde{\delta}$ is defined on p-form φ by

$$\tilde{\delta}\varphi = (-1)^p \tilde{*} d \tilde{*} \varphi.$$

THEOREM 3.4.23.

(i) $\tilde{\delta}^2 = 0.$

(ii) $d\Lambda - \Lambda d = -\tilde{\delta}.$

(*iii*) $\tilde{\delta}L - L\tilde{\delta} = -d.$

For a Kähler manifold the codifferential is defined by

$$\delta = (-1)^p *^{-1} d *.$$

One easily checks that δ and $\tilde{\delta}$ are related by

$$\tilde{\delta} = -C^{-1}\delta C.$$

The Laplacian on p-forms is defined by $\Delta = d\delta + \delta d$. If $\mathscr{H}^p = \{p\text{-form } \varphi \mid \Delta\varphi = 0\}$ we have the Hodge isomorphism

$$H^p{}_{\mathrm{dR}} \simeq \mathscr{H}^p.$$

THEOREM 3.4.24. Let (M, Ω) be a symplectic manifold. If $X = p(\varphi)$ then $\mathscr{L}(X)\Omega = -\mathrm{d}\varphi$.

Proof. Since $\mathrm{d}\Omega = 0$ and $i(X)\Omega = -\varphi$, we have $\mathscr{L}(X)\Omega = i(X)\mathrm{d}\Omega + \mathrm{d}i(X)\Omega = -\mathrm{d}\varphi$.

THEOREM 3.4.25. If $Z^1(M)$ denotes the R-module of all closed 1-forms on the symplectic manifold (M, Ω), then p gives a bijection between the R-modules Z^1 and symp.

Proof. p is a bijection and from the last theorem X is in symp iff $\mathrm{d}\varphi = 0$.

The map μ allows us to define a Lie algebra structure on $Z^1(M)$; viz.

$$[\varphi, \psi] = \mu[p(\varphi), p(\psi)],$$

or equivalently by setting $X = p(\varphi)$, $Y = p(\psi)$ we have

$$[\varphi, \psi] = i[X, Y]\Omega.$$

By 3.1.4 we have $i[X,Y]\Omega=\mathscr{L}(X)i(Y)\Omega=\mathrm{d}i(X)i(Y)\Omega$. Thus Z^1 has the structure of a Lie algebra; and the space of exact 1-forms $\mathrm{d}A^0$, is an ideal with

$$[Z^1, Z^1]\subset \mathrm{d}A^0.$$

From the bijection given by p we see that if φ is exact, say $\varphi=\mathrm{d}f$, then we have a naturally associated vector field $p(\varphi)=X_f$ to f. Of course, if the first Betti number, b_1 (M), is zero, then φ is always exact.

DEFINITION 3.4.26. X_f is called the *Hamiltonian vector field* associated to f in $A(M)$.

Clearly X_f is in symp (M,Ω).

DEFINITION 3.4.27. Let ham $(M,\Omega)=\{X_f \mid f \text{ in } A(M)\}$.

THEOREM 3.4.28. [symp (M), symp (M)] $\subset$ ham (M).

Let $B=p(Z^1(M))$ and $B^*=p(\mathrm{d}A(M))$. Then by deRham's theorem we have

THEOREM 3.4.29. $B/B^*\simeq H^1(M,R)$.

DEFINITION 3.4.30. Let B^c denote the Lie algebra of *conformal symplectic transformations* – i.e. X such that $\mathscr{L}(X)\Omega=-K_X\Omega$, where K_X is a constant.

B and B^* are ideals of B^c and we have $K_X\Omega=-\mathrm{d}\mu(X)$. Thus if Ω is not exact (e.g. if M is compact) $K_X=0$ for every X in B^c so $B^c=B$.

If Ω is exact, set $\Omega=\mathrm{d}\omega(X_0)$ for X_0 in B^c with $K_{X_0}=1$. Then B^c is the direct sum $B^c=B+C_0$ where C_0 is the one dimensional subspace generated by X_0. Thus we have

THEOREM 3.4.31. If Ω is exact, $B^c=B+C_0$ and $\dim B^c/B=1$.

DEFINITION 3.4.32. A Lie algebra (of finite of infinite dimension) is called *semisimple* if it admits no nonzero abelian ideals.

THEOREM 3.4.33. B^c, B, and B^* are semisimple.

DEFINITION 3.4.34. A *derivation of a Lie algebra* $\mathfrak{g}$ is a linear map

$$D:\mathfrak{g}\to\mathfrak{g} \text{ satisfying } D[u,v]=[Du,v]+[u,Dv].$$

D is said to be *inner* if there exists an X in L such that $DY = [X, Y]$ for all Y in $\mathfrak{g}$.

THEOREM 3.4.35. All derivations of B^c, B, and B^* are infinitesimal conformal symplectic transformations of the form $Y \to [X, Y]$ where X in B^c.

We briefly review the concepts of Lie algebra cohomology which we need for the following results. Let $\mathfrak{g}$ be a Lie algebra and let $X \to \rho(X)$ be a representation of $\mathfrak{g}$ on $GL(E)$. An *n-cochain* is a skew symmetric multilinear map

$$\mathfrak{g} \times \cdots \times \mathfrak{g} \to E$$
$$X_1, \ldots, X_n \to \omega(X_1, \ldots, X_n) \in E.$$

The vector space of n-cochains is denoted $C^n(\mathfrak{g}, E)$; here $C^0 = E$. The coboundary operator is defined by

$$\mathrm{d}\omega(X_1, \ldots, X_{n+1}) = \sum_{i=1}^{n+1} (-1)^{n+1-i} \rho(X_i)(\omega(X_1, \ldots, \hat{X}_i, \ldots, X_{n+1}))$$
$$+ \sum_{1 \leqq i, j \leqq n} (-1)^{i+j} \omega(X_1, \ldots, \hat{X}_i, \ldots \hat{X}_j \ldots, [X_i, X_j])$$

where $\hat{}$ means that argument is omitted.

E.g. for $n = 0$ we have $\mathrm{d}\omega(X) = \rho(X)(\omega)$. For $n = 1$ we have

$$\mathrm{d}\omega(X_1, X_2) = \rho(X_1)(\omega(X_2)) - \rho(X_2)(\omega(X_1) - \omega([X_1, X_2])).$$

For $\omega \in C^n$ we have the contraction map i: $C^n \to C^{n-1}$ given by

$$i(X)\omega(X, \ldots, X_{n-1}) = \omega(X, X_1, \ldots, X_{n-1}).$$

The Lie derivative of ω by X is

$$\mathscr{L}(X)\omega(X_1, \ldots, X_n) = \rho(X)(\omega(X_1, \ldots, X_n)) -$$
$$- \omega([X, X_1], X_2, \ldots, X_n) - \omega(X_1, \ldots [X, X_n]).$$

We leave it to the reader to check that

THEOREM 3.4.36.

(1) $\mathscr{L}(X)\omega = i(X)\mathrm{d}\omega + \mathrm{d}(i(X(\omega)))$.

(2) $\mathrm{d}(\mathrm{d}\omega) = 0$ for ω in C^n.

As usual ω in C^n is said to be a *cocycle* if $\mathrm{d}\omega = 0$; the vector space of cocycles is denoted Z^n. ω is a *coboundary* if $\omega \in \mathrm{d}(C^{n-1})$; the vector space of coboundaries is denoted B^n. We set $C^{-1} = 0$. Clearly $B^n \subset Z^n$. The quotient vector space $H^n = Z^n/B^n$ is defined to be the nth *cohomology group* of $(\mathfrak{g}, \rho)$.

The elements ω in Z^0 satisfy $0 = \mathrm{d}\omega(X) = \rho(X)(\omega)$ for all X in $\mathfrak{g}$. Thus $H^0 = Z^0$ is the set of vectors in E that are annihilated by $\rho(\mathfrak{g})$.

Consider now the case $E = \mathfrak{g}$; then

$$\mathrm{d}\omega(X_0, \ldots, X_p) = \sum_{k=0}^{p} (-1)^k [X_k, \omega(X_0, \ldots, \hat{X}_k, \ldots X_p)] +$$

$$+ \sum_{k<l} (-1)^{k+l} \omega([X_k, X_l], X_0, \ldots, \hat{X}_k, \ldots, \hat{X}_l, \ldots, X_p).$$

If $\omega = \mathrm{d}Y$ for Y in $\mathfrak{g}$, then $\omega(X) = [X, Y]$. If ω is a 1-cochain then

$$\mathrm{d}\omega(X, Y) = [X, \omega(Y)] + [\omega(X), Y] - \omega([X, Y]).$$

Thus the space of closed 1-cochains of $\mathfrak{g}$ coincide with the space of derivations of $\mathfrak{g}$. The exact cochains are precisely the inner derivations. From the results above we see that the space of derivations of B is B^c and the space of inner derivations is B. Thus $H^1(B) = B^c/B$. By deRham's theorem $H^1(M; R)$ is isomorphic B/B^* (Theorem 3.4.29). These facts combine to give the following two results:

COROLLARY 3.4.36. $H^1(B^c) = 0$ and $H^1(B) = B^c/B$.

THEOREM 3.4.37. If Ω is not exact (in particular if M is compact) $\dim H^1(B) = 0$ and $\dim H^1(B^*) = b_1(M)$; while if Ω is exact, $\dim H^1(B) = 1$ and $\dim H^1(B^*) = b_1(M) + 1$.

THEOREM (Darboux) 3.4.38. For any point m in symplectic manifold (M, Ω) there is an open neighborhood U of m in M and local coordinates $(q_1, \ldots, q_n, p, \ldots, p^n)$ on U such that

$$\Omega | U = \sum \mathrm{d}p^i \wedge \mathrm{d}q_i.$$

In the Darboux coordinates we have

$$P(f)g = \{f, g\} = \sum \frac{\partial f}{\partial p^i}\frac{\partial g}{\partial q_i} - \frac{\partial g}{\partial p^i}\frac{\partial f}{\partial q^i}.$$

3.5. Classical mechanics and symmetry groups

Definition 3.5.1. Let Y be a vector field on symplectic manifold (M, Ω, H). Y is said to be an *infinitesimal symmetry* if

$$\mathscr{L}(Y)\Omega = 0 \quad \text{and} \quad \mathscr{L}(Y)(X_H) = [Y, X_H] = 0.$$

Let φ_t be the 1-parameter group associated to the infinitesimal symmetry Y. Then clearly $\varphi_t^* \Omega = \Omega$ and $\varphi_t(X_H) = X_H$.

Definition 3.5.2. If φ_t satisfies these conditions we say that φ_t is a *symmetry* of the Hamiltonian system.

Let h_Y be the local Hamiltonian function associated to Y. Then $\mu([X_H, Y]) = \mathrm{d}(\{H, h_Y\}) = 0$.

Definition 3.5.3. h is an *integral of motion* if $\{h, H\} = 0$.

Clearly then the energy H is an integral of motion.

If we take the Hilbert space $L^2(M, \tilde{v})$, then the volume preserving flow g_t on M induces a 1-parameter group $U_t f(m) = f(g_{-t} m)$ of unitary operators. And by Stone's theorem there is a self-adjoint operator H such that $U = \exp(-itH)$.

Similarly in quantum mechanics the dynamics is given by a 1-parameter group $V(t) = \exp(-itH)$. And other observables are specified by self adjoint operators A.

Definition 3.5.4. A quantum observable A is said to be an *integral of motion* if $V(t)\exp(-isA) = \exp(-isA)V(t)$ for all s, t.

Taking the derivative of this expression we have formally

Theorem 3.5.5. If A is an integral of motion, $[H, A] = 0$.

Corollary 3.5.6. $[H, H] = 0$ – i.e. energy is always conserved during motion.

We can relate this discussion to unitary representations as follows. Given a unitary representation U of G then U induces a representation of the Lie algebra $\mathfrak{g}$ of G by

$$\dot{U}(X) = \frac{\mathrm{d}U}{\mathrm{d}t}(\exp tX)|_{t=0}$$

for X in $\mathfrak{g}$. Then G invariance under U of the Hamiltonian implies that

$$U(\exp tX)H = HU(\exp tX).$$

Upon differentiation we have formally

$$[\dot{U}(X), H] = 0.$$

Thus formally $\dot{U}(X)$ for X in $\mathfrak{g}$ are integrals of motion. E.g. if H is invariant under $SO(3)$, then the angular momentum operators are integrals of motion. Finer analysis is required to make this rigorous, esp. regarding the various domains. See Helgason H11.

3.6. Homogeneous symplectic manifolds

Definition 3.6.1. Let G be a Lie group acting on a symplectic manifold (M, Ω) by symplectomorphisms. In this case we say (G, M, Ω) is a *symplectic G-space*. If G acts transitively, it is called a *homogeneous symplectic G-space*.

We define the map $\sigma\colon \mathfrak{g} \to V(M)$ by

$$(\sigma(X)f(m)) = \frac{\mathrm{d}f}{\mathrm{d}t}(\exp tX.m)|_{t=0}.$$

Definition 3.6.2. A symplectic G-space is called *strongly symplectic* if $\sigma(\mathfrak{g}) \subset \mathrm{ham}\,(M, \Omega)$.

Definition 3.6.3. If (M, Ω) is a strongly symplectic G-space then a *Lie* (resp. *smooth*) *lift* of σ is a Lie algebra homomorphism (resp. smooth map) $\lambda\colon \mathfrak{g} \to A(M)$ such that

$$\begin{array}{ccccccccc} 0 & \to & R & \to & A(M) & \xrightarrow{p} & \mathrm{ham}\,(M, \Omega) & \to & 0 \\ & & & & {\scriptstyle\lambda}\nwarrow & & \uparrow{\scriptstyle -\sigma} & & \\ & & & & & & \mathfrak{g} & & \end{array}$$

commutes.

Consider the sequence

$$\begin{array}{ccccccc} H^0(M, R) & \xrightarrow{p_0} & A(M) & \xrightarrow{p} & \mathrm{symp}\,(M, \Omega) & \xrightarrow{p_1} & H^1(M, R) \to 0. \\ & & {\scriptstyle\lambda}\nwarrow & & \uparrow{\scriptstyle -\sigma} & & \\ & & & & \mathfrak{g} & & \end{array} \qquad (*)$$

Since $[\mathrm{symp}, \mathrm{symp}] \subset p_0(H^0(M, R))$, p_1 is a Lie algebra homomorphism by taking the abelian Lie algebra on $H^1(M, R)$. And taking the abelian Lie algebra structure on $H^0(M, R)$, p_0 becomes a Lie algebra homomorphism. Thus $(*)$ is an exact sequence of Lie algebras.

The smooth lift λ in $(*)$ exists if $-\sigma(\mathfrak{g}) \subset p(A(M))$, i.e. when $p_1 \cdot p = 0$.

However, $p_1 \cdot p$ is a Lie algebra homomorphism and $H^1(M, R)$ is Abelian so $p_1 \cdot p$ annihilates $[\mathfrak{g}, \mathfrak{g}]$. If $\bar{p}$ denotes the factored map $\mathfrak{g}/[\mathfrak{g}, \mathfrak{g}] \to H^1(M, R)$, the vanishing of $\bar{p}$ is necessary and sufficient for the smooth lifting of σ. Clearly (a) if $\mathfrak{g}$ is semisimple, a smooth lift exists; (b) if $\Omega = \mathrm{d}\omega$ and ω is invariant under $\mathfrak{g}(\mathscr{L}\sigma(\mathfrak{g})\omega = 0)$, then there is a smooth lift $\lambda(X) = -\omega(X)$. Finally (c) if $H^1(M, R) = 0$, σ admits a smooth lift.

EXAMPLE 3.6.4. Consider the Euclidean group E(2). The Lie algebra of E(2) is the set of matrices

$$r(a, b, c) = \begin{pmatrix} 0 & -c & a \\ c & 0 & b \\ 0 & 0 & 0 \end{pmatrix}$$

for a, b, c, in R. The Lie algebra has a basis $I_1 = r(1,0,0)$, $I_2 = r(0,1,0)$ and $I_3 = r(0, 0, 1)$ with commutation relations $[I_1, I_2] = 0$, $[I_3, I_1] = I_2$ and $[I_3, I_2] = -I_1$. Consider the 1-parameter subgroups $g_k(t) = \exp(tI_k)$, $k = 1, 2, 3$. Then under the symplectic form $\Omega = \mathrm{d}p \wedge \mathrm{d}q$, $g_k(t)$ generate Hamiltonian vector fields since $g_k(t)^*\Omega = \Omega$. Viz., $\sigma(I_1) = -\partial/\partial p$, $\sigma(I_2) = -\partial/\partial q$, and $\sigma(I_3) = p\partial/\partial q - q\partial/\partial p$. Clearly $\mathscr{L}(\sigma(I_k))\Omega = 0$. Noting that $\Omega = \mathrm{d}\omega$ we see that there is a map λ given by $\lambda(X) = -\omega(\sigma(X))$. Taking $\omega = \frac{1}{2}(p\mathrm{d}q - q\mathrm{d}p)$, then we have $\lambda(I_1) = q/2, \lambda(I_2) = -p/2$ and $\lambda(I_3) = H = \frac{1}{2}(p^2 + q^2)$.

To analyze when the lift is a Lie algebra homomorphism we use the concept of *generalized momentum* due to Souriau.

DEFINITION 3.4.5. Let (M, Ω) be a strongly symplectic G-space with a smooth lift λ. Define the map $s: M \to \mathfrak{g}^*$ by $s(m)(X) = \lambda(X)(m)$ for X in $\mathfrak{g}$.

The map $s(X_1, X_2) = s([X_1, X_2]) - \{s(X_1), s(X_2)\}$ is a skew-symmetric bilinear map from $\mathfrak{g} \times \mathfrak{g}$ into $A(M)$ with $p \cdot s = 0$. s also satisfies $\sum_{\text{cyclic}} s([X, Y], Z) = 0$ i.e. s is a 2-cocycle.

THEOREM 3.6.6. There is a Lie lift to σ iff s is a coboundary.

The obstruction to this lift lies in $H^2(\mathfrak{g}, R) \otimes H^0(M, R)$. One immediately checks that the three cases (a), (b), and (c) cited above also have Lie lifts.

DEFINITION 3.6.7. If (M, Ω) is a strongly symplectic G-space with a Lie lift σ, then (M, Ω, G, λ) is called a *Hamiltonian* G-space.

For X, Y in ham (M) with $\mu(X) = \mathrm{d}f$ and $\mu(Y) = \mathrm{d}g$ we set

$A(X, Y) = \int fg\tilde{v}$. Since

$$A([Z,X],Y) + A(X,[Z,Y]) = \int_M \Lambda\{(\varphi \wedge \mathrm{d}f)g + (\varphi \wedge \mathrm{d}g)f\}\tilde{v}$$

$$= \int_M \Lambda\{\varphi \wedge \mathrm{d}(fg)\}\tilde{v} = -\int_M \Lambda\{\mathrm{d}(fg\varphi)\}\tilde{v}.$$

Thus we have

THEOREM 3.6.8. $A([Z,X],Y) + A(X,[Z,Y]) = 0$, i.e. A is invariant by ad(ham).

If $M = G/H$ is a compact symplectic manifold we set $\mathfrak{g}^0 = \mathfrak{g} \cap \mathrm{ham}$. We assume that $\mathfrak{g}$ is finite dimensional. Let S be a subspace of $\mathfrak{g}^0$ invariant by $\mathrm{ad}(\mathfrak{g})$. Its ortho complement with respect to A is also invariant. Thus the representation defined by $\mathrm{ad}(\mathfrak{g})$ on $\mathfrak{g}^0$ is completely reducible. Thus we have

THEOREM 3.6.9. In the case above, $\mathfrak{g}^0$ is reductive – i.e. $\mathfrak{g}^0 = \mathfrak{a} + [\mathfrak{g}^0, \mathfrak{g}^0]$ where $\mathfrak{a}$ is an abelian ideal.

If $X \in \mathfrak{g}$ is an element in the centralizer of $\mathfrak{g}^0$ in $\mathfrak{g}$ then there is a 1-form $\varphi = \mu(X)$ and an element Y in $\mathfrak{g}^0$ such that

$$-i(Y)i(X)\Omega = -\Lambda(\varphi \wedge \mathrm{d}g) = \text{constant},$$

where $\mu(Y) = \mathrm{d}g$. Since $\int_M \Lambda(\varphi \wedge \mathrm{d}g)\tilde{v} = 0$ we see that the constant is zero. Thus $i(Y)\varphi = 0$ for all Y. However, G is transitive so $\varphi = 0$ and $X = 0$. Thus the centralizer of $\mathfrak{g}^0$ is zero – i.e. $\mathfrak{g}^0$ is semisimple. Thus we have

THEOREM 3.6.10. If $M = G/H$ is a compact homogeneous symplectic manifold with $b_1(M) = 0$, then G (compact or noncompact) is semisimple.

The classification theorem of Wang shows that $M = G/H$ is a homogeneous symplectic space with G compact semisimple iff H is the centralizer of a torus of G. And when this is true M is Kählerian, $\chi(M) > 0$ and M is simply connected. This result has been generalized by Kostant which we develop in Chapter 7 on orbit space.

We note here that if $M = G/H$ is a homogeneous symplectic space where G is not necessarily compact, but the fundamental group of M is finite, then G is semisimple. Then by a theorem of Montgomery, the maximal compact subgroup G^u of G acts transitively on M and $M = G^u/H^u$, $H^u = G^u \cap H$, is of the form discussed in the last paragraph. We will see later that spaces of this form are Hodge manifolds and hence are algebraic varieties.

PROBLEMS

EXERCISE 3.1. Show that the Poisson bracket can be represented by $\{f, g\} = \Lambda\,(\mathrm{d}f \wedge \mathrm{d}g)$.

EXERCISE 3.2. Show that if $M = G/H$ is a homogeneous symplectic manifold with G compact, then the associated almost Kähler structure is invariant by G and $\mathfrak{g} = [\mathfrak{g}, \mathfrak{g}] + \mathfrak{c}$ where $[\mathfrak{g}, \mathfrak{g}] \subset \mathrm{ham}(M)$ and the center $\mathfrak{c}$ is generated by the inverse image by p of the harmonic 1-forms.

EXERCISE 3.3. Consider the spherical rotator introduced in Section 0.2.2 with coordinates $q = (\vartheta, \varphi, \psi)$ and metric $g_{11} = g_{22} = g_{33} = I$, $g_{23} = g_{32} = I\cos\vartheta$ and all other $g_{ij} = 0$. Show that the curvature tensor satisfies $R_{ijkl} = (4I)^{-1}(g_{ik}g_{jl} - g_{il}g_{jk})$. Recall that a linear form in momenta p_i, $L = \sum v^i p_i$, is a constant of motion iff the components v_i satisfy the Killing equation $\nabla_i v_j + \nabla_j v_i = 0$ where ∇_i is the covariant derivative with respect to q^i. Show that the solution to the Killing equations are

$$v_1 = -a_1 \sin\varphi + a_2 \cos\varphi - b_1 \sin\psi + b_2 \cos\psi,$$
$$v_2 = a_3 + (b_1 \cos\psi + b_2 \sin\psi)\sin\varphi + b_3 \cos\vartheta,$$
$$v_3 = (a_1 \cos\varphi + a_2 \cos\varphi)\sin\vartheta + a_3 \cos\vartheta + b_3,$$

where a_i, b_i $i = 1, 2, 3$ are arbitrary constants. Show that the Killing components define six operators $-\sqrt{-1}\, v^i(\partial/\partial q_i)$ which represent a constant of motion in quantum mechanics. Let

$$M_1(\varphi, \psi) = \frac{1}{\sqrt{-1}}\left[-\sin\varphi\frac{\partial}{\partial\vartheta} - \cot\vartheta\cos\varphi\frac{\partial}{\partial\vartheta} + \operatorname{cosec}\vartheta\cos\varphi\frac{\partial}{\partial\psi}\right],$$
$$M_2(\varphi, \psi) = \frac{1}{\sqrt{-1}}\left[\cos\varphi\frac{\partial}{\partial\vartheta} - \cos\vartheta\sin\varphi\frac{\partial}{\partial\vartheta} + \operatorname{cosec}\vartheta\sin\varphi\frac{\partial}{\partial\psi}\right],$$
$$M_3(\varphi, \psi) = \frac{1}{\sqrt{-1}}\frac{\partial}{\partial\varphi},$$

and $N_i(\varphi, \psi) = M_i(\psi, \varphi)$, $i = 1, 2, 3$. The reader should identify that M_i, N_i represent angular momenta. Show $[M_i, N_j] = 0$, $[M_i, M_j]\sqrt{-1}\,\varepsilon_{ijk}$, $[N_i, N_j] = \sqrt{-1}\,\varepsilon_{ijk}N_k$. Thus M_i, N_i form two commuting sets of operators, each one generating the Lie algebra of $SO(3)$; and together they generate the Lie algebra of $SO(3) \times SO(3) = SO(4)$. $SO(4)$ is said to be the kinematical symmetry group of the freely rotating spherical rotator.

Chapter 4

Geometry of Contact Manifolds

4.1. Contact Manifolds

Definition 4.1.1. A $2n+1$ dimensional manifold M is said to be a *contact manifold* if there is a global 1-form ω in $A^1(M)$ which satisfies $\omega \wedge (\mathrm{d}\omega)^n \neq 0$ at every point of M.

Definition 4.1.2. A $2n+1$ dimensional manifold M is called an *almost contact manifold* if there is a global 1-form ω in $A^1(M)$ and a global 2-form π in $A^2(M)$ which satisfy $\omega \wedge (\pi)^n \neq 0$ at every point of M.

Clearly a contact manifold is an almost contact manifold.

Definition 4.1.3. Given an almost contact manifold (M, ω, π) define the global vector field V_ω in $V(M)$ and maps

$$l: A^1 \to V$$
$$L: A(M) \to V$$

by

$$V_\omega(f)\omega \wedge (\pi)^n = \mathrm{d}f \wedge (\pi)^n$$
$$l(\varphi) f \omega \wedge (\pi)^n = n\varphi \wedge \mathrm{d}f \wedge \omega \wedge (\pi)^{n-1}$$
$$L(f) g \omega \wedge (\pi)^n = n \mathrm{d}f \wedge \mathrm{d}g \wedge \omega \wedge (\pi)^{n-1}$$

for f, g in $A(M)$ and φ in A^1. V_ω is called the *canonical field* and l, L are called *Lagrange brackets.*

$V_\omega(f)$, $l(\varphi)f$, and $L(f)g$ as vector fields are uniquely determined since the $2n+1$ form $\omega \wedge (\pi)^n$ gives a base for the *V-module* $A^{2n+1}(M)$. The following relationships are easily verified:

Theorem 4.1.4. Let l, L be the Lagrange brackets of an almost contact manifold. Then

(i) $L(f) = l(\mathrm{d}f)$.

(ii) l is A-linear.

(iii) L is R-linear and an A-derivation.
(iv) $L(f)g = -L(g)f$.

THEOREM 4.1.5. Let V_ω, l, L be the canonical field and Lagrange brackets of an almost contact manifold. Then V_ω, l, L are uniquely characterized by

(i) $X = V_\omega$ iff $i(X)\omega = 1$ and $i(x)\pi = 0$.
(ii) $X = l(\varphi)$ iff $i(X)\omega = 0$ and $i(X)\pi = \varphi(V_\omega)\omega - \varphi$.
(iii) $X = L(f)$ iff $i(X)\omega = 0$ and $i(X)\pi = V_\omega(f)\omega - \mathrm{d}f$.

Proof. The proof is left as an exercise.

DEFINITION 4.1.6. Define the map $K: A(M) \to V(M)$ for an almost contact manifold M by $K(f) = fV_\omega + L(f)$.

It follows quickly from the above theorems that

THEOREM 4.1.7.

(i) K is R-linear.
(ii) $K(fg) = fK(g) + fK(f) - fgV_\omega$.
(iii) $X = K(f)$ iff $i(X)\omega = f$ and $i(X)\pi = V_\omega(f)\omega - \mathrm{d}f$.

COROLLARY 4.1.8. $K: A \to V$ is an injection with left inverse $\omega: V \to A$.

Proof. $\omega(K(f)) = i(K(f))\omega = f$. Thus $\omega \circ K: V \to V$ is the identity.

DEFINITION 4.1.9. X in V is called *horizontal* if $\omega(X) = 0$. Let W denote the space of *horizontal vector fields.*

DEFINITION 4.1.10. $\varphi \in A^p(M)$ is called *basic* if $i(V_\omega)\varphi = 0$. Let $B^1(M)$ denote the space of basic 1-forms.

THEOREM 4.1.11. The following sequences are exact:

$$0 \to W \to V \xrightarrow{\omega} A \to 0,$$

$$0 \to B^1 \to A^1 \xrightarrow{i(V_\omega)} A \to 0,$$

with splittings $r(V_\omega): A \to V$ where $r(V_\omega)f = fV_\omega$ and $r(\omega): A \to A^1$: $r(\omega)f = f\omega$.

Let $\tilde{\pi}$: $V \to A^1$ denote also the map $\tilde{\pi}(X) = i(X)\pi$.

THEOREM 4.1.12. $\tilde{\pi}$ gives a bijection between the V-modules W and B^1 with inverse -1. Thus we have

$$\begin{array}{ccccccccc} 0 & \to & W & \to & V & \to & A & \to & 0, \\ & & \uparrow\downarrow & & \uparrow\downarrow\alpha & & \| & & \\ 0 & \to & B^1 & \to & A^1 & \to & A & \to & 0, \end{array}$$

where $\alpha = \tilde{\pi} + r(\omega)\omega$.

EXAMPLE 4.1.13. R^{2n+1} with coordinates $(x^1, \dots x^n, y^1 \dots y^n, z)$ is a contact manifold with contact form $\omega = \mathrm{d}z - \sum y^k \,\mathrm{d}x^k$.

EXAMPLE 4.1.14. Under fairly obvious conditions on regularly immersed manifolds M, $2n+1 = \dim(M)$ in R^{2n+2} or more generally of $T(N)$ or $T^*(N)$, M will be a contact manifold. In particular the tangent space of M cannot pass through the origin. Making this more precise (which is left as an exercise) we see that S^{2n+1} given by $\sum_{i=1}^{2n+2} (x^i)^2 = 1$ is a contact manifold with contact form $\omega = i^*\beta$ where $\beta = x^1\,\mathrm{d}x^2 - x^2\,\mathrm{d}x^1 + \cdots - x^{2n+2}\mathrm{d}x^{2n+1}$ and i is the injection $i: M \to R^{2n+2}$.

Since the contact form just defined is invariant under the antipodal map $x^i \to -x^i$, we see that

EXAMPLE 4.1.15. The real projective spaces $RP(2n+1)$ are contact manifolds with contact from ω as above.

Contact manifolds arise naturally in classical mechanics as follows. Consider the geodesic equations

$$\dot{x}^r = X^r$$
$$\dot{X}^r = -\Gamma^r_{jk} X^j X^k, \qquad r = 1, \dots, n$$

On an open set U in TN with coordinates $(x^1, \dots, x^n, X^1, \dots, X^n)$ where N is a Riemannian manifold with metric g and Riemannian covariant derivative ∇. By uniqueness of the solutions of differential equations, the geodesic equations integrate to give one and only one geodesic from $p(0)$ to $p(t)$ in $T(N)$. The map $T_t: p(0) \to p(t)$ is a diffeomorphism of $T(N)$ for every t. The set T_t (t in R) is an abelian group called the geodesic *flow* of N with trajectory $p(t)$.

The vector field on $T(N)$ given locally by $(X^i, -\Gamma^i_{jk} X^j X^k)$ on $U \times R^n$ is called the *geodesic vector field*. $T(N)$ has a natural Riemannian metric if N is a Riemannian manifold, viz. $\mathrm{d}s^2 = g(\mathrm{d}x, \mathrm{d}x) + g(\nabla X, \nabla X)$.

THEOREM (Liouville) 4.1.16. The geodesic vector field leaves invariant the Riemannian metric $\mathrm{d}s^2$ on $T(N)$.

By conservation of energy the geodesic flow T_t lives on the unit tangent bundle $T_1(N)$. $T_1(N)$ is a fiber bundle with fiber S^n and group $O(n)$. Inducing the Riemannian metric from $T(N)$, the unit tangent bundle is a $2n-1$ dimensional Riemannian submanifold of $T(N)$. The geodesic field induces a geodesic flow on $T_1(N)$. This gives

EXAMPLE 4.1.17. $T_1(N)$ is a contact manifold with contact form $\omega = = g_{ij}X^j dx^i$. Thus $d\omega = g_{ij}\nabla X^j \wedge dx^i$ and so $\omega \wedge (d\omega)^{n-1} \neq 0$. We leave it to the reader to check that the geodesic vector field X satisfies $i(X)d\omega = 0$ and $i(X)\omega = 1$.

Using the injection i: $T_1(N) \to T(N)$ we pull back ω to the cosphere bundle $T^*(N)$ to give

EXAMPLE 4.1.18. The cosphere bundle is a contact manifold.

Particular examples of the sphere and cosphere bundles are presented next. First we recall that the sphere bundles $T_1(S^n)$ over the n-spheres S^n are the Stiefel manifolds $V(n+1, 2)$, where $V(n, k) = SO(n)/SO(n-k)$. The first example is the case $n = 2$.

EXAMPLE 4.1.19. Consider $N = S^2$; then the tangent sphere bundle $T_1(S^2)$ is diffeomorphic to $RP(3)$. Namely for p in $T_1(S^2)$ consider the unit vector $e_1(p)$ from 0 to p in S^2. Then the diffeomorphism is just $f: T_1(S^2) \to SO(3)$; $p \to (e_1(p), e_2(p) = p, e_1(p) \times e_2(p))$ where $\times$ is the vector cross product in R^3. As is well known $SO(3)$ is diffeomorphic to $RP(3)$. If the geodesic is parametrized locally by $(x(s), y(s))$, $s =$ arc length of the geodesic, then the geodesic equations of $T_1(S^2)$ for x and y are

$$x'' = -by + ay',$$
$$y'' = cy,$$

where $' = d/ds$ and $a = (x', y)$, $b = (x', y')$, and $c = -(y', y')$.

EXAMPLE 4.1.20. An example of a unit cosphere bundle is given by compact Riemannian manifolds $M = \Gamma \backslash \mathscr{P}$ where $\mathscr{P}$ is the Poincaré upper half plane $\{z \in C \mid \mathrm{Im}(z) > 0\}$. It is easily checked that $\mathscr{P} = SL(2, R)/SO(2)$ as a homogeneous space. We want to show that $T_1^*(M)$ is diffeomorphic to $SL(2, R)/\Gamma$. To see this one checks that the lifted action of $PL(2, R)$ on $T_1^*(\mathscr{P})$ is transitive and free. Thus $T_1^*(\mathscr{P}) = PL(2, R)$. Factoring by Γ gives the result.

The Lie algebra $s1(2, R)$ of 2×2 real matrices of trace zero has a basis

$$E_\alpha = \begin{pmatrix} 0 & 1 \\ 0 & 0 \end{pmatrix}, \quad E_{-\alpha} = \begin{pmatrix} 0 & 0 \\ 1 & 0 \end{pmatrix} \text{ and } H_\alpha = \tfrac{1}{2}\begin{pmatrix} 1 & 0 \\ 0 & -1 \end{pmatrix}.$$

Note that $[E_\alpha, E_{-\alpha}] = 2H_\alpha$. And $[H_\alpha, E_{\pm\alpha}] = E_{\pm\alpha}$. Exponentiating gives

$$\exp(tH_\alpha) = \begin{pmatrix} \exp(t/2) & 0 \\ 0 & \exp(-t/2) \end{pmatrix},$$

$$\exp\left(\frac{\vartheta}{2}(E_\alpha \quad - E_{-\alpha})\right) = \begin{pmatrix} \cos(\vartheta/2) & \sin(\vartheta/2) \\ -\sin(\vartheta/2) & \cos(\vartheta/2) \end{pmatrix}$$

and

$$\exp(t(E_\alpha + E_{-\alpha})) = \begin{pmatrix} \mathrm{ch}(t/2) & \mathrm{sh}(t/2) \\ \mathrm{sh}(t/2) & \mathrm{ch}(t/2) \end{pmatrix}.$$

Let $X_\pm$, Y denote the vector fields corresponding to $E_{\pm\alpha}$, H_α on $N = SL(2, R)/\Gamma$. $X_\pm$, Y then satisfy the same bracket relationships as $E_{\pm\alpha}, H_\alpha$. There is then a unique 1-form ω on N which satisfies $i(Y)\omega = 1$ and $i(X_\pm)\omega = 0$. Since $\mathscr{L}([X_\pm, X_-])\omega = \mathrm{d}\omega(X_+, X_-) = 2$ we see that $\omega \wedge \mathrm{d}\omega \neq 0$. There is then a unique 1-form ω on N which satisfies $i(Y)\omega = 1$ and $i(X_+)\omega = 0$. Therefore ω defines a contact structure on N. Since $[Y, X_\pm] = X_\pm$ we see that $(\exp(tY))_* X_\pm = \exp(^\pm t)X_\pm$. Thus $\exp(tY)$ preserves the contact structure. We leave it to the reader to check that the 1-form is equivalent to the canonical contact form on $T_1^*(\mathscr{P}/\Gamma)$.

Theorem 4.1.21. If M is a compact orientable 3-dimensional manifold then there exists a 1-form ω on M such that (M, ω) is a contact manifold.

We have already seen that S^3 in R^4 admits a contact structure

$$\omega_1 = i^*(-x_2\mathrm{d}x_1 + x_1\mathrm{d}x_2 + x_3\mathrm{d}x_4 - x_4 - \mathrm{d}x_3),$$

where $i: S^3 \to R^4$ is the natural injection. Under the involution τ in $SO(4)$, $\tau: S^3 \to S^3: (x_1, x_2, x_3, x_4) \to (x_1, x_2, x_3, -x_4)$ we see that $\tau^*\omega_1 = \omega_{-1}$ where

$$\omega_{-1} = i^*(-x_2\,\mathrm{d}x_1 + x_1\,\mathrm{d}x_2 - x_3\,\mathrm{d}x_4 + x_4\,\mathrm{d}x_3).$$

Definition 4.1.22. The *contact distribution* of a contact manifold (M, ω) is the subbundle Σ_ω of the tangent bundle defined by $\omega = 0$.

It is easily checked that $\sum_\omega$ is integrable – i.e. write $d\omega = \sum_{i=1}^n \alpha^i \wedge \alpha^{n+i}$ and let X_i be dual to α^i. Then X_i are linearly independent and $\omega[X_i, X_j]) = -2d\omega(X_i, X_j) \neq 0$ iff $j \neq n+1 \pmod{2n}$.

THEOREM 4.1.23. If Σ_ω is a contact distribution on M, then (M, ω) is a contact manifold.

DEFINITION 4.1.24. Two contact distributions Σ, Σ' are called *isomorphic* if there is a diffeomorphism f such that $f^*_m\Sigma_m = \Sigma'_{f(m)}$ for all m in M.

DEFINITION 4.1.25. Two contact structures on M are called *conjugate* if there is a diffeomorphism $f\colon M \to M$ such that $\omega' = g f^*(\omega)$ where g is in $A(M)$.

THEOREM 4.1.26. ω and ω' are conjugate iff the contact distributions Σ and Σ' are isomorphic.

We have already seen that on $S^3 \omega_1$ and ω_{-1} are conjugate contact forms.

DEFINITION 4.1.27. Two contact structures on M are called *isomorphic* if their contact distributions are isomorphic.

THEOREM 4.1.28. Let M be a compact orientable 3-manifold. Then there exists an infinity of nonisomorphic contact structures on M.

COROLLARY 4.1.29. In particular the theorem holds for S^3 and S^3/Γ where Γ is a finite group operating properly without fixed point.

Thus we conclude that there is an infinity of nonisomorphic contact structures on $SO(3) = T_1(S^2) = RP(3)$.

Another example covered in this theorem is the lens space $L(p, q)$. If we set $x = (x_1, x_2, x_3, x_4)$ in R^4 and let $z = x_1 + ix_4$ and $w = x_3 + ix_4$ then S^3 is the subspace of C^2 given by $|z|^2 + |w|^2 = 1$. Define the map $\gamma\colon S^3 \to S^3$ by $\gamma(z, w) = (\exp(2\pi i/p)z, \exp(2\pi iq/p)w)$. Then clearly $\gamma^p = I$. Set $L(p, q) = S^3/\Gamma$, where $\Gamma = \{\gamma | \gamma$ as above with γ in $SO(4)\}$. $L(p, q)$ is the *lens space.* It is a compact orientable 3-manifolds with fundamental group Z_p. If $\pi\colon S^3 \to$ $\to L(p, q)$ is the quotient map we see that a form ω on S^3 is the image by π^* of a form ω' on $L(p, q)$ if $\gamma^*\omega = \omega$. Since $\omega_{\pm 1}$ on S^3 are invariant by γ we see that $L(p, q)$ is a contact manifold. Of course $L(2, 1) = RP(3)$.

The analogue of Darboux' theorem is due to Cartan.

THEOREM 4.1.30. If (M,ω) is a contact manifold, $\dim M = 2n+1$, then we can always find variables (q^k, p^k, z) such that locally $\omega = \mathrm{d}z - \sum p^k \mathrm{d}q^k$.

4.2. ALMOST CONTACT METRIC MANIFOLDS

If M is an $2n+1$ dimensional manifold almost contact manifold, then the structure group $SO(2n+1)$ of the tangent bundle reduces to $U(n) \times I$, and the almost contact structure is also specified by a tensor field ϕ of type $(1,1)$, a contravariant vector field V_ω, and a covariant field ω which satisfy $\omega(V_\omega) = 1$ and $\phi^2 = -1 + V_\omega . \omega$.

THEOREM 4.2.1. (ϕ, V_ω, ω) also satisfy $\phi V_\omega = 0$, $\omega \cdot \phi = 0$, $\phi^3 + \phi = 0$ and rank $\phi = 2n$.

Define the Riemannian metric g on M by $g(X, V_\omega) = \omega(X)$ and $g(\phi X, \phi Y) = g(X, Y) - \omega(X)\omega(Y)$. g is called the *associated Riemannian metric* of the almost contact manifold. We define $\tilde{\Omega}$ by $\tilde{\Omega}(X,Y) = g(X, \phi Y)$. Then $\tilde{\Omega}$ is a 2-form of rank $2n$.

THEOREM 4.2.2. If M admits a 2-form Ω of rank $2n$, then M admits an almost contact metric structure $(\phi, V_\omega, \omega, g)$ for which $\Omega(X,Y) = g(X, \phi Y)$.

Clearly if (M,ω) is a contact manifold then (M,ω) admits an almost contact metric structure.

The map ϕ can be extended to a linear map of the complex tangent space $T^C_m(M)$ at m. The eigenvalues of this map are 0, $\pm i$ with eigenspaces V_0, $V_i V_{-i}$ of dimension 1, n, n resp. Let D_0, $D_{\pm i}$ denote the corresponding distributions spanned by V_0, $V_{\pm i}$ at m in M.

DEFINITION 4.2.3. The almost contact structure is called *normal* if D_i and $D_i + D_0$ are completely integrable.

THEOREM 4.2.4. (ϕ, V_ω, ω) is normal iff ω is invariant under the local group of transformations generated by V_ω.

THEOREM 4.2.5. If (M,ω) is a contact manifold, all trajectories of V_ω are geodesics of the associated Riemannian metric. And if (M,ω) is a normal almost contact manifold V_ω is a Killing vector field of the Riemannian metric, i.e. $\mathscr{L}(V_\omega)g = 0$.

THEOREM 4.2.6. Let G be a reductive Lie group of odd dimension. Then G admits a left invariant normal almost contact structure.

Proof. G reductive means that the Lie algebra of G has the form $\mathfrak{g} = \mathfrak{s} + \mathfrak{a}$ where $\mathfrak{s}$ is semisimple and $\mathfrak{a}$ is Abelian. Let J be a linear map in $\mathrm{End}(\mathfrak{g})$. Assume $J(V) = 1$ in $\mathfrak{a}$. Writing X in $\mathfrak{g}$ as $X = X' + aV$ for a in R, X' in $\mathfrak{s}$ then we define $\phi(X) = J(X')$ and $\omega(X) = a$. Then the structure (ϕ, V, ω) is easily checked to be as desired.

COROLLARY 4.2.7. Every odd dimensional compact connected Lie group has a left invariant normal almost contact structure.

4.3. DYNAMICAL SYSTEMS AND CONTACT MANIFOLDS

Reeb began the study of the relationship of contact manifolds and dynamical systems by noting that if X is a smooth vector field on M then the dynamical system $\dot{x} = X(x)$ on M has naturally attached a contact structure. If B is the space of orbits, then the fibration $M \to B$, generated by the involutive distribution associated with X was studied by Reeb. Reeb noted early on that examples of such systems were given by geodesic flows on S^n, $RP(n)$, $CP(n)$, and $QP(n)$. Reeb was able to prove, using results that we will cover, that the product of different odd dimensional spheres do not admit this structure.

We study the geometry of the orbit space defined by a non-zero vector field X.

DEFINITION 4.3.1. X is called *proper* if it generates a group of diffeomorphisms of M.

DEFINITION 4.3.2. X is called *regular* if it is regular in the sense of Palais.

The Palais theorem shows that

THEOREM 4.3.3. If X is proper and regular, then the orbit space, B, of X, with the quotient topology is a $2n$ dimensional smooth manifold with smooth projection $p: M \to B$.

DEFINITION 4.3.4. $P_X(m) = \inf(t \mid t > 0, \varphi_t(m) = m$ in M where $\varphi_t = \exp(tX))$ is called the *period function.*

THEOREM 4.3.5. If X is proper and regular and satisfies $\mathscr{L}(X)\omega = 0$ and $i(X)\omega = 1$, then P_X is constant on M.

DEFINITION 4.3.6. If (M, ω) is a contact manifold with associated vector field X and if X is regular, we say (M, ω) is a *regular contact manifold.*

THEOREM 4.3.7. If (M, ω) is a connected compact regular contact manifold, then there is a map $\omega' = s\omega$ such that X' associated to ω' has associated Lie group, a 1-dimensional compact Lie group acting freely on M.

Proof. Since P_X is constant we substitute $\omega' = (1/P)\omega$; thus $X' = P_X X$. Since $P_{X'} = 1$, the group generated by X' depends only on t mod 1.

THEOREM (Boothby-Wang). 4.3.8. Let (M, ω) be a compact regular contact manifold. Then $S^1 \to M \xrightarrow{p} B$ is a principal circle bundle with connection $\tilde{\omega}$ defined by ω; and (B, Ω) is a symplectic manifold with Ω defining the curvature of the connection – i.e. $d\tilde{\omega} = p{*}\Omega$. Finally Ω determines an integral cohomology class.

Proof. By the last result we can modify ω so that X associated to ω generates S^1. We leave it to the reader to show that $p: M \to B$ is a principal circle bundle. Clearly we have $\mathscr{L}(X)\omega = 0 = \mathscr{L}(X)\,d\omega$. Again the reader may check that ω defines a s^1-valued 1-form $\tilde{\omega}$ on M which is right invariant and satisfies $\tilde{\omega}(X) = 1$.

Since S^1 is Abelian, $[\tilde{\omega}(X), \tilde{\omega}(Y)] = 0$; we have

$$d\tilde{\omega}(X, Y) = -\tfrac{1}{2}[\tilde{\omega}(X), \tilde{\omega}(X)] + \tilde{\Omega}(X, Y)$$

so $d\tilde{\omega} = \tilde{\Omega}$, where $\tilde{\Omega}$ is the curvature form. Since $R(g)\tilde{\Omega} = \tilde{\Omega}$ for g in S^1 and $i(X)\tilde{\Omega} = 0$, there is a unique 2-form Ω on B such that $\tilde{\Omega} = p{*}\Omega$. By the isomorphism between forms on B and horizontal S^1 invariant forms on M we have $p{*}d\Omega = dp{*}\Omega = dd\Omega = 0$. And $p{*}(\Omega^n) = (p{*}\Omega)^n = (d\omega)^n \neq 0$. Thus (B, Ω) is symplectic. The reader may check that Ω is integral.

From the exact sequence of abelian groups

$$0 \to Z \to R \to S^1 \to 0$$

we have an exact sequence of sheaves of abelian groups

$$0 \to Z \to \mathbf{R} \to \mathbf{S}^1 \to 0.$$

Since the sheaf $\mathbf{R}$ is fine, we have a bijection

$$H^1(B, \mathbf{S}^1) \to H^2(B, Z)$$

$$L \to \mathfrak{x}(L)$$

mapping the line bundle L to its Euler–Poincaré class. From the cohomology sequence

$$H^1(B, S^1) \to H^2(B, Z) \to H^2(B, R)$$

we map $\mathfrak{x}(L) \to d \cdot R(\Omega)$. This provides the basis for

THEOREM 4.3.9. Let (B, Ω) be a symplectic manifold with Ω determining an integral cohomology class. Then there is a principal circle bundle over B with a connection $\tilde{\omega}$ such that ω defined by $\tilde{\omega}$ is a contact form with associated vector field which generates right translations of the structure group S^1.

The proof is left to the reader.

Finally, we relate the normal contact metric structure to the present result:

THEOREM 4.3.10. A compact regular contact metric manifold is normal iff the base manifold under the Boothby–Wang theorem is a Hodge manifold (i.e. a compact Kähler manifold (B, Ω) with Ω integral).

EXAMPLE 4.3.11. Consider the 2 spheres $S^2 = \{x \in R^3 \mid \|x\|^2 = 1\}$. It is closed compact 2-dimensional smooth manifold. Identifying S^2 with $CP(1)$ we see that S^2 is a Kähler manifold. We will examine the symplectic structure in complex coordinates. S^2 has two neighborhoods $U_j = S^2 - \{(0, 0, j)\}$ where $j = \pm 1$, such that $S^2 = U_{-1} \cup U_{+1}$. If x is in U_j we set $z_j = (x_1 \pm ix_2)(1 \pm x_3)^{-1}$. This allows us to identify U_j with C. On $U_1 \cap U_{-1}$ we have $z_1 z_{-1} = 1$. We define the line bundle over S^2 by $c_{jj} = 1$ on U_j, while $c_{12} = z_1^n$ and $c_{21} = z_{-1}^n$ on $U_1 \cap U_{-1}$.

We take for the symplectic form on S^2, $\Omega = 2i(1 + z_j\bar{z}_j)^{-2} \mathrm{d}\bar{z}_j \wedge \mathrm{d}z_j$. We want two one forms α_j such that $(\Omega - \mathrm{d}\alpha_j)|U_j = 0$. The natural choice is then $\alpha_j = 2iv\bar{z}_j(1 + z_j\bar{z}_j)\,\mathrm{d}z_j$ on U_j. Note that $\alpha_1 - \alpha_{-1} = -2iv\mathrm{d}z_1/z_1$. But $\mathrm{d}c_{12}/2\pi i c_{12} = -in\mathrm{d}z/2\pi z$. Thus α_j defines a connection α on L if $2v = n/2\pi$. The curvature of α is Ω where $\Omega_j^{(n)} = 2iv(1 + z_j\bar{z}_j)^{-2}\ \mathrm{d}\bar{z}_j \wedge \mathrm{d}z_j$. Now $H^2(S^2, R)$ is isomorphic to R by the map $d \cdot R.\ (\beta) \to \int_{S^2} \beta$. In our case

$$\int_{S^2} \Omega = 4\pi v = n \in Z.$$

Thus (S^2, Ω) is a Hodge manifold. (L, ω), where $\omega_j = (1/2\pi)\,\mathrm{d}\log(z_j) + \alpha_j$, is the contact manifold over S^2. We define the polarization (see Section 6.1) on S^2 by $\mathscr{F}_x = C\,\partial/\partial\bar{z}_j$ for x in U_j. Thus $(S, \Omega^{(n)}, \mathscr{F})$ is a Kähler manifold for

$n \neq 0$. We leave it to the reader to verify that $SU(2)$ acting on S^2 leaves $\mathscr{F}$ and Ω invariant. Thus $(S^2, \Omega, \mathscr{F})$ is an admissably polarized strongly symplectic manifold (see Section 6.2). Also the reader should find the maps λ, σ so that

$$\begin{array}{ccccccccc} 0 & \to & R & \to & A(S^2) & \to & \mathrm{ham}(S^2, \Omega) & \to & 0 \\ & & & & {\scriptstyle\lambda}\nwarrow & & \uparrow{\scriptstyle -\sigma} & & \\ & & & & & su(2) & & & \end{array}$$

commutes, thus showing that S^2 is a Hamiltonian $SU(2)$-space.

Definition 4.3.12. A *Hermitian structure* h on a line bundle is the smooth assignment $x \to h_x$ of a positive definite inner product to L_x.

Definition 4.3.13. If L has a connection ∇ and Hermitian structure, then h is called ∇-*invariant* if

$$Xh(f,g) = h(\nabla f, g) + h(f, \nabla g).$$

Theorem 4.3.14. If $H^1(M, R) = 0$ then there is a ∇-invariant Hermitian structure in $L \to M$ iff the curvature of ∇, $(1/2\pi i)\,\mathrm{curv}\,(L, \nabla)$, is a real 2-form.

If $L_c(B, \Omega)$ denotes the set of all line bundles L over B with ∇-invariant Hermitian structure such that $\mathrm{curv}\,(L, \nabla) = \Omega$, then we have

Theorem 4.3.15. $L_c(B, \Omega)$ is nonempty iff $[\Omega] \in H^2(B, R)$ is integral. In this case $L_c(B, \Omega)$ is the inverse image of Ω under the map $H^2(B, Z) \to H^2(B, R)$, i.e. $c_1(L) = \Omega$.

Definition 4.3.16. Let $P_i(z_0, \ldots, z_n) = \sum_{j=0}^{n} \alpha_{ij} z_j^{a_{ij}}$, $i = 1, \ldots, m$ be a set of m polynomials of n variables where $\alpha_{ij} \in R$ and a_{ij} are positive integers. Let $V = \{\mathbf{z} \in C^{n+1} (P_i(z) = 0, 1 \leqq i \leqq m\}$. If $S(\varepsilon) =$ the hypersphere of radius ε at the origin, $\Sigma(\varepsilon) = V \cap S(\varepsilon)$ is a *generalized Brieskorn manifold.*

Theorem 4.3.17. $\Sigma(\varepsilon)$ admits a 1-parameter family of normal contact structures $(\phi(t), V_\omega(t), \omega(t)) - \infty < t < \infty$. These structures are in general nonregular.

Example 4.3.18. Let $P(z) = z_0 + \cdots + z_{m-1} + z_m^l$. Since V has no singular point, Σ is diffeomorphic to S^{2m-1}. The action of S^1 given by

$$t(z_0, \ldots, z_m) = (\exp{(2\pi l t_i)} z_0, \ldots, \exp{(2\pi l t_i)} z_{m-1}, \exp{(2\pi t_i z_m)}$$

has only one isotropy group, Z_l, other than $\{e\}$. Thus if $k \neq l$, $Z_k \cap Z_l \neq \{e\}$ and the Z_k-action on Σ does not have any fixed point. Consider the manifold $M = \Sigma/Z_k$. The normal contact structure (ϕ, V, ω) on Σ induces one on M, $(\bar{\phi}, \bar{V}, \bar{\omega})$. We claim that $\bar{\omega}$ is nonregular unless $l = 1$. When $l = 1$, the contact foliation associated to $\bar{\omega}$ is just the Hopf fibration and M has a regular contact structure with $\pi_1(M) = k$. In particular for $k = 2$ we have the real projective space as M.

EXAMPLE 4.3.19. Let

$$P(z) = z_0^{q_1 \cdots q_{n-1}} + \cdots + z_{n-1}^{q_1 \cdots q_{n-2} q_{n-1}} + z_n^{q_1 \cdots q_{n-1}}$$

be a Brieskorn polynomial. Then Σ is diffeomorphic to the generalized lens space $L(p, q_1, \ldots, q_{n-1}) = S^{2n-1}/\Gamma$ where $\Gamma = Z_p$ acting on z in C^{n-1} by $t(z) = (\exp(2\pi it) z_0, \exp(2\pi it) z_1, \ldots, \exp((2\pi i q_{n-1} t) z_{n-1})$. One can show that Σ has normal contact structure which is in general nonregular.

DEFINITION 4.3.20. We say that we contact structures ω and ω' on M are *strictly conjugate* if there is a diffeomorphism f of M with $f*\omega = \omega'$.

THEOREM 4.3.21. Let $\Sigma^{(a_0, \ldots, a_n)}$ and $\Sigma^{(b_0, \ldots, b_n)}$ be two Brieskorn manifolds with normal contact structures ω_a and ω_b. Then ω_a is not strictly conjugate to ω_b if their respective slice diagrams (v. JS1) do not coincide.

EXAMPLE 4.3.22. Consider $P_q(z) = z_0 + z_1 + z_2^q + \ldots + z_n^q$, $q > 0$. Again Σ_q is diffeomorphic to S^{2n-1}. The C-action is given by $t(z) = (\exp(2\pi qt) z_0, \exp(2\pi qt) z_1, \ldots, \exp(2\pi t) z_n)$. If $\Delta(S^1, \Sigma_q)$ denotes the slice diagram, then one finds that $\Delta(S^1, \Sigma_q) \neq \Delta(S^1, \Sigma_r)$ for $q \neq r$. Thus the contact structures ω_q, $q = 1, 2, \ldots$ are all distinct in the sense of strict conjugation on S^{2n-1}. Of course ω_1 is the normal contact structure given by the Hopf fibration

$$S^1 \to S^{2n-1} \to CP(n-1).$$

4.4. Topology of regular contact manifolds

The topological properties of regular contact manifolds do not depend on their underlying Riemannian metric.

THEOREM 4.4.1. Let M be a compact regular contact manifold with fibration

$S^1 \to M \xrightarrow{p} B$. Then there is an almost complex structure J and almost Hermitian metric h on B satisfying $d\omega = 2p^*\Omega$ and $\Omega(X, Y) = h(X, JY)$.

COROLLARY 4.4.2. The almost Kählerain structure on B is Kählerian iff the contact metric structure is normal.

THEOREM 4.4.3. If M is a contact regular Riemannian manifold, then $b_1(M) = b_1(B)$.

Proof. This theorem follows from the Gysin sequence for the circle bundle $S^1 \to M \to B$:

$$0 \to H^1(B, R) \xrightarrow{p^*} H^1(M, R) \to H^0(B, R) \to H^0(B, R) \xrightarrow{L_0} H^2(B, R)$$
$$\dots \xrightarrow{L_{p-2}} H^p(B, R) \xrightarrow{p^*} H^p(M, R),$$

where L_p: $\alpha \to \Omega \wedge \alpha \in H^{p+2}(B, R)$ for α in $H^p(B, R)$. Since L_0 is an isomorphism, the theorem follows.

THEOREM 4.4.4. If M is as above, if φ is a harmonic 1-form on B, then $p^*\varphi$ is harmonic and if ψ is a harmonic 1-form on M then $\psi = p^*\varphi$ for some harmonic φ on B.

Noting that if B is Kählerian that L_p is an into isomorphism for $p \leqq (m-3)/2$, $m = \dim M$ and p^* is onto we have

THEOREM 4.4.5. If M is a normal regular compact contact Riemannian manifold, then $b_1(M) = b_1(M)$, $b_p(M) = b_p(B) - b_{p-2}(B)$ for $2 \leqq p \leqq$ $\leqq (m-1)/2$ and $b_p(M) = b_{p-1}(B) - b_{p+1}(B)$ for $(m+1)/2 \leqq p < m$.

Since $H^0(B, R) \simeq R$ we have

COROLLARY 4.4.6. $b_2(M) = b_2(B) - 1$.

Since $b_p(B)$ is even if p is odd we have

COROLLARY 4.4.7. $b_p(M)$ is even or zero if p is odd and $\leqq (m-1)/2$; and $b_p(M)$ is even or zero if p is even and $\geqq (m+1)/2$.

COROLLARY 4.4.8. For any harmonic r-form ψ on M there exists a harmonic r-form φ on B such that $\psi = p^*\varphi$ for $r \leqq (m-1)/2$.

4.5. INFINITESIMAL CONTACT TRANSFORMATIONS

For a contact manifold (M, ω) since $\omega, d\omega$ defines an almost contact structure on M we have the structure (V_ω, l, L, K)

Theorem 4.5.1. If (M, ω) is a contact manifold:

(i) $X = V_\omega$ iff $i(X)\omega = 1$ and $i(X)\,\mathrm{d}\omega = 0$.

(ii) $X = l(\varphi)$ iff $i(X)\omega = 0$ and $i(X)\,\mathrm{d}\omega = \varphi(\mathrm{V}_\omega)\omega - \varphi$.

(iii) $X = L(f)$ iff $i(X)\omega = 0$ and $i(X)\,\mathrm{d}\omega = (V_\omega f)_\omega - \mathrm{d}f$.

(iv) $X = K(f)$ iff $i(X)\omega = f$ and $i(X)\,\mathrm{d}\omega = (V_\omega f)\omega - \mathrm{d}f$.

Definition 4.5.2. A vector field X on a contact manifold (M, ω) is called an *infinitesimal contact transformation* if there is a function k in $A(M)$ such that $\mathscr{L}(X)\omega = k\omega$. Let $\mathrm{cont}\,(M, \omega)$ denote this set of vector fields.

Definition 4.5.3. X in $V(M)$ is an *infinitesimal automorphism of the contact structure if* $\mathscr{L}(X)\omega = 0$. Let $\mathrm{cont}_0(M, \omega)$ denote this set.

Since $\mathscr{L}([X, Y]) = [\mathscr{L}(X), \mathscr{L}(Y)]$, the R-modules cont and cont_0 are Lie algebras.

Theorem 4.5.4. If X is in cont, then the k in $A(M)$ such that $\mathscr{L}(X)\omega = k\omega$ is given by $k = V_\omega \omega(X)$.

Proof. Since $\mathscr{L}(X)\omega = \mathrm{d}\omega(X) + i(X)\,\mathrm{d}\omega = k\omega$, taking the inner product $i(V_\omega)$ gives the result.

Theorem 4.5.5. If (M, ω) is a contact manifold, $K: A(M) \to V(M)$ gives a bijection $K: A \to \mathrm{cont}$ with inverse $\omega|_{\mathrm{cont}}$.

Corollary 4.5.6. $V_\omega = K(1) \in \mathrm{cont}_0$.

In summmary we have an exact commutative diagram of R-modules

$$\begin{array}{cccccc} 0 \to & A_\emptyset & \to & A & \to & A \\ & \uparrow\downarrow & & \alpha\uparrow\downarrow K & & \| \\ 0 \to & \mathrm{cont}_0 & \to & \mathrm{cont} & \to & A \end{array}$$

where $A_\emptyset$ is the R-module of all first integrals of V_ω and $\alpha: X \to k_x$ where $\mathscr{L}(X)\omega = k_X\omega$.

Definition 4.5.7. If $X \to K(f)$ and $Y = K(g)$, then the *Jacobi bracket* $[f, g]$ is given by $[f, g] = \omega([X, Y])$.

Clearly then

Theorem 4.5.8. $K([f, g]) = [K(f), K(g)]$.

THEOREM 4.5.9. The Lie algebra cont is isomorphic to the Lie algebra $(A(M), [,])$ where $[,]$ are the Jacobi brackets.

We summarize two elementary facts in

THEOREM 4.5.10. If $X = K(f)$ and $Y = K(g)$ for f, g in $A(M)$, then

(i) $L(f)g = \mathrm{d}\omega(X, Y)$.

(ii) $[f,g] = L(f)g + fV_\omega g - gV_\omega f$.

COROLLARY 4.5.11. If X is in cont_0 (i.e. f is basic) then $[f,g] = K(f)g$; and if X, Y are in cont_0, then $[f,g] = L(f)g$.

Several other properties of V_ω, K, L hold:

THEOREM 4.5.12.

(i) $V_\omega(L(f)g) = L(V_\omega(f))g + L(f)V_\omega(g)$.

(ii) $[fV_\omega, gV_\omega] = (fV_\omega(g) - gV_\omega(f))V_\omega$.

(iii) $L(f)L(g)h + L(g)L(h)f + L(h)L(g)f = V_\omega(f)L(g)h + {} + V_\omega(g)L(h)f + V_\omega(h)L(f)g$.

(iv) $[L(f), fV_\omega] = L(f)gV_\omega - gL(V_\omega(f))$.

(v) $[L(f), L(g)] - L(L(f)g)) = L(f)gV_\omega + {} + V_\omega(f)L(g) = V_\omega(g)L(f)$.

If locally

$$\omega = \mathrm{d}z - \textstyle\sum p^j \mathrm{d}q^j,$$

then

$$V_\omega = \partial/\partial z$$

and

$$L(f)g = \textstyle\sum \partial g/\partial p^j(\partial f/\partial q^j + p^j \partial f/\partial z) - {} - \textstyle\sum \partial f/\partial p^j(\partial f/\partial q^j + p^j \partial g/\partial z).$$

Thus

$$[f,g] = \{f,g\} - \partial f/\partial z(g + \textstyle\sum p^j \partial g/\partial p^j) + {} + \partial g/\partial z(f + \textstyle\sum p^j \partial f/\partial p^j).$$

In particular if f is basic,

$$K(f)g = (f - \textstyle\sum p^j \partial f/\partial p^j)\partial g/\partial z + \{f,g\}.$$

THEOREM 4.5.13. $0 \to \mathrm{cont}_0(M, \omega) \to \mathrm{cont}(M, \omega) \to V(M) \to H^1(M, \mathbf{cont}_0) \to 0$ and $H^q(M, \mathbf{cont}_0) = 0$ for $q \geqq 2$; here $\mathbf{cont}_0$ is the sheaf of germs of infinitesimal contact automorphisms.

Let M be a regular contact manifold which is a G-bundle over B where $G = S^1$ or R, defined by vector field V_ω.

DEFINITION 4.5.14. If φ is a p-form on M, $0 \leqq p \leqq 2n+1$, φ is called *invariant* if $\mathscr{L}(V_\omega)\varphi = 0$.

DEFINITION 4.5.15. φ is called *dynamic* if φ is invariant and $i(V_\omega)\varphi = 0$ for $1 \leqq p \leqq 2n+1$. If $p = 0$ and φ is invariant, φ is called *dynamic* or a *first integral* for V_ω. Sometimes basic forms are said to be *absolute integral invariants* for V_ω.

DEFINITION 4.5.16. If X is a vector field which satisfies $i(X)\omega = 0$ and $\mathscr{L}(X)\omega = 0$, then X is called a *characteristic vector field* for ω.

THEOREM 4.5.17. If M is a regular contact manifold over B, $p: M \to B$, then a p-form φ on M is dynamic iff there is a p-form ψ on B such that $\varphi = p^*\psi$.

Proof. By $i(V_\omega)\varphi = 0$ we see that φ is specified for horizontal vector fields $X_1, \ldots, X_p$ at m in M. However, $\mathscr{L}(V_\omega)\varphi = 0$ implies that φ is invariant under right translation $R(g)$ so that $\varphi(R(g)X_1, \ldots, R(g)X_p) = R_g^*\varphi(X_1, \ldots, X_p) = \varphi(X_1, \ldots, X_p)$. Thus there is a p-form ψ at $p(m)$ with $\varphi = p^*\psi$. And conversely if $\varphi = p^*\psi$ we have $R_g^*\varphi = \varphi$ and $i(V_\omega)\varphi = 0$; so φ is dynamic.

COROLLARY 4.5.18. If $A_\phi^p(M)$ is the set of dynamic p-forms on M, then $p^*: A^p(B) \to A_\phi^p(M)$ is an isomorphism.

Clearly the set of invariant and basic forms on M forms a ring with differential operator d. One may show that

THEOREM 4.5.19. If M is an compact regular contact manifolds, $H_{\mathrm{dR}}^*(B, R)$ is the deRham cohomology and $H_0^*(M, R)$ is the ring of closed invariant forms on M, then $H_{\mathrm{dR}}^*(B, R)$ is isomorphic to $H_0^*(M, R)$.

Clearly every invariant 0-form u on M can be decomposed as $u = u_1 \wedge \omega + u_0$ with u_1 and u_0 dynamic – viz. take $u = u_1 = u_0$. This generalizes to$H_0^*(M, R)$.

THEOREM 4.5.20. Every invariant p-form φ on a regular contact manifold M can be decomposed as $u = u_1 \wedge \omega + u_0$ where u_1 and u_0 are dynamic $(p-1)$ resp. p-forms on M; and if u is closed, then u_1 is closed.

COROLLARY 4.5.21. Every invariant 1-form over a regular contact manifold is a dynamic form.

DEFINITION 4.5.22. Let $K^p = \{\text{closed } p\text{-forms } \varphi \text{ on } B \mid i(\varphi)\Omega = 0\}$. Let $\beta_p = \dim K^p$. Let $\beta_p = 0$ for $p < 0$.

THEOREM 4.5.23. Let M be a compact regular contact manifold over B. Then $b_p(M) = b_p(B) - b_{p-2}(B) + \beta_{p-1} - \beta_{p-2}$ for $p \geqq 0$.

4.6. HOMOGENEOUS CONTACT MANIFOLDS

DEFINITION 4.6.1. A contact manifold (M, ω) is called *homogeneous* if there is a transitive Lie group of strict contact transformations acting on M.

If (M, ω) is a homogeneous contact manifold and M is compact and simply connected, then the Lie group acting transitively on M has a compact semisimple subgroup which acts transitively. Furthermore, in this case we may assume that this group is simply connected.

We return to the general set-up of an arbitrary homogeneous contact manifold. Let $M = G/K$ where K is the isotropy subgroup at point x_0. We define the usual map $\pi: G \to M$ by $\pi(g) = gx_0$ for each g in G. Then $\pi^{-1}(x_0) = K$.

By definition ω is invariant under G, and so V_ω is invariant under G. Thus the orbits of V_ω are transformed transitively by G. And if an orbit is regular (or closed), then every other orbit is regular (or closed).

Let ω be a G-invariant contact form on M. Then setting $\tilde{\omega} = \pi{*}\omega$ we define

$$H = \{g \in G \mid \mathrm{ad}(g){*}\,\tilde{\omega} = \tilde{\omega}\}.$$

One easily checks that H is a subgroup of G which contains K. The Lie algebra of H is

$$\mathfrak{h} = \{X \in \mathfrak{g} \mid \mathrm{d}\tilde{\omega}(X, Y) = 0 \text{ for all } Y \text{ in } \mathfrak{g}\}.$$

Finally $\dim \mathfrak{h} = \dim \mathfrak{k} + 1$. This follows since rank $\mathrm{d}\tilde{\omega} = \text{rank } \mathrm{d}\omega = 2n$ and rank $\mathrm{d}\tilde{\omega} = \dim \mathfrak{g} - \dim \mathfrak{h}$ and rank $\mathrm{d}\omega = \dim(\mathfrak{g}/\mathfrak{k}) - 1 = \dim \mathfrak{g} - \dim \mathfrak{k} - 1$. These results lead to

THEOREM (Boothby–Wang) 4.6.2. Let G be a connected Lie group and let $(M = G/K, \omega)$ be a homogeneous contact manifold. Then ω is regular and the integral curves of V_ω are fibers of the bundle $G/K \to G/H^0K$ where H^0 is the identity subgroup of H. Thus the integral curves, which are homeomorphic to H_0K/K, are either simple closed curves or open arcs.

As we mentioned if M is compact and simply connected we may assume that G is semisimple. Consider this case when $K = \{e\}$.

THEOREM 4.6.3. If G is connected semisimple Lie group with left invariant contact form ω, then G is locally isomorphic to either $SO(3)$ or $SL(2, R)$.
Proof. Let B be the Killing form on $\mathfrak{g}$. B is nondegenerate so there is a unique element A is $\mathfrak{g}$ such that $B(A, X) = \omega(X)$ for every X in $\mathfrak{g}$. If $(\operatorname{ad} g)^*\omega = \omega$, then since $\omega(\operatorname{ad} gX) = \omega(X)$ we see that $B(A, X) = \omega(X) = \omega(\operatorname{ad} gX) = B(A, \operatorname{ad} gX)$. However, $\operatorname{ad}(g^{-1})$ is an automorphism of $\mathfrak{g}$ and it leaves B invariant. Thus, we have

$$B(A, X) = B(\operatorname{ad}(g^{-1})A, X).$$

Since B is nondegenerate and X is arbitrary we have $\operatorname{ad} gA = A$. And conversely.

The centralizer of a subset $N \subset G$ is

$$C(N) = \{g \text{ in } G \mid gng^{-1} = n \text{ for any } n \text{ in } N\}.$$

Thus, we see that H is the centralizer of the 1-parameter subgroup of G generated by A.

Since $M = G$, i.e. $K = \{e\}$ we have $\dim H = 1$ from the remarks preceding Theorem 4.6.2. But $\tilde{\omega}$ is now just ω and any maximal abelian subgroup of G belongs to H. Maximal abelian subgroups are mutually conjugate and have common dimension the rank of G. Thus

$$\operatorname{rank} G \leqq \dim H = 1.$$

Thus, G must be simple. From the classification of Lie algebras we have the result.

We have already seen the contact structures on $SO(3) = RP(3)$ and on $SL(2, R)$.

EXAMPLE 4.6.4. As we saw in Example 4.1.14 S^{2n+1} inherits a contact structure from R^{2n+2}. If M is an odd dimensional complete connected Riemannian manifold of constant curvature $K > 0$, then $M = S^{2n+1}/\Gamma$ is a contact manifold with contact structure inherited from S^{2n+1}. Here Γ is a finite group of matrices either of the form λI_r, $\lambda \in C$, $|\lambda| = 1$ where $(2n+1) = 2r - 1$ (r odd) or of the form ρI_r, $\rho \in Q$ (quaternions) $|\rho| = 1$, where $(2n+1) = 4r - 1$.

If, in addition to being a complete connected Riemannian manifold of constant positive curvature, M is required to be a homogeneous contact manifold, then M is of the above form. However, the converse is not the case in general.

4.7. Contact structures in the sense of Spencer

Definition 4.7.1. Let $\{U_\alpha\}$ be an open covering of a $(2n+1)$ dimensional manifold M. If there is a system of 1-forms $\{\omega_\alpha\}$, $\omega_\alpha \in A^1(U_\alpha)$ with $\omega_\alpha \wedge (d\omega_\alpha)^n \neq 0$ and a system of functions $\{\bar{g}_{\alpha\beta}\}$ with $\omega_\alpha = g_{\alpha\beta}\omega_\beta$ on $U_\alpha \cap U_\beta$ for $g_{\alpha\beta}$ in $A(U_\alpha \cap U_\beta)$ then M is said to be a *contact manifold in the sense of Spencer*.

Theorem 4.7.2. Let M be a contact manifold in the sense of Spencer. Then:

(i) If n is odd, M is orientable.

(ii) If n is even, $w_1(L) = w_1(M)$, where L is the line bundle defined by $\{g_{\alpha\beta}\}$.

(iii) If n is even and M is orientable, then the Spencer contact structure is given a global contact form.

Proof. From the exact sequence

$$0 \to R^+ \to R^* \overset{j}{\to} Z_2 \to 0$$

we examine the sheaf sequence

$$0 \to \mathbf{R}^+ \to \mathbf{R}^* \to Z_2 \to 0.$$

Since A is a finel sheaf and log: $\mathbf{R}^+ \to A$ is bijective, we see that $\mathbf{R}^+$ is a fine sheaf. Therefore

$$0 \to H^1(M, \mathbf{R}^*) \overset{j^*}{\to} H^1(M, Z_2) \to 0$$
$$L \mapsto w_1(L)$$

is exact where j^* maps the class of the smooth R^*-bundle L to the Stiefel–Whitney class $w_1(L)$.

Let L be the line bundle defined by $\{g_{\alpha\beta}\}$. Since

$$d\omega_\alpha \wedge (d\omega_\alpha)^n = g_{\alpha\beta}^{n+1}\omega_\beta \wedge (d\omega_\beta)^n$$

it follows that $L^{-(n+1)} \in H^1(M, \mathbf{R}^*)$ is the canonical line bundle – i.e. the line bundle on M consisting of $(2n+1)$-forms. Thus $j^*(L^{n+1}) \in H^1(M, Z_2)$ gives $w_1(M)$. If we set $a = j^*(L) = w_1(L)$ we have $(n+1)a = w_1(M)$. The theorem follows from this equality.

Spencer contact structure extends to the complex analytic case when M is a complex analytic manifold and $g_{\alpha\beta}$ is a non-vanishing complex analytic function on nonempty intersection $U_\alpha \cap U_\beta$.

THEOREM 4.7.3. A complex analytic contact manifold in the sense of Spencer is a (restricted) contact manifold iff $c_1(M) = 0$.

Proof. From the exact sequence of abelian groups

$$0 \to Z \to C \to C^* \to 0$$

we get an exact sequence of sheaves

$$0 \to Z \to \mathcal{O} \to \mathcal{O}^* \to 0$$

over M. By the cohomology sequence we have the homomorphism

$$\to H^1(M, \mathcal{O}^*) \xrightarrow{\delta^*} H^2(M, Z)$$

$$L \to c_1(L).$$

Using the holomorphic C^*-bundle $L = \{g_{\alpha\beta}\} \in H^1(M, \mathcal{O}^*)$ and setting $\alpha = c_1(L)$ we have, following the proof of Theorem 4.7.2.

$$(n+1)\alpha = c_1(M).$$

The theorem follows.

4.8. HOMOGENEOUS COMPLEX CONTACT MANIFOLDS

The homogeneous complex contact manifolds with first Chern class nonzero have been classified by Boothby using the Wang classification of homogeneous Kählerian manifolds.

Consider a homogeneous complex, compact simply connected contact manifold M. M has the form $M = G/L$ where G is complex semisimple and L is a closed complex subgroup with positive Euler characteristic. L has a closed complex normal subgroup L_1 with $L/L_1 = C^*$. The line bundle defined by the contact structure is $L/L_1 \to G/L_1 \to G/L$.

Let B be the Killing form on $\mathfrak{g}$. Then

THEOREM 4.8.1. There is a vector Z in $\mathfrak{g}$ such that:

(i) $B(Y, Z) = 0$ for all Y in $\mathfrak{l}$ = Lie algebra of L.

(ii) $\mathfrak{l}_1 = \mathfrak{c}(Z) = \{X \in \mathfrak{g} \mid [X, Z] = 0\}$ = Lie algebra of L_1.

(iii) $[\mathfrak{l}, Z] = \{Z\}$.

Using this result and the Wang classification we have

THEOREM 4.8.2. M is a compact simply connected complex homogeneous contact manifold with positive Euler characteristic iff M is a Kähler manifold, $M = G/L$ where the Lie algebra of L has the form $\mathfrak{l} = \mathfrak{c}(h'_\rho) + V_\Theta$, where ρ is a maximum root of $\mathfrak{g}$, $h'_\rho \in$ Cartan subalgebra $\mathfrak{h} \subset \mathfrak{l}$ with $B(h'_\rho, Y) = \rho(Y)$ for Y in $\mathfrak{h}$, and V'_Θ is spanned by the root vectors e_σ, $\sigma > 0$, which are not in the centralizer $\mathfrak{l}(h'_\rho)$.

We have not reviewed Lie algebras yet, so we are not in a position to prove this result. We refer the reader to Boothby B12.

COROLLARY 4.8.3. If M is as above and M contains more than one point, then $c_1(M) \neq 0$ and there is exactly one such manifold for each of the classes of simple Lie subgroups A_n, B_n, C_n, and D_n and the five exceptional simple groups. No other manifolds satisfying these hypotheses exist.

For the proof see Boothby B13.

For each class of complex simple Lie groups the homogeneous complex contact manifold are given as follows:

$$SU(n+1)/SU(n-1) \times T^2, SO(2n+1)/SO(2n-3) \times SO(3) \times T^1,$$
$$Sp(n)/SP(n-1) \times T^1, \ SO(2n)/SO(2n-4) \times SO(3) \times T^1,$$
$$G_2/SO(3) \times T^1, F_4/Sp(3) \times T^1, E_6/SU(6) \times T^1, E_7/SO(2) \times T^1,$$

and

$$E_8/E_7 \times T^1.$$

In the A_n-case if we let

$$M = SU(n+1)/SU(n-1) \times T^1,$$

and if we let

$$B = SU(n+1)/SU(n) \times T^1$$

and

$$F = SU(n) \times T^1/SU(n-1),$$

then

$$F \to M \to B = CP(n)$$

is a fiber bundle with fiber $F = CP(n-1)$. This is precisely cotangent bundle on B. To see this we note that $SU(n+1)$ acts transitively on $T(CP(n))$ with isotropy group $SU(n-1) \times T^2$. Taking the dual we have $F \to M \to B$; or in other words M is homeomorphic to the bundle of complex codirections over a complex analytic manifold. However, this is the exception rather than the rule as we note in the following theorem.

THEOREM 4.8.4. Other than the A_n-case, none of the manifolds in 4.8.3 can be homeomorphic to a bundle of complex codirections over a complex manifold.

Proof. If $M = G/L$ is a bundle of complex codirections over a complex analytic manifold B of dimension $n+1$ then the fiber F would be a complex projective space of dimension n. M is Kähler from the last theorem and F is Kähler; thus B is Kähler by general results on Kähler manifolds. Since $H^1(F) = 0$ we find by transgression that $H^1(F) \to H^2(B)$ vanishes. Since F and M are simply connected, we have $\pi_1(B) = 0$. By a result of Blanchard the real cohomology of M is isomorphic to that of $B \times F$. In particular for the Poincaré polynomials of M, F, B we have

$$p_M(t) = p_F(t)p_B(t) = (1 + t^2 \cdots + t^n)(1 + at^2 + \cdots + t^{n+1}),$$

where $a \geqq 1$ since B is Kählerian. Thus, $b_2(M) \geqq 2$.

Since $\pi_2(G) = 0$ and

$$0 \to \pi_2(G/L) \to \pi_1(L) \to \pi_1(G) \to 0$$

we have rank $\pi_2(G/L) <$ rank $\pi_1(L)$. Using the maximal compact subgroups G^u and L^u of G and we have $L^u = S \times T^n$ where S is semisimple. Thus, rank $(\pi_1(L)) = r$. Since $2 \leqq b_2(M) \leqq r$ we must have $r \geqq 2$. But this only occurs in the case A_n where $G^u = SU(n+1)$, $L^u = SU(n-1) \times T^2$.

PROBLEMS

EXERCISE 4.1. Show that if f_1 and f_2 are basic, then $\{K(f_1), K(f_2)\} =$ $= K(\{f_1, f_1\})$.

EXERCISE 4.2. Show that cont(M, ω) is a commutative ring for product $XY = K(fg)$ where $X = K(f)$, $Y = K(g)$. Show that $XY =$ $= fK(g) + gK(f) - fgV_\omega$.

EXERCISE 4.3. Show that $A(M)$ is a semisimple Lie algebra.

EXERCISE 4.4. Show that $D = \mathscr{L}(X) + a$ is a derivation of Lie algebra $A(M)$

iff X is in cont(M, ω). Thus show that every derivation of Lie algebra $A(M)$ is inner; hence $H^1(A) = 0$.

EXERCISE 4.5. Consider R^{2n+1} with contact structure $\omega = -\mathrm{d}s + \sum p_j \mathrm{d}q_j$. Show that every strict contact transformation is of the form $(p, q) \to (p', q')$, a symplectomorphism and $s \to s' = s + \pi(p, q)$ where $\mathrm{d}\pi = \sum (p'_j \mathrm{d}q'_j - p_j \mathrm{d}q_j)$.

EXERCISE 4.6. Let G denote the group of global strict contact transformations of the last exercise. Let $\underset{\sim}{C}$ be that group of transformations $(s, p, q) \to (s + r, p, q)$ for r in R. Show that $\underset{\sim}{C}$ is an invariant subgroup of G; in fact, the centre of G. Show that $G/\underset{\sim}{C}$ is the group of global symplectomorphisms.

Let T denote the subgroup formed by $(s, p, q) \to (s + \sum \alpha_j q_j + r, p_j + \alpha_j, q_j + \beta_j)$ where α_j, β_j, r are in R. Then $\underset{\sim}{C} \subset T$ and $T/\underset{\sim}{C}$ is the abelian group of translations.

Let L be the group of transformations

$$s' = s + \sum_{i,j,k} (\tfrac{1}{2} a_{ij} c_{ik} p_j p_k + \tfrac{1}{2} b_{ij} d_{ik} q_j q_k + b_{ij} c_{ij} q_i p_j) + \\ + \sum_{jk} d_{jk} a_j q_k + \sigma,$$

$$p' = \sum (a_{jk} p_k + b_{jk} q_k) + \alpha_j,$$

$$q' = \sum (c_{jk} p_h + d_{jk} q_k) + \beta_j,$$

where

$$\begin{pmatrix} \mathrm{A} & \mathrm{B} \\ \mathrm{C} & \mathrm{D} \end{pmatrix}$$

is symplectic. Show that L is a subgroup of G with $L/T = Sp(2n, R)$.

EXERCISE 4.6. Let G be the Galilean group where $g = (a, b, v, R)$ for $a \in R^3$, $b \in R$, $v \in R^3$, $R \in SO(3)$. If $(t, x) \in R \times R^3$ we let $M = T^*R^4$ have coordinates $(t, x, -h, p)$ and G acts on M by $(t^1 = t + b, x^1 = Rx + vt + a, p^1 = Rp + mv$, $h^1 = h + (Rp, v) + \frac{1}{2} mv^2$. However, M is not a homogeneous space. Let Y be the homogeneous space defined by $h - p^2/2m = w \in R$. Let $\omega_Y = \sum p_i \mathrm{d}x^i - ((p^2/2m) + w)\mathrm{d}t$. Show that (Y, ω_Y) is a 7-dimensional contact manifold with canonical vector field $V_\omega = p_i(\partial/\partial x) + m(\partial/\partial t)$. Show that the space of orbits $\mathcal{O}$ for V_ω is strongly G-homogeneous symplectic manifold with coordinates (q, p) such that $q^i = x^i - tp_i/m$ where $(x, p, t) \in Y$, G acts on (q, p) by $q^1 = Rq - b(R(p/m) + v) + a$ and $p^1 = Rp + mv$. Show that there is a unique G-invariant polarization given by $\{\partial/\partial q^i\}$ and show that $\sigma: \mathfrak{g} \to V(\mathcal{O})$ does not admit a lift.

EXERCISE 4.7. Let Y be a compact Riemannian manifold of dimension m with Laplacian Δ. Let $A = \sqrt{-\Delta}$. (A is then an elliptic pseudo differential operator of degree one; we can define a symbol for A, $a = \sigma(A)$; v. G13). Let λ denote the 1-form $\lambda = \sum \eta_j \, \mathrm{d}y_j$ where $\mathrm{d}\lambda$ is the canonical symplectic form on T^*Y. Let $X \subset T^*Y$ be the hypersurface defined by $\sigma(A) = 1$ and set $\alpha = \lambda | X$. Let $\Xi_f = \{f, \cdot\}$. Then Ξ_a is the infinitesimal generator of the $U(1)$ action on T^*Y and X.

Assume that X is a principal fiber bundle for $U(1)$ and set $B = X/U(1)$. Since $\langle \Xi_a, \alpha \rangle = a = 1$ on X, show that α is the connection form for this principal bundle. Show that $\mathrm{d}\alpha$ induces a symplectic structure on B, which form Ω in turn is the curvature form for the fibration: $X \to B$. (Let c_1 denote the Chern class of this fibration (i.e. $\Omega/2\pi$). Let $\mathscr{T}_d$ denote the Todd class of M. The spectra of A is contained in the union of intervals $I_n = [n + (\mu/4) + \sigma - (c_1/n),\ n + (\mu/4) + \sigma + (c_1/n)$ where μ is the Arnold–Maslov index. Colin de Verdiere has shown that the number $P(n)$ of eigenvalues of A contained in I_n is a polynomial in n for n large enough. It has been conjectured that number $P(n)$ is given by an analogue of the Riemann–Roch theorem $P(n) = e^{nc_1} \mathscr{T}_d[M]$; v. L. Boutet de Monvel, Sem. Bourbaki (1978/79) # 532.)

Chapter 5

The Dirac Problem

5.0. Derivations of Lie algebras

Definition 5.0.1. A *derivation* of a Lie algebra $(L, [\ ,\ ])$ is a linear map $D: L \to L$ which satisfies $D[X, Y] = [DX, Y] + [X, DY]$ for all X, Y in L. A derivation D is said to be *inner* if there exists an X in L such that $DY = [X, Y]$ for all Y in L i.e. $D = \mathrm{ad}X$; otherwise D is said to be *outer*.

Note that if D is an inner derivation then D vanishes on the center of L whereas outer derivations need not so vanish.

The set of all derivations $D(L)$ of L forms a Lie algebra under the Lie bracket $[D_1, D_2]X = D_1(D_2 X) - D_2(D_1 X)$. The set of inner derivations forms an ideal $\mathrm{ad}(L)$ in $D(L)$.

The reader can also check that every derivation is a skew symmetric transformation with respect to the Killing form B, i.e. $B(X, DY) + B(DX, Y) = 0$.

Clearly if L is commutative each of its inner derivations is the zero map and any nonzero linear transformation is an outer derivation. At the other extreme we have

Theorem (Zassenhaus) 5.0.2. If L is a semisimple Lie algebra (i.e. a nondegenerate Killing form) then every derivation D of L is inner.

In the next section we want to compare the properties of the Lie algebra of smooth functions on a manifold and the Lie algebra given by linear operators on a Hilbert space. We begin by studying the Lie algebra given by the set of complex polynomials in real variables p, q which forms a Lie algebra P under the Poisson bracket $\{f, g\}$.

Theorem (Wollenberg) 5.0.3. Every derivation D of P is of the form

$$Df = \{a_\alpha, f\} + \beta\left(f - \alpha p\frac{\partial f}{\partial p}\right) - (1-\alpha)q\left(\frac{\partial f}{\partial q}\right)$$

with a_α in P, α, β in C.

Thus every derivation of P is a sum of an inner derivation $\{a_\alpha, f\}$ and an explicitly determined outer derivation. However this decomposition is not unique. For further details, see Joseph J1.

Consider now an associative and distributive algebra Q over C generated by finite linear combinations and finite powers elements q, p where $qp - pq = 1$. Q is then a Lie algebra with respect to the standard Lie bracket $[f, g] = fg - gf$ for f, g in Q. Let ad$X$ denote the linear transformation of Q into Q given by ad$X(Y) = [X, Y]$ for X and Y in Q. The reader should check that

$$\operatorname{ad} X^m = \sum_{k=1}^{m} \binom{m}{k} (-1)^{k-1} X^{m-k} \operatorname{ad}^k X.$$

In particular we have $[q, p^n] = np^{n-1}$. Similarly for p. Thus if an element in Q commutes with p or q, then it is independent of p or q – i.e. the center of Q is formed of constant multiples of 1.

THEOREM 5.0.4. All the derivations of Q are inner

DEFINITION 5.0.5. A derivation D of Lie algebra L_1 into Lie algebra L_2 where L_1 is a subalgebra of L_2 is a linear map $D: L_1 \to L_2$ for which $D[X, Y] = [DX, Y] + [X, DY]$ for all X, Y in L_1. In this case D is said to be inner if there is an X in L such that $DY = [X, Y]$ for all Y in L_1.

In the next section we consider the question of the existence of a 'Dirac map' $\mathbf{q}$ which maps $C^\infty(M)$ to linear operators on a Hilbert space with the properties that $\mathbf{q}(1) = 1$ and $[\mathbf{q}f, \mathbf{q}g] = \mathbf{q}\{f, g\}$ – i.e. $\mathbf{q}$ gives an isomorphism between the two Lie algebras. However, if two Lie algebras are isomorphic then so are their derivation algebras. And if one Lie algebra has outer derivations and the other not, then they cannot be isomorphic. In particular let D be an outer derivation of P into $C^\infty(R^2)$ with $D(1) = 1$; e.g. $Df = f - \frac{1}{2}q(\partial f/\partial q) - \frac{1}{2}p(\partial f/\partial p)$. A Dirac map $\mathbf{q}$ is then given by $(qf)(\varphi) = \{f, \varphi\} + (Df)\varphi$ for φ in $C_0(R^2)$. The problem arises by requiring that the Dirac map is irreducible – i.e. $[\mathbf{q}(q), \mathbf{q}(p)] = 1$, i.e. the only operators commuting with both $\mathbf{q}(q)$ and $\mathbf{q}(p)$ are constant multiples of 1. We see immediately that $\mathbf{q}$ cannot form an isomorphism of the Lie algebra P and the Lie algebra Q for one algebra admits outer derivations while the other does not.

The reader should check that the subalgebra of Q formed by the linear span of $(1, q, p, q^2, p^2, qp)$ has outer derivations given by $D(q^m p^n + p^n q^m) = (\alpha/2)(2 - m - n)(q^m p^n + p^n q^m)$ with $n = 1, 2$.

5.1. GEOMETRIC QUANTIZATION: AN INTRODUCTION

Quantization unfortunately was never codified by the founding fathers. There was never a precise definition. Some interpretations would have a map q from a subalgebra of the algebra of smooth functions on a symplectic manifold to the algebra of self adjoint operators of a Hilbert space H such that:

(q1) $q(f_1+f_2)=q(f_1)+q(f_2)$;
(q2) $q(af)=aq(f), a$ in R;
(q3) $q(\{f_1,f_2\})=(1/i)[q(f_1),q(f_2)]$;
(q4) $q(1)=I_H$;
(q5) $q(x^j)$, $q(p_k)$ are unitarily equivalent to M_{x^j} and $(1/i)(\partial/\partial x^k)$.

There are many problems with this philosophy, and it is unclear that the founding fathers actually believed quantization would work this simply. Before developing the faults with this theory, let us formulate a process to start with functions or vector fields and associate to them self adjoint operators.

THEOREM (van Hove) 5.1.1. Let $t \to g_X(t)$ be a 1-parameter group of diffeomorphisms of a smooth manifold M associated to vector field X. And assume there is a Borel measure v which is invariant by $g_X(t)$. Then the operator $B_X: f \to -iX(f)$ for f in $A_0(M)$ is a symmetric operator. The closure (also denoted B_X) of B_X is self adjoint in $H=L^2(M,v)$. And $\exp(-itB_X)\psi(m)=\psi(g_X(t)^{-1}m)$ for ψ in $L^2(M,v)$.

If we assume that $K(f)$ are complete vector fields leaving invariant a Borel measure v on M, then we have a map q from $A_\phi(M)\simeq A(B)$ to self-adjoint operators on $L^2(B,v)$ which by Theorems 4.1.7 and 4.5.8 satisfies (q1), (q2), and (q3). However, $K(1)=\partial/\partial s$. Applying a Fourier transform in the s-variable would resolve (q4). However, (q5) is the real problem which problem is delineated in another theorem due to van Hove.

To state this result of van Hove, let us specialize to quantization on Euclidean spaces $M=R^{2n+1}$. As we know from Exercise 4.5 the group G of strict contact transformations on M are then of the form

$$(s,p,q)\to(s+\pi_g(p,q), g(p,q)),$$

where $\pi_{g_1g_2}(y)=\pi_{g_2}(y)+\pi_{g_1}(g_2y)$.

THEOREM 5.1.2. The left regular representation $(U, L^2(M))$ of G is unitary representation with infinitesimal generator $(U(g_t(f))=\exp(itH(f))$ given by $H(f)\varphi=K(f)\varphi$ for φ in $A_c(M)$.

THEOREM 5.1.3. The representation of G given by

$$U^{(a)}(g)f(y)=\exp(ia\pi_g(g^{-1}y))f(g^{-1}y)$$

defined on $L^2(R^{2n})$ is unitary with infinitesimal generator $U^{(a)}(g_t(f)) = \exp(itH^{(a)}(f))$ given by $H^{(a)}(f)\varphi = a(f - \sum p_j \partial f/\partial p_j) + i\{f, \varphi\}$ for φ in $A_c(R^{2n})$. For $a \neq 0\, U^{(a)}$ is an irreducible unitary representation. And U is unitarily equivalent (by Fourier transform) to a continuous direct sum of representations $U^{(a)}$.

The relationship between $H(f)$ and $H^{(a)}(f)$ is given by the Fourier transform as noted. More precisely the map $A_k : \varphi(p, q) \to \sqrt{(a/2\pi)} \times \exp(-aks)\varphi(p, q)$ is a unitary transform on $L^2(R^{2n})$ for $k = 0, 1, 2, \ldots$. We are able to decompose $\varphi(s, y) = \sum A_k \sqrt{(a/2\pi)} \int_0^{2\pi/a} \exp(iaku)\varphi(u, y)du$. Then we find that $A_k^{-1} H(f) A_k \varphi = H^{(ka)}(f)\varphi$ for φ in $A_c(R^{2n})$. For details of the proof see Van Hove H29.

In summary from the commutative diagram

$$\begin{array}{ccccc} 0 \to A_\phi(M) & \to & A(M) & \to & A(M) \\ \uparrow\downarrow & & \uparrow\downarrow & & \| \\ 0 \to \text{cont}_0 & \to & \text{cont} & \to & A(M) \end{array}$$

we associate infinitesimal strict contact transformations to functions in $A_\phi(M) \simeq A(B)$. Now not every such f produces a complete vector field. Furthermore, the vector field associated to $f_1 + f_2$ is not necessarily complete even if it were so for f_1 and f_2. Examples are given in van Hove H29.

DEFINITION 5.1.4. Let $\tilde{A}_0$ denote the functions $f(p, q)$ with associated complete vector fields in cont_0.

THEOREM 5.1.5. For f in $\tilde{A}_0, H(f), H^{(a)}(f)$ are essentially self adjoint operators on their domains. If f_1, f_2 are in $\tilde{A}_0$ and $a_1 f_1 + a_2 f_2$ and $\{f_1, f_2\}$ are in $\tilde{A}_0$, then

$$a_1 H(f_1) + a_2 H(f_2), a_1 H^{(a)}(f_1) + a_2 H^{(a)}(f_2), [H(f_1), H(f_2)]$$

and $[H^{(a)}(f_1), H^{(a)}(f_2)]$ are essentially self adjoint on their domains with self adjoint extensions $H(a_1 f_1 + a_2 f_2)$, $H^{(a)}(a_1 f_1 + a_2 f_2)$, $H(\{f_1, f_2\})$ and $H^{(a)}(\{f_1, f_2\})$, respectively.

Thus using $H^{(a)}$ we have a map $q : \tilde{A}_0 \to$ self adjoint operators on $L(R^{2n})$ which satisfies (q1)–(q4).

However, (q5) as noted above will cause the difficulty. If we let $P_j = (1/a) H^{(a)}(P_j)$ and $Q_k = (1/a) H^{(a)}(q)$ then P_j, Q_k have self-adjoint extensions also denoted by P_j, Q_k which satisfy $[Q_j, Q_k] = 0$, $[P_j, P_k] = 0$ and

$[Q_j, P_k] = (i/a^2)\, H^{(a)}(\{q_j, p_k\}) = (i/a^2)\, H^{(a)}(\delta_{jk}) = (i\delta_{jk}/a)$. Thus we have the canonical commutation relations holding on $L^2(R^{2n})$.

As we know from any elementary course on quantum theory, the wave functions are functions of only one half the degrees of freedom – usually position $(q_1, q_2, \ldots, q_n)$ or momentum $(p_1, \ldots, p_n)$. Admitting this constraint we examine what happens. First we recall the Stone–von Neumann theorem.

Let E be a real vector space with dual space E^*.

DEFINITION 5.1.6. A pair of unitary representations (U, V) of the additive groups E, E^* are said to satisfy *the Weyl commutation relations* if $U(x)V(f) = \exp(if(x))V(f)U(x)$ for x in E and f in E^*.

Let $P(x)$ and $Q(f)$ denote the infinitesimal generators of U and V respectively. Consider the 1-dimensional case for illustration: $\exp(itp)\exp(isq) = \exp(ist)\exp(isq)\exp(itp)$. This is just the Weyl form of the canonical commutation relations $[p, q] = 1/i$. As we know there is a distinguished representation – the *Schrödinger representation* where $U(tp)\varphi(x) = \varphi(x+t)$ and $U(tq)\varphi(x) = \exp(itx)\varphi(x)$ on $L^2(R)$; i.e. $p = (1/i)\partial/\partial x$ and $q = M_x$. This generalizes to give:

DEFINITION 5.1.7. Let $H = L^2(E)$. Then the *Schrödinger representation* of the Weyl relations is $U(x)\psi(y) = \psi(y+x)$ and $V(f)\psi(y) = \exp(if(y))\psi(y)$, where ψ is in H.

THEOREM (Stone–von Neumann) 5.1.8. Any Weyl system (U, V, E) is unitarily equivalent to a direct sum of copies of the Schrödinger system.

If we write $\mathscr{E} = E + E^*$, then $T(z) = U(x)V(f)$ where $z = (x, f)$ satisfies

$$T(z)\,T(z') = \exp(iB(z, z')/2)\,T(z + z'),$$

where $B(z, z') = f'(x) - f(x')$ is a skew symmetric bilinear form – i.e. a symplectic form. The space $(\mathscr{E}, B)$ is a symplectic vector space. And T is a projective representation of $\mathscr{E}$. The infinitesimal generator of $T(tz) = \exp(itA(z))$ provides a map from z in $(\mathscr{E}, B)$ to self adjoint operators on a Hilbert space. If $\{(e_i, f_i)\}$ is the canonical basis of $(\mathscr{E}, B)$ – i.e. $B(e_i, e_j) = B(f_i, f_j) = 0$ and $B(e_i, f_j) = \delta_{ij}$, then $(A(e_i), A(f_j))$ are pairs of canonically conjugate dynamical operators.

There is a general technique of passing from projective representations of a group G to standard representations of an associated group:

THEOREM 5.1.9. For every projective representation V of G with multiplier μ there is a standard representation $U(t,g) = tV(g)$ of the associated group $G^{\mu} = \{t, g \mid t \text{ in } T^1, g \text{ in } G\}$ with multiplication to be defined below. The correspondence $V \to U$ is one to one and U is a representation such that $(t,e) \to U(t,e)$ is a multiple of the character $(t,e) \to t$.

DEFINITION 5.1.10. The *associated group* G^{μ} is the set of pairs (t,g) as above under the product rule $(t,g)(t',g') = (tt'/\mu(g,g'), gg')$. If G has a Borel structure, then under the product structure G is a separable locally compact topological group.

The associated group to the projective representation $(T, \mathscr{E})$ is the set $\tilde{\mathscr{E}}$ of $\mathscr{E} \times T^1$ under the product $(t,z)(t,z') = (tt' \exp(iB(z,z')), z+z')$. The identity is $(1, 0)$ and the inverse is $(t^{-1}, -z)$. Clearly $(t,z)\,(t',z') = (t',z')\,(t,z)$ for all (t',z') iff $B(z,z') = 0$ for all z', i.e. $z = 0$. Thus the center of $\tilde{\mathscr{E}}$ is isomorphic to T^1. The homomorphism $(t,z) \to z\colon \tilde{\mathscr{E}} \to \mathscr{E}$ has kernel T^1. Thus

THEOREM 5.1.11. $\tilde{\mathscr{E}}$ is the central extension of $\mathscr{E}$ by T^1.

DEFINITION 5.1.12. $\tilde{\mathscr{E}}$ is called the *Heisenberg group*.

THEOREM 5.1.13. $\tilde{\mathscr{E}}$ is a locally compact real connected 2-step nilpotent Lie group with maximal compact subgroup and center T^1.

Proof. We check only the 2-step nilpotent part, which follows from examining the central subgroups:

$$\begin{aligned} C^0(\tilde{\mathscr{E}}) &= \{(0,0)\} \\ C^1(\tilde{\mathscr{E}}) &= \{(t,z) \mid ((t,z)(t',z')^{-1}(t',z')^{-1} = (0,0)\} \\ &= \{(t,z) \mid B(z,z') = 0 \bmod 2\pi\} \\ &= \{(t,0)\} = T^1 \\ C^2(\tilde{\mathscr{E}}) &= \{(t,z) \mid (t,z)(t',z')(t,z)^{-1}(t',z')^{-1} \text{ in } T^1\} \\ &= \{(t,z) \mid (2B(z,z'),0) \text{ in } T^1\} \\ &= \tilde{\mathscr{E}}. \end{aligned}$$

The Lie algebra $\tilde{\varepsilon}$ of $\tilde{\mathscr{E}}$ is the space $\{(ix,z) \mid x \text{ in } R,\ z \text{ in } \mathscr{E}$ with $[(ix,z),(iy,z')] = (2\pi i B(z,z'),0)$.

5.2. THE DIRAC PROBLEM

DEFINITION 5.2.1. The *Dirac problem of quantization* is stated to be a map from some subalgebra of $A(B)$ to the self adjoint operators on $L^2(R^n)$ which satisfies (q1)–(q5).

However, this does not work in general as was shown by van Hove:

THEOREM 5.2.2. For every real a there does not exist a map $f \to A(f)$ from $\tilde{A}_0$ to self adjoint operators on a Hilbert space $L^2(R^n)$ which satisfies:

(Q1) there exists a common domain in H for $A(f)$ for f in $\tilde{A}_0$ which is invariant for $A(f)$ and $\exp(iatA(f))$;

(Q2) if $a_1 f_1 + a_1 f_2$ and $\{f_1, f_2\}$ are in $\tilde{A}_0$ for f_1, f_2 in $\tilde{A}_0$ then $a_1 A(f_1) + a_2 A(f_2) = A\left(\sum a_i f_i\right)$ and $[A(f_1), A(f_2)] = A(\{f_1, f_2\})$;

(Q3) if $g_s(f_3) = g_t(f_1) g_s(f_2) g_{-t}(f_1)$ then $\exp(iasA(f_3)) = \exp(iatA(f_1))$ $\exp(iasA(f_2)) \exp(-iatA(f_1))$;

(Q4) the operators $A(p_j)$ and $A(q_j)$ are equivalent to $(1/ai)(\partial/\partial x_j)$ and M_{x_j}, $j = 1, \ldots, n$.

On the positive side van Hove showed that if we restricted G to the subgroup L (cf. Exercise 4.5) then the corresponding subalgebra of strict contact transformations of L is the family of polynomials of degree 0, 1, 2, in p_j, q_k. We denote this set by A_0^L.

THEOREM 5.2.3. There does exist a quantization map from $A_0^L \to L^2(R^n)$ which satisfies (Q1)–(Q4).

Rather than proving the negative or positive result we demonstrate the truth of the positive result of van Hove by a construction.

First we note that $H^{(a)}(f)$ in terms of the symplectic geometry is just $H^{(a)}(f) = a(f + \omega_0(p(f)) + ip(\mathrm{d}f)$ where $\Omega = \mathrm{d}\omega_0$. As we know from the last chapter $p(\mathrm{d}f) = P(f) = \{f, \cdot\}$. Taking $\omega_0 = \frac{1}{2}\sum(p_j \mathrm{d}q_j - q_j \mathrm{d}p_j)$ we have $\omega_0(p(f)) = \frac{1}{2}\sum(q_i(\partial f/\partial q_i) + p_i(\partial f/\partial p_i))$.

EXAMPLE (Harmonic Oscillator) 5.2.4. The harmonic oscillator has Hamiltonian $H = \frac{1}{2}(q^2 + p^2)$. As we see from van Hove's positive result this system admits a solution to Dirac's version of quantization. Taking $\hat{f} = H^{(1)}(f)$ and ω_0 as above we have $\hat{q} = \frac{1}{2}q + i(\partial/\partial p), \hat{p} = \frac{1}{2}p - i(\partial/\partial q)$, and $\hat{H} = i(q(\partial/\partial p) - p(\partial/\partial q))$.

Now consider the functions $T_1 = -q$, $T_2 = p$ and $M = \frac{1}{2}(p^2 + q^2)$. Although the classical invariant $\mathcal{S} = T_1 + T_2 - 2M$ vanishes, the image under the van Hove map, $\hat{\mathcal{S}}$, of $\mathcal{S}$ is not a scalar multiple of the identity on $L^2(R^2)$. An invariant subspace of $L^2(R^2)$ must be selected on which $\hat{p}$, $\hat{q}$ act irreducibly – viz. look for a subspace corresponding to a fixed eigenvalue of $\hat{\mathcal{S}}$. Checking that $2\hat{\mathcal{S}} + 1 = (i\bar{z} - 2i(\partial/\partial z))(iz + 2i(\partial/\partial \bar{z}))$ with coordinate $z = p + iq$, it follows that $(2\hat{\mathcal{S}} + 1)(\exp(-\frac{1}{4}|z|^2)\psi(z)) = 0$ for every holomorphic function $\psi(z)$. Clearly $\hat{H}$, $\hat{p}$, $\hat{q}$ leave the subspace of holomorphic functions invariant and the subspace selected is the Fock space $\mathscr{F}$ of holomorphic functions with the scalar product

$$\langle\psi_1|\psi_2\rangle = \frac{1}{\pi}\int_C \exp(-\tfrac{1}{2}|z|^2)\overline{\psi_1(z)}\,\psi_2(z)\frac{\overline{\mathrm{d}z}\wedge \mathrm{d}z}{2i}$$

which we saw before in Section 2.2.

The symplectic manifold $(R^2, \mathrm{d}p\wedge \mathrm{d}q)$ has the natural structure of a Kähler manifold which is being used in this example. To see this, we note that $\omega = \mathrm{d}s + \frac{1}{2}(p\mathrm{d}q = q\mathrm{d}p) = \mathrm{d}s + (1/4i)(\bar{z}\mathrm{d}z - z\mathrm{d}\bar{z})$. Thus $\Omega = \mathrm{d}\omega = = (1/2i)\,\overline{\mathrm{d}z}\wedge \mathrm{d}z$ is the Kähler (1, 1) form. Noting that Ω can be rewritten as $\Omega = i\,\partial\bar{\partial}(\frac{1}{2}\ln|z|^2)$, this suggests the following generalization due to Onofri. Let N be a contact manifold over a Kähler manifold M with $\omega = = \mathrm{d}\mathfrak{z} - i\partial f(z,\bar{z})$ where $\mathrm{Im}(\mathfrak{z}) = \frac{1}{2}f(z,\bar{z})$. As above we have $\Omega = i\partial\bar{\partial}f(z,z)$. Let G be a group of holomorphic automorphisms of ω. One can check (exercise for the reader) that in this case $f(g(z), g(\bar{z})) = f(z,\bar{z}) + h_g(z) + h_g(\bar{z})$ for g in G where $h_g(z)$ are holomorphic functions. Then we form the unitary representation of G on the space of holomorphic functions which are square integrable with respect to

$$\langle\psi_1|\psi_2\rangle = \int \exp(-f(z,z))\overline{\psi_2(z)}\,\psi_2(z)\Omega$$

given by $(U(g)\psi)(x) = (\exp(h_g)\psi)(g^{-1}z)$.

By a theorem due to Kobayashi K 10 this representation U is known to be either trivial or irreducible.

In the case of the harmonic oscillator the group G is the group of Euclidean transformations $g(z) = \exp(it)z + c$.

DEFINITION 5.2.5. A Lie algebra $\mathfrak{g}$ is called *nilpotent* if $[\mathfrak{g},[\mathfrak{g},\ldots[\mathfrak{g},]\ldots]] = 0$.

Clearly we have

THEOREM 5.2.6. The set $(\hat{q}, \hat{p}, I)$ forms a nilpotent Lie algebra.

This is just the Heisenberg Lie algebra.

DEFINITION 5.2.7. A Lie algebra $\mathfrak{g}$ is *solvable* if setting $\mathfrak{g}_2 = [\mathfrak{g}, \mathfrak{g}]$, $\mathfrak{g}_3 = [\mathfrak{g}_2, \mathfrak{g}_2]$, etc., we have $\mathfrak{g}_k = \{0\}$ for some k.

One may check that the set $\hat{p}, \hat{q}, \hat{H}, I$ forms a Lie algebra. This Lie algebra is called the *oscillator Lie algebra.* And since $[\mathfrak{g},\mathfrak{g}]$ = Heisenberg Lie algebra we see that $\mathfrak{g}_3 = I$ and $\mathfrak{g}_4 = \{0\}$. Thus

THEOREM 5.2.8. The oscillator Lie algebra is a solvable Lie algebra.

5.3. KOSTANT AND SOURIAU APPROACH

Kostant and Souriau generalized the van Hove construction by realizing that the van Hove q-map is just the map

$$\nabla_X \varphi + 2\pi i \varphi,$$

where ∇_X is the covariant derivative $\nabla_X f = Xf + 2\pi i\, i(X)\vartheta f$ associated to the connection $\alpha = \vartheta + (1/2\pi i)(dz/z)$ in the line bundle $L = B \times C^* \to B$. More specifically the generalization goes as follows. Let $L_c(B, \Omega)$ be the set of line bundles with connection ω over a symplectic manifold (B, Ω) having Ω as the curvature of the connection ω. Let S denote the space of smooth sections of the line bundle. Now the covariant derivative associated to a connection ω is a linear map ∇: $V(B) \to \operatorname{End} S$ which satisfies

(i) $\nabla_{fX} = f\nabla_X$;

(ii) $\nabla_X(fs) = X(f)s + f\nabla_X s$;

for f in $A(B)$ and s in S. Thus ∇ is related to the connection 1-form ω by noting that $X \to (1/2\pi i)\nabla_X(s)/s$ is C-linear from $V(B) \to A(B)$. Thus, there is a unique 1-form $\omega = \omega(s)$ such that $\nabla_X s = 2\pi i\ i(X)\omega s$.

THEOREM (Kostant-Souriau) 5.3.1. The generalized van Hove map q is q: $A(B) \to S$ given by $q(f)s = (\nabla_{X_f} s + 2\pi i f s)$ for f in $A(B)$ and s in S. In particular q is a representation of $A(B)$ on S – i.e. $q(\{f_1, f_2\}) = [q(f_1), q(f_2)]$.

However, as we saw in even the simplest dynamical system, the harmonic oscillator, the only hope of achieving some form of Dirac quantization involves selecting out a further sub-algebra from $A(B)$. The Kostant–Souriau approach involves the theory of polarization to which we turn next.

PROBLEMS

EXERCISE 5.3.1. Let G be a locally compact Abelian group with character group $\hat{G}$. Let $\sigma: G \to \hat{G}$ be given by $\sigma[s](\alpha) = (s, \alpha)$ for s in G and α in G. Let R: $\alpha \to R_\alpha$ and M: $a \to M_a$ be representations of G and $\hat{G}$ in $H = L^2(\hat{G})$ where $R_\alpha \hat{f}(\tau) = \hat{f}(\tau\alpha)$ and $M_u \hat{f}(\tau) = z(\tau)\hat{f}(\tau)$ for f in $L^2(\hat{G})$. Show that M, σ provide a representation M: $s \to M_{\sigma(s)}$ of G in H. Show that M, R are weakly continuous unitary representations of $G, \hat{G}$ respectively and satisfy $M_s R_\alpha = (s, \alpha) R_a M_s$ for all s in G and α in G. Thus by Mackey's version of the

Stone–von Neumann theorem there exists a linear isometry S from H to a direct sum of n-fold copies of $L^2(G)$ such that

$$SM_sS^{-1}\{f_1(t),\ldots,f_k(t),\ldots\}=\{f_1(ts),\ldots,f_k(ts),\ldots\}$$
$$SR_\alpha S^{-1}\{f_1(t),\ldots,f_k(t),\ldots\}=\{\alpha(t)f_1(t),\ldots,\alpha(t)f_k(t),\ldots\}$$

where $f_i(t)\in L^2(G)$. Use this fact to show that G is topologically isomorphic to $\hat{G}$ (i.e. Pontryagin's theorem).

EXERCISE 5.3.2. Let G and $\hat{G}$ be as in the last exercise. For a Borel subset S of G and the characteristic function $c_S(g)$ define $E(S)f(g)=c_S(g)f(g)$. This defines a spectral measure dE on $L^2(G)$ such that $U(g)E(S)=E(gS)U(g)$ where $U(g)f(.)=f(g.)$ is the usual representation of G on $L^2(G)$. Define the representation of $\hat{G}$ by $V(\mathfrak{X})=\int_G\overline{\mathfrak{X}(g)}\,dE(g)$ where $\mathfrak{X}\in\hat{G}$. Then the reader can check that $U(g)V(\mathfrak{X})=\mathfrak{X}(g)V(\mathfrak{X})U(g)$. Now reverse the roles of U, V and G, $\hat{G}$; call these representations U' and V'. Then by Mackey's theorem there is an isometry $T\colon L^2(G)\to L^2(\hat{G})$ such that $U'T=TU$ and $V'T=TV$. Use this to show that there is a map $\mathcal{F}:\varphi\in L^2(G)\to(1/c)\int\mathfrak{X}(g)\,\varphi(g)\,dg$ such that $\int|\mathcal{F}\varphi(\mathfrak{X})|^2\,d\mathfrak{X}=\int|\varphi(g)|^2\,dg$, $\int\mathcal{F}\varphi(\mathfrak{X})\overline{\mathcal{F}\psi(\mathfrak{X})}$, $d\mathfrak{X}=\int\varphi(g)\psi(g)\,dg$ and $\mathcal{F}^{-1}(f)(g)=(1/c)\int\mathfrak{X}(g)f(\mathfrak{X})\,d\mathfrak{X}$ for f in $L^1(G)\cap L^2(G)$.

EXERCISE 5.3.2. Let G be a compact Hausdorff group, let $R(G)$ be the representation algebra of G and let S be the compact group of homomorphisms from $R(G)$ into C. Let R_g be the right regular representation of S and let $\hat{}: G\to S$ be the map $\hat{x}(f)=f(x)$ for f in $R(G)$. Show that $\hat{}$ is a continuous group homomorphism. Let $R(g)=R_{\hat{g}}$ define a weakly continuous representation of G on $L^2(S)$. Let $T\colon g\to T(g)$ be the multiplication representation of $L^\infty(S)$ on $L^2(S)$ given by $T(g)k(t)=g(t)k(t)$. Let ${}^0: R(G)\to R(S)$ be the map $f^0(s)=s(f)$ and set $T(f)=T(f^0)$. Show that

$$R(g)R(f)=T(f_x)R(g)$$

where $(f_x)(y)=f(yx)$.

Mackey's generalization of the Stone–von Neumann theorem states that for G any locally compact group (not necessarily abelian) and $B(G)$ the *-algebra of measurable functions on G, then if U is a weakly continuous representation of G on H and V is a *-representation of $B(G)$ as bounded operators on H and $U_xV_n=V_n\,U_x$, then there exists a linear isometry $A\colon H$ to a direct sum of n-copies of $L^2(G)$ such that $AU_xA^{-1}=\sum\oplus T(x)$ and $AV_nA^{-1}=\sum\oplus R(n)$. Show that this implies that G is topologically isomorphic to S (i.e. the Tannaka duality theorem).

Chapter 6

Geometry of Polarizations

6.1. Polarizations

Let M be a real manifold of dimension $2n$. Then an *almost complex structure* on M is a tensor field of type $(1,1)$ $j: m \to j_m$ in End $(T_m M)$ for m in M such that (i) $j_m^2 = -1$ and (ii) j is smooth. j thus defines a complex distribution $F: m \to F_m \subset T_m^C$ such that $T_m^C = F_m \oplus \bar{F}_m$. The almost complex manifold will be denoted (M, j). And the smooth functions f in $A(M)$ which satisfy $X(f) = 0$ for all X in $V_F(M) = \{X \in V^C(M) \mid X_m \text{ in } F_m, m \text{ in } M\}$ is the algebra of holomorphic functions.

Definition 6.1.1. The almost complex manifold (M, j) is called *complex* if F is involutive – i.e. $[X, Y] \in V_F$ for all X, Y in V_F.

Definition 6.1.2. A *Kähler manifold* is a symplectic manifold (M, Ω) with complex (F, j) structure such that $\Omega_m|_{F_m} = 0$ for all m in M.

Let E be a real vector space and B be an alternating bilinear form on E. If $E(B)$ is the kernel of B, i.e. $E(B) = \{x \mid B(x, y) = 0 \text{ for all } y \text{ in } E\}$, then B defines a nondegenerate alternating form on $E/E(B)$; thus dim $(E/E(B))$ is even. If W is a vector subspace of E, $W(B)$ is the orthogonal complement of W with respect to B on E.

Definition 6.1.3. W is said to be a *Lagrangian* or *totally isotropic subspace* if $B|W$ is identically zero.

Theorem 6.1.4. W is a maximal Lagrangian subspace if $W = W(B)$ or if W is a Lagrangian subspace and dim $W = \frac{1}{2}(\dim E + \dim(E(B)))$.

Definition 6.1.5. If dim $E = 2n$ and B is nondegenerate, then the set of all Lagrangian subspaces of dimension n is denoted $\Lambda(n)$ and is called the *Lagrangian Grassmannian.*

Theorem 6.1.6. $\Lambda(n) = U(n)/0(n)$.

If E^C is the complexification of E and B is nondegenerate, then B extends to a form on E^C and defines a Hermitian form H on E^C by $H(x, y) = 2iB(x, y)$.

DEFINITION 6.1.7. Maximal Lagrangian subspaces W of E^C for which H is positive – i.e. $iB(w, w) \geqq 0$ for all w in W – are said to be *positive.*

DEFINITION 6.1.8. A *polarization* of a symplectic manifold (M, Ω) of dimension $2n$ is a smooth distribution $F: m \to F_m$ in T_m^C of dimension n such that (i) $\Omega_m|_{F_m} = 0$ for all m in M (i.e. F_m is a complex Lagrangian subspace of T_m^C) and (ii) F is involutive.

If $X \to \bar{X}$ denotes the complex conjugation, then the conjugate $\bar{F}: m \to \bar{F}_m$ to a polarization F is a polarization.

DEFINITION 6.1.9. Let (M, Ω, F) be a polarized manifold. F is said to be *real* if $F = \bar{F}$; and F is said to be *Kählerian* if $F \cap \bar{F} = \{0\}$(i.e. $T_m^C = F_m \oplus \bar{F}_m$).

DEFINITION 6.1.10. Let (M, Ω, F) be a polarized manifold. F is said to be *admissable* if

(i) $m \to (F_m \cap \bar{F}_m)$ is an involutive distribution of dimension k and
(ii) $m \to F_m + \bar{F}_m$ is an involutive distribution of dimension $(2n - k)$.

Thus if F is admissable, there are two distributions

$$D: m \to D_m = F_m \cap T_m = \bar{F}_m \cap T_m$$
$$E: m \to E_m = (F_m + \bar{F}_m) \cap T_m.$$

Clearly $\{0\} \subset D_m \subset E_m \subset T_m$ and $\dim D_m = k$, $\dim (T_m/E_m) = k$ and $\dim E_m/D_m = 2(n - k)$. Also F is real iff $D = E$; and F is Kählerian iff $D = 0$.

DEFINITION 6.1.11. Let (M,Ω,B) be a polarized manifold and let $V_F(M) = \{X | X_m \in F_m$ for all m in $M\}$. Let $A_F(M) = \{f$ in $A(M) | Xf = 0$ for all X in $V_F\}$. A_F is called the *algebra* of *F-holomorphic functions.*

Clearly $A_F(U)$ for U open in M defines a sheaf, $\mathscr{A}_F$, of germs of F-holomorphic functions. Two elementary facts follow regarding A_F.

THEOREM 6.1.12. (i) For f in $A(M)$, f is in A_F iff $p(f)$ is in V_F where p is the map in $0 \to C \to A(M) \overset{p}{\to} V(M) \to 0$.

(ii) A_F is an abelian Lie subalgebra of $A(M)$.

DEFINITION 6.1.13. Let (M, Ω, F) be a polarized manifold. Set Set $C_F^{(k)} =$

$= \{f \text{ in } A(M) | \{f_1, \ldots \{f_{k+1}, f\} \ldots\}|_U = 0$ for functions f_i in $A_F(U)$ for all opens U in $M\}$.

THEOREM 6.1.4.

(i) $C_F^{(k)} \subset C_F^{(k+1)}$;

(ii) $\{C_F^{(k)}, A_F\} \subset C_F^{(k-1)}$;

(iii) if (M, F) is an admissably polarized manifold $C_F^{(0)} = A_F$ and $\{C_F^{(k)}, C_F^{(l)}\} \subset C_F^{(k+l-1)}$ where $C_F^{(-1)} = 0$;

(iv) if f in $A(M)$ is such that $p(f)$ is globally integrable with 1-parameter group $g(t)$ and (M, F) is admissably polarized, then $g(t)_* F = F$ for all t in R iff f is in C_F^1.

COROLLARY 6.1.15. If (M, F) is as in (iii) then $\{C_F^{(1)}, C_F^{(k)}\} \subset C_F^{(k)}$ for all k.

From this Corollary we see that $C_F^{(1)}$ is a Lie subalgebra of A and A_F is an abelian ideal of $C_F^{(1)}$.

EXAMPLE 6.1.16. The standard example of a polarized manifold is $M = T^*N$ with local coordinates (p_i, q_i). Let F_m be spanned by $\{(\partial/\partial p_i)_m$, $i = 1, \ldots, n$. Then $A_F = \{f$ in $A(M)$ with $f(p, q) = f(q)\}$ and $C_F^{(1)} = \{f$ in $A(M) | f(p, q) = \sum p_i \hat{f}^i(q) + \hat{f}(q)\}$.

Let (M, Ω, F) be an admissably polarized manifold. Let (L, α) be a fiber bundle over M with connection α such that $d\alpha = \Omega$. The sections of L constant along F are defined by

DEFINITION 6.1.17. $S_F = \{f$ in $S | \nabla_X f = 0$ for all X in $V_F\}$.

EXAMPLE 6.1.18. If (M, Ω, F) is a Kählerian manifold, then S_F is the space of holomorphic sections of (L, α).

THEOREM 6.1.19. If q is the Kostant–Souriau map, then S_F is $q(C_F^{(1)})$ stable.

Thus we have a Lie algebra representation $q: C_F^{(1)} \to S_F$. Note that if f is in $A_F \subset C_F^{(1)}$, then $q(f)s = 2\pi$ ifs for s in S.

DEFINITION 6.1.20. A polarization F on a G-symplectic manifold is G-*invariant* if $\sigma_*(g) F_m = F_{g,m}$ for all g in G and m in M.

Representations of G on S follow if we consider a polarized G-manifold (M, Ω, F) with $\sigma: \mathfrak{g} \to V(M)$ admitting a lifting $\lambda: \mathfrak{g} \to A(M)$ such that

$\lambda(\mathfrak{g}) \subset C_F^{(1)}$. If F is G-invariant, then S_F is G-stable and we have the natural homomorphism $T_F: G \to \mathrm{Aut}(S_F)$.

6.2. RIEMANN-ROCH FOR POLARIZATIONS

Let $F^0 \subset (T^*M)^C$ be the subcotangent bundle of covectors which vanish on F. Thus f is in $A_F(U)$ over open U in M iff $\mathrm{d}f$ is a section of F^0 on U. If there are functions $f_1, \ldots, f_m$ in $A_F(U)$ with $(\mathrm{d}f_1, \ldots, \mathrm{d}f_m)$ forming a frame of F^0 at each point of U, then $(f_1, \ldots, f_m)$ is called an *A_F-coordinate system* and U is an *A_F-coordinate neighborhood. F* is *integrable* if M can be covered by A_F-coordinate neighborhoods. And if F is integrable then F is involutive. For the converse we need:

THEOREM (Frobenius-Nirenberg) 6.2.1. If F is involutive and if either
(i) $\dim_C (F_m \cap \bar{F}_m)$ is constant and $F + \bar{F}$ is involutive, or
(ii) M is a real analytic manifold and F is an analytic subbundle, then F is integrable.

Let $\Omega_F^p(U)$ denote the sections of $\Lambda^p F^*$ over open U in M. Thus $\Omega_F^0(U) = A(U)$. Viewing sections of $\Lambda^p F^*$ as alternating $A(U)$-multilinear maps of $V_F(U)$ into $A(U)$ for the case that F is involutive, we can define a differential $\mathrm{d}_F: \Omega_F^p(U) \to \Omega_F^{p+1}(U)$ by

$$(\mathrm{d}_F\alpha)(X_1, \ldots, X_{p+1}) = \sum_{i=1}^{p+1} (-1)^{i+1} X_i[\alpha(X_1, \ldots, X_{p+1})] +$$

$$+ \sum_{i<j} \alpha([X_i, X_j], X_1, \ldots, \hat{X}_i, \ldots, \hat{X}_j, \ldots, X_{p+1})$$

for X_i in $V_F(U)$, $i = 1, \ldots, p+1$. One quickly checks that $\mathrm{d}_F^2 = 0$ and $\mathrm{d}_F f = \mathrm{d}f|_F$ for f in $A(M)$. Thus we have $A_F(U) = \{f \text{ in } A(M) | \mathrm{d}_F f = 0\}$.

The differential complex

$$\Omega_F^0(M) \xrightarrow{\mathrm{d}_F} \Omega_F^1(M) \to \ldots \xrightarrow{\mathrm{d}_F} \Omega_F^n(M) \to 0$$

has cohomology groups denoted by $H^p(\Omega_F^*(M))$.

If $\mathscr{A}_F^p$ is the sheaf of germs of local smooth sections of $\Lambda^p F^*$ we have the induced sequence

$$0 \to \mathscr{A}_F \to \mathscr{A}_F^0 \xrightarrow{\mathrm{d}_F} \ldots \xrightarrow{\mathrm{d}_F} \mathscr{A}_F^n \xrightarrow{\mathrm{d}_F} 0. \qquad (*)$$

We now state the generalized deRham theorem:

THEOREM 6.2.2. If condition (i) of Theorem 6.2.1 holds for involutive F then $H^p(M, \mathscr{A}_F) \simeq H^p(\Omega_F^*(M))$ for all p and (*) is a fine resolution of $\mathscr{A}_F$.

EXAMPLE 6.2.3. In the case $F = TM^C$ then this theorem is just the usual deRham theorem because $\mathscr{A}_F$ is just the constant sheaf C.

EXAMPLE 6.2.4. In the case $TM^C = F \oplus \bar{F}$ and J in End TM has $J = -i$ on F and $J = i$ on $\bar{F}$, then M is a complex manifold; A_F-neighbourhoods are holomorphic coordinate systems; $\Omega_F^p(M) \simeq \Omega^{0,p}(M)$ and (*) is the $\bar{\partial}$-resolution of the sheaf of germs of holomorphic functions $\mathscr{A}_F$.

REMARK. For the rest of this section we assume F is involutive and satisfies (i) of Theorem 6.2.1.

By selecting a direct summand say F^1 of F with respect to some Hermitian structure on TM^C – i.e. $TM^C = F^1 \oplus F$, we can develop a double grading $\Lambda^k(T^*M)^C = \oplus_{p+q=k}(\Lambda^pF^1 \otimes \Lambda^qF)^*$ on the sheaf of germs of C-valued smooth differential forms on M. Then $\Omega_F^k = \oplus \Omega_F^{p,q}$ where $\Omega_F^{p,q}$ is the sheaf of germs of sections of $(\Lambda^pF^1 \otimes \Lambda^qF)^*$. The sequence (*) is now of the form

$$0 \to \mathscr{A}_F^p \to \Omega_F^{p,0}(C) \xrightarrow{d_F} \Omega_F^{p,1}(C) \to \ldots \Omega_F^{p,n}(C) \to 0.$$

where $n =$ rank of F.

For complex vector space E we set $\Omega_F^{p,q}(E) =$ the sheaf of E-valued forms. We restate Theorem 6.2.2 as

THEOREM 6.2.5. $H^q(M, \mathscr{A}_F^p(E)) = H^q(\Omega_F^{p,q}(M))$.

COROLLARY 6.2.6. If $q > n = \text{rank}(F)$, then $H^q(M, \mathscr{A}_F^p(E)) = 0$.

DEFINITION 6.2.7. If $\mathbf{E} \to M$ is a complex vector bundle over M with fiber E, then $\mathbf{E}$ is called *F-holomorphic* if there is a vector bundle atlas (U_α) on M such that the 1-cocycle $(g_{\alpha\beta})$ defining $\mathbf{E}$ for this atlas satisfies $d_F g_{\alpha\beta} = 0$.

Let s in $S(\mathbf{E})$ be a section represented by $s_\alpha : U_\alpha \to \mathbf{E}$ where $\mathbf{E}|U_\alpha = U_\alpha \times E$. Then $s_\beta = g_{\beta\alpha}s_\alpha$. Applying d_F we have $d_F s_\beta = (d_F g_{\beta\alpha})s_\alpha + g_{\beta\alpha} d_F s_\alpha = = g_{\beta\alpha} d_F s_\alpha$. Thus $d_F s_\alpha$ represents a global section of $\mathbf{E}$ which we denote $d_F s$. We have a sheaf complex $(\Omega^{p,q}(\mathbf{E}), d_F)$ defined on $\mathbf{E}$-valued forms of type (p, q). So by Theorem 6.2.2 we have the generalized Dolbeault–Serre theorem:

THEOREM 6.2.8. $(\Omega_F^p(\mathbf{E}), d_F)$ is a fine resolution of $\mathscr{A}_F^p(\mathbf{E})$ and $H^q(M,$ $\mathscr{A}_F^p(\mathbf{E})) = H^q(\Omega^p(\mathbf{E}))$.

Let $L(\mathbf{E}_1, \mathbf{E}_2)$ denote the bundle of linear maps $\mathbf{E}_1 \to \mathbf{E}_2$. Thus we can view $\Omega^{p,q}(\mathbf{E})$ as the sheaf of sections of $L(\Lambda^p F^1 \otimes \Lambda^q F, \mathbf{E}) = D^{p,q}(\mathbf{E})$. If $\mathbf{E}$ is F-holomorphic we have a sequence.

$$D^{p,0}(\mathbf{E}) \overset{d_F}{\to} D^{p,1}(\mathbf{E}) \overset{d_F}{\to} \cdots \to D^{p,n}(\mathbf{E}) \to 0.$$

DEFINITION 6.2.8.   The symbol $\sigma(d_F)$ is given by $\sigma(d_F)(m)(\alpha, \beta) =$ $= (d_F(f - f(m)s)(m)$ where $\alpha \in T^*M$ (real cotangent space), $\beta \in D^{p,q}(\mathbf{E})_m$, f is chosen such that $df(m) = \alpha$, s is a section of $D^{p,q}(\mathbf{E})$ such that $s(m) = \beta$.

H. Fischer pointed out that the associated symbol sequence is exact iff $F + \bar{F} = T(M)^C$. Thus we have

THEOREM 6.2.9. $F + \bar{F} = T(M)^C$ holds iff for $p \geqq 0$ and $\mathbf{E}$ F-holomorphic $(D^{p,q}(\mathbf{E}), d_F)$ is elliptic.

COROLLARY 6.2.10.   In particular, under this condition the cohomologies $H^*(M, \mathscr{A}^p(E))$ are finite dimensional.

COROLLARY 6.2.11.   If M is a polarized symplectic manifold, ellipticity holds iff M is Kähler.

Selecting Hermitian structures on bundles $D^{p,q}(\mathbf{E})$ allows us to define the formal adjoint d_F^* to d_F and hence the generalized Laplacian $\Delta_F = = d_F * d_F + d_F d_F^*$. Checking the symbol $\sigma(\Delta_F)$ we see Δ_F will be an elliptic operator of order 2.

DEFINITION 6.2.12.   Let $\mathscr{H}_F^{p,q}(\mathbf{E})$ be the *space of Δ_F-harmonic sections* of $D^{p,q}(\mathbf{E})$.

The Hodge isomorphism holds: i.e. $H^q(\Omega^{P, \cdot}(\mathbf{E})) = \mathscr{H}_F^{p,q}(E)$.

Consider now the case that $p = 0$. The Euler–Poincaré characteristic for F is given by $\mathfrak{x}_F(M, \mathbf{E}) = \mathfrak{x}(M, \mathscr{A}_F(\mathbf{E})) = \sum (-1)^q \dim (\mathscr{H}_F^{0,q}(\mathbf{E}))$ where $\mathscr{A}_F(\mathbf{E})$ is the sheaf of F-holomorphic sections of the F-holomorphic complex bundle $\mathbf{E}$.

The bundle $D^{0,q}(\mathbf{E}) = L(\Lambda^q F, \mathbf{E})$ is simply equal to $\Lambda^q F^* \otimes \mathbf{E}$. The operator $d_F : D^{0,q}(E) \to D^{0,q+1}(E)$ has formal adjoint d_F^* and we set $D_F = = d_F + d_F^*$; thus $\Delta_F = D_F^* D_F$.

THEOREM 6.2.12. D_F is elliptic and $\mathfrak{x}_F(M, \mathbf{E}) = \mathrm{ind}_\alpha(D_F)$, the analytic index of D_F.

The next result requires some knowledge of characteristic classes. Space does not permit an adequate development of this topic. The reader is referred to Hirzebruch H22. Several exercises regarding the Todd class are included for the benefit of the student.

Basically for any $GL(q, C)$-bundle ξ over a compact almost complex manifold M, the Chern classes are elements $c_i \in H^{2i}(M, Z)$. The total Todd class is defined by the formal factorization

$$\sum_{j=0}^{\infty} c_j x^j = \prod_{i=1}^{\infty} (1 + \gamma_i x), \text{ viz. } \text{Td}(\xi) = \prod_{i=1}^{q} \frac{\gamma_i}{1 - e^{-\gamma_i}}.$$

The Chern character is defined by $\text{ch}(\xi) = \sum_{i=1}^{q} e^{\gamma_i}$.

Using these results Fischer and Williams have demonstrated the following generalized version of the Riemann–Roch theorem

THEOREM (Riemann–Roch) 6.2.13. Let M be a closed even dimensional oriented manifold and assume F is elliptic. If the foliation defined by $D = = F \cap \bar{F} \cap T(M)$ defines a fiber bundle $M \to M/D$, then M/D is a complex manifold with complex structure J_D induced by F. And for any F-holomorphic bundle $\mathbf{E}$ over M

$$\mathfrak{x}_F(M, \mathbf{E}) = \text{ch}(\mathbf{E}) e(D) \mathcal{T}(T(M)/D)[M],$$

where $\mathcal{T}(\cdot)$ is the Todd class of a complex vector bundle, $\text{ch}(\cdot)$ is the exponential Chern character of a complex vector bundle and $e(\cdot)$ is the Euler class of a real vector bundle.

The proof here is very straight forward. We refer the reader to Fischer–Williams F3.

EXAMPLE 6.2.13. If F is real and $E = 1$, the trivial line bundle, we have $d_F = d$, $T(M)/D = 0$, $\mathscr{A}_F(1) = C$ and $\mathfrak{x}(M) = e(T(M))[M]$.

EXAMPLE 6.2.14. If $D = 0$, i.e. M is a complex manifold, $d_F = \bar{\partial}$, $T(M)/D = = T(M)$ and

$$\mathfrak{x}(M, \mathbf{E}) = \text{ch}(E) \mathcal{T}(M)[M]$$

is the Hirzebruch–Riemann–Roch theorem.

Finally, Fischer and Williams proved the generalized version of Serre duality. Viz. let $k = \text{rank}(D)$, let $G = D_C^{\perp} \cap F$ and let $g = \text{rank}(G)$. The canonical line bundle is defined by $K_M = \Lambda^g \bar{G}^*$.

THEOREM (Serre duality) 6.2.15. $H^q(M, \mathscr{A}_F^q(K_M^* \otimes \mathbf{E}))$ has dual space $H_c^{n-q}(M, \mathscr{A}_F(K_M \times \mathbf{E}^*))$ where $H_c(\cdot)$ denotes cohomology with compact support.

COROLLARY 6.2.16. If $\mathbf{E} = M \times C$ and M is compact we have $H^q(M, \mathscr{A}_F)^* \simeq H^{n-q}(M, \mathscr{A}_F(K_M))$, where $n = \text{rank}(F)$.

6.3. LIE ALGEBRA POLARIZATIONS

Let $\mathfrak{g}$ be a finite dimensional real Lie algebra and let f be in $\mathfrak{g}^*$. Extend f to be a linear form on $\mathfrak{g}^C$. Then f defines a bilinear form B_f by $B_f(X, Y) = f([X, Y])$ whose kernel is $\mathfrak{g}(f)$.

DEFINITION 6.3.1. A Lie subalgebra $\mathfrak{h}$ of $\mathfrak{g}$ is said to be a *Lie algebra polarization* at f if:

(i) the vector subspace of $\mathfrak{h}$ is maximal totally isotropic for B_f; thus $\dim \mathfrak{h} = \frac{1}{2}(\dim \mathfrak{h} + \dim \mathfrak{g}(f))$. And

(ii) $\mathfrak{h} + \bar{\mathfrak{h}}$ is a subalgebra of $\mathfrak{g}$.

DEFINITION 6.3.2. The polarization $\mathfrak{h}$ at f is called *real* if $\mathfrak{h} = \bar{\mathfrak{h}}$.

B_f defines a Hermitian form H_f on $\mathfrak{g}^C$ by $H_f(X, Y) = 2iB_f(X, Y) = 2if([X, Y])$.

DEFINITION 6.3.3. A polarization $\mathfrak{h}$ at f is said to be *positive* if $H_f|\mathfrak{h}$ is positive, i.e. if $f([X, \bar{X}]) \geqq 0$ for all X in $\mathfrak{h}$.

Clearly if $\mathfrak{h}$ is a real polarization, then $\mathfrak{h}$ is positive since $H_f|\mathfrak{h} = 0$.

Two subalgebras of $\mathfrak{g}$ are associated to a polarization $\mathfrak{h}$:

$$\mathscr{E} = (\mathfrak{h} + \bar{\mathfrak{h}}) \cap \mathfrak{g} \quad \text{or} \quad \mathscr{E}^C = \mathfrak{h} + \bar{\mathfrak{h}}$$

and

$$\mathscr{A} = \mathfrak{h} \cap \mathfrak{g} \qquad \text{or} \quad \mathscr{A}^C = \mathfrak{h} \cap \bar{\mathfrak{h}}.$$

We note that $\mathscr{D}^C$ is the orthocomplement to $\mathscr{E}^C$ with respect to B_f; and the real form B_f defines an alternating bilinear form on $\mathscr{E}/\mathscr{D}$. Thus $\dim(\mathscr{E}/\mathscr{D})$ is even.

THEOREM 6.3.3. If $\mathfrak{h}$ is a positive polarization at f, then $\mathscr{D} = \{X \in \mathfrak{h} \mid f([X, \bar{X}]) = 0\}$.

EXAMPLE 6.3.4. Let $\mathfrak{n}$ be the Heisenberg Lie algebra $\{P, Q, I\}$ with

$[P, Q] = I$. If $f \in \mathfrak{n}^*$ is such that $f|\mathfrak{z} = 0$ where $\mathfrak{z}$ is the center of $\mathfrak{n}$ then $f|[\mathfrak{n}, \mathfrak{n}] = 0$; so $\mathfrak{g}(f) = \mathfrak{n}$ is the only polarization.

If $f|\mathfrak{z} \neq 0$, then $\mathfrak{h}_0 = CP \oplus CI$ and $\mathfrak{h}_1 = C(P + iQ) \oplus C(P - iQ) \oplus CI$ are polarizations. Clearly $\mathfrak{h}_0$ is real and $\mathfrak{h}_1$ is such that $\mathfrak{h}_1 + \bar{\mathfrak{h}}_1 = \mathfrak{n}^C$ and it is positive.

The center $\mathfrak{z}$ is isomorphic to R and can be mapped in the Heisenberg group N by $i(t) = \exp(tz)$. Call the image Z. Then the characters of Z are of the form $\mathfrak{X}_\mu(i(t)) = \mathbf{e}(\mu t)$ where $\mathbf{e}(\cdot) = \exp(2\pi i.)$. Consider the direct product ZH where $H = \exp(\mathfrak{h}_0)$. Then there is a unique character of ZH inducing $\mathfrak{X}_\mu$ on Z – viz. $\mathfrak{X}_f(i(t)e^P) = \mathbf{e}(\mu t)$, where we select f in η^* so that $f(X) = \mu t$ for X in η. We can view the elements of N as having the form $g = i(t)e^{pP}e^{qQ}$. Clearly we have $N/ZH \simeq R$. Then the reader may check that the representation induced from $\mathfrak{X}_f$ acting on $L^2(N/ZH)$ is given by $(U_f(g)\varphi)(q') = \mathbf{e}(\mu t)\mathbf{e}(\mu pq')\varphi(q' + q)$ (i.e. the Schrödinger representation). And by the Stone–von Neumann theorem every irreducible unitary representation such that $U(i(t)) = \mathfrak{X}_\mu(i(t))$ is unitarily equivalent to this representation. Thus every representation of this type is formed by inducing from a Lie algebra polarization.

We will relate the Lie algebra polarizations to the manifold polarizations in Chapter 7.

6.4. Spin Structures, Metaplectic Structures and Square Root Bundles

The Clifford algebra A_n is generated by R^n with a basis $e_1, \ldots, e_n$ with the relations $e_i^2 = 1, e_ie_j + e_je_i = 0$ for $i \neq j$. The group $\text{Pin}(n)$ is the group of g in A_n which admit an inverse g^{-1} and such that x in $R^n \to gxg^{-1}$ in R^n. Since $\|gxg^{-1}\|^2 = \|x\|^2$ we see that $\alpha_g : x \to gxg^{-1}$ is an orthogonal transformation of R^n. Thus we have $\text{Pin}(n) \to 0(n) : g \to \alpha_g$. The inverse image of $SO(n)$ in $\text{Pin}(n)$ is called the $\text{Spin}(n)$ group. Then we have

$$0 \to Z_2 \to \text{Spin}(n) \to SO(n) \to 0$$

i.e. $\text{Spin}(n)$ is the universal covering group of $SO(n)$. Let γ in $H^1(SO(n), Z_2)$ be the corresponding cohomology class.

If M is a Riemannian manifold, there is a principal bundle over M, $SO(n) \to P \to M$, the bundle of orthonormal frames.

Definition 6.4.1. A *Spin structure* on M is a double covering $Q \to P$; so we have

$$\begin{array}{ccccc} \mathrm{Spin}(n) & \to & Q & \to & M \\ \downarrow & & \downarrow & & \| \\ SO(n) & \to & P & \to & M \end{array} \qquad (*)$$

DEFINITION 6.4.2. Given a central extension of Lie groups

$$0 \to C \to \tilde{G} \xrightarrow{p} G \to 0 \qquad (\substack{*\\*})$$

and a principal G-bundle P then a ρ-lifting of P is a $\tilde{G}$-bundle $\tilde{P}$ and a ρ-equivalent bundle morphism $\tilde{\rho}: P \to \tilde{P}$.

DEFINITION 6.4.3. Given two p-lifts $(\tilde{P}', \tilde{\rho}')$ and $(\tilde{P}, \tilde{\rho})$ they are called *equivalent* if there is a principal bundle morphism ϕ such that

$$\begin{array}{ccc} \tilde{P} & \xrightarrow{\phi} & \tilde{P}' \\ & {\scriptstyle\tilde{\rho}}\searrow \quad \swarrow{\scriptstyle\tilde{\rho}'} & \\ & P & \end{array}$$

is commutative. The set of isomorphism classes of principal G-bundles is equivalent to $H^1(M, \mathbf{G})$. Corresponding to the exact sequence $(\substack{*\\*})$ there is a cohomology sequence

$$\to H^1(M, \mathbf{C}) \to H^1(M, \tilde{\mathbf{G}}) \to H^1(M, \mathbf{G}) \xrightarrow{\delta^1} H^2(M, \mathbf{C}) \to .$$

THEOREM 6.4.4. P admits a ρ-lifting iff δ^1 $([P])$ vanishes.

THEOREM 6.4.5. $H^1(M, \mathbf{C})$ operates in a simply transitive manner on $H^1(M, \mathbf{G})$. Thus this set is either empty or in bijection with $H^1(M, \mathbf{C})$.

From the sequence $(*)$ we have an exact sequence

$$0 \to H^1(M, Z_2) \to H^1(P, Z_2) \to H^1(SO(n), Z_2) \xrightarrow{\delta} H^2(M, Z).$$

Clearly we have

THEOREM 6.4.6. A spin structure exists on M iff $\delta(\gamma) = w_2(M) = 0$. And if $\delta(\gamma) = 0$, the spin structures are classified by a coset of $H^1(M, Z_2)$ in $H^1(P, Z_2)$.

If M admits an almost complex structure then the $SO(2n)$-bundle P reduces to a $U(n)$-structure.

Theorem 6.4.7. If M is almost complex, the set of spin structures on M is in bijective correspondence with the double coverings of the bundle $U(1) \to \det(P) \to M$ which restrict to the squaring map, mapping $U(1) \to U(1)$, on each fiber.

Corollary 6.4.8. If M is complex, the set of spin structures on M is in bijective correspondence to isomorphism classes of continuous line bundles (L^2, α) where $\alpha : L^2 \to K$ (the canonical line bundle) is a continuous isomorphism.

However, given a pair (L, α), K induces a holomorphic structure on L making α a holomorphic isomorphism. Thus we have

Theorem 6.4.9. The set of spin structures on a compact complex manifold correspond bijectively to the isomorphism classes of holomorphic line bundles L with $L^2 \simeq K$.

Consider now the symplectic manifold (M, Ω). Its structure group is $\mathrm{Sp}(2n, R)$, which admits a 2-fold covering group

$$0 \to Z_2 \to \mathrm{Mp}(2n, R) \to \mathrm{Sp}(2n, R) + 0.$$

$\mathrm{Mp}(2n, R)$ is called the *metaplectic group*. The ρ-lift of $Sp(2n) \to P \to M$ is the *metaplectic bundle*. Since $U(n)$ is the maximal compact subgroup of $Sp(2n, R)$, we have a $U(n)$-bundle contained in $P \to M$; i.e. (M, Ω) admits an almost complex structure j. If $T_m^C M$ is the complexification of $T_m M$ and if $F_m = \{X \in T_m^C M \mid jX = -iX\}$ then the fiber $L_m = \Lambda^n F_m$ defines a line bundle L over M and $c_1(L) =$ the first Chern class of M, c_1. (Thus c_1 is independent of j.)

Theorem 6.4.10. The symplectic manifold (M, Ω) admits a metaplectic structure iff there is a c in $H^2(M, Z)$ with $2c = c_1$ – i.e. iff there is a line bundle $\tilde{L}$ on M such that $\tilde{L}^2 = L$.

Taking the exact sequence

$$0 \to Z \xrightarrow{2} Z_2 \to 0$$

then α induces a map $\alpha : H^2(M, Z) \to H(M, Z_2)$. Let $\bar{c} = \alpha(c_1)$.

Theorem 6.4.11. A metaplectic structure exists over (M, Ω) iff $\bar{c} = 0$.

And by theorem 6.4.5 we have

THEOREM 6.4.12. If $\bar{c}=0$, the set of equivalence classes of metaplectic structures is given by $H^1(M,Z_2)$. In particular, if M is simply connected, there is a unique metaplectic structure.

Finally if M admits a real polarization F, then $\Lambda^n F_m$ defines a real line bundle on M – i.e. an element τ in $H^1(M,Z_2)$.

THEOREM 6.4.13. $\bar{c}=\tau^2$ and hence a metaplectic structure exists iff $\tau^2=0$.

EXAMPLE 6.4.14. If $M=T^*N$ and F is the polarization defined by cotangent spaces, τ is the pullback of $w_2(N)$ to the tangent bundle. Thus, a metaplectic structure exists on M iff $w_2(N)^2=0$. In particular, if N is orientable, $w_2(N)=0$; and thus a metaplectic structure exists.

PROBLEMS

EXERCISE 6.1. Show that $CP(n)$ has a metaplectic frame bundle iff n is even.

EXERCISE 6.2. Let E, F be complex vector bundles over M. Let $c_j(E\in H^{2j}(M)$ denote the jth Chern class. Recall from Hirzebruch H22 that the Todd cohomology class $\mathscr{T}(E,F)\in H^{**}(M)$ is given by

$$\mathscr{T}(E,F)=\mathrm{ch}(F)\sum_{j=1}^{\infty} T_j(c_1(E),\ldots,c_j(E)),$$

where $\mathrm{ch}(F)$ is the Chern character. If F is a complex line bundle with $d\in H^2(M)$ as its first Chern class set $\mathscr{T}(E,d)=\mathscr{T}(E,F)$. Finally the Todd genus is given by $T(M,d)=\mathscr{T}(M,d)[M]$.

Let G be a compact connected Lie group and T a maximal torus. The homogeneous space G/T has 2^m invariant almost complex structures, which determines the first Chern class $c_1(G/T)\in H^2(G/T,Z)$ (v. Theorem 9.1.20). Show that $\mathscr{T}(G/T,d)=\exp(\mathrm{d}+\frac{1}{2}c_1)$ and $m!\ T(G/T,d)=((c_1/2)+d)^m[G/T]$. Show that $T(G/T,d)=0$ if $d+c_1/2$ is a singular weight and $T(G/T,d)=\pm$ degree (λ) when d is a nonsingular weight and λ is a suitable irreducible representation of G.

Chapter 7

Geometry of Orbits

7.1. Orbit Theory

Let G be a real connected Lie group and let $\mathfrak{g}$ be its Lie algebra. G acts on $\mathfrak{g}$ by the adjoint representation. Thus, if $f \in \mathfrak{g}^*$ and g is in G, then the *coadjoint representation* gf is defined by

$$\langle g.f, Y \rangle = \langle f, \mathrm{Ad}(g)^{-1} Y \rangle$$

for all Y in $\mathfrak{g}$. Similarly, there is a linear representation of $\mathfrak{g}$ in $\mathfrak{g}^*$; viz. if $X \in \mathfrak{g}$ then $X.f$ in $\mathfrak{g}^*$ is defined by

$$\langle X.f, Y \rangle = \langle f, [X, Y] \rangle$$

for all Y in $\mathfrak{g}$.

Let G act on f by the coadjoint action. The isotropy subgroup in G of f is denoted $G(f) = \{g \text{ in } G | g.f = f\}$. The Lie algebra of $G(f)$ is $\mathfrak{g}(f) = \{X | X.f = 0\}$. Under the orbit map $g \to g.f$ of $G \to \mathfrak{g}^*$, we see that the orbit $\mathcal{O}_f = Gf$ can be identified with $G/G(f)$ and $\mathcal{O}_f$ is a smooth manifold.

The vector field $\sigma(\mathcal{O}, X)$ on $\mathcal{O}$ is defined by

$$\sigma(\mathcal{O}, X)_f \varphi = \frac{\mathrm{d}}{\mathrm{d}t} \varphi(\exp tX.f)|_{t=0},$$

where φ is in $A(\mathcal{O})$ and $f \in \mathcal{O}$. If φ is a smooth function on $\mathfrak{g}^*$ we denote the vector field defined similarly by $\sigma(X)_f$.

Define the differential $\mathrm{d}_f\varphi$, φ a smooth function on $\mathfrak{g}^*$ given in a neighborhood of f, as

$$\langle f_1, \mathrm{d}_f \varphi \rangle = \frac{\mathrm{d}}{\mathrm{d}t} \varphi(f + tf_1)|_{t=0}$$

for f_1 in $\mathfrak{g}^*$. Since $\exp tX.f = f + f X.f \bmod t^2$ we have

$$\sigma(X)_f \varphi = \langle X.f, \mathrm{d}_f \varphi \rangle.$$

Definition 7.1.1. For Y in $\mathfrak{g}$ define the function Ψ^Y on $\mathfrak{g}^*$ by $\langle \Psi^Y, f_1 \rangle = \langle f_1, Y \rangle$.

THEOREM 7.1.2. $\sigma(X)\Psi^Y = -\Psi^{[X,Y]}$.

Proof. $\sigma(X)\Psi^Y(f) = (d/dt)\Psi^Y(\exp - tX.f) = (d/dt)\langle \exp tX.f, Y\rangle = = \langle X.f, Y\rangle = -\langle f,[X,Y]\rangle$.

THEOREM 7.1.3. For X in $\mathfrak{g}(f)$, $d_f(\sigma(X)\varphi) = -[X, d_f\varphi]$.

COROLLARY 7.1.4. If φ is locally invariant by G – i.e. $\sigma_f(X)\varphi = 0$ for X in $\mathfrak{g}$, then $d_f\varphi$ is in the center of $\mathfrak{g}(f)$.

Proof. $\langle X.f, d_f\varphi\rangle = \sigma(X)_f\varphi = 0$. Thus $d_f\varphi$ is in $\mathfrak{g}(f)$ and by the theorem $d_f\varphi$ is in the center of $\mathfrak{g}(f)$.

Since $\sigma(X)_f \in T_f(\mathcal{O}_f)$ we view σ_f as the vector field

$$\mathfrak{g} \to T_f(\mathcal{O}_f): X \to \sigma(X)_f.$$

THEOREM 7.1.5. $0 \to \mathfrak{g}(f) \to \mathfrak{g} \overset{\sigma_f}{\to} T_f(\mathcal{O}_f) \to 0$ is exact.

Thus we see that $T_f(\mathcal{O}) = \{\sigma(\mathcal{O}, \mathfrak{g})_X\} = \mathfrak{g}/\mathfrak{g}(f)$.

The orbit $\mathcal{O}$ is a homogeneous symplectic manifold under the 2-form

$$\Omega(\sigma(\mathcal{O}, X)_f, \sigma(\mathcal{O}, Y)_f) = \langle f, [X, Y]\rangle.$$

THEOREM 7.1.6. $(\mathcal{O}, \Omega)$ is a Hamiltonian G-space.

Proof. Let B_f be the alternating 2-form on $\mathfrak{g}$ defined by $B_f(Y, X) = = \langle f, [X, Y]\rangle = \langle -X.f, Y\rangle$. Then B_f induces a nonsingular alternating form Ω on $\mathfrak{g}/\mathfrak{g}(f)$. Let $i_{\mathcal{O}}: \mathcal{O} \to \mathfrak{g}^*$ be the natural injection and consider the 1-form $i^*(d\psi)$. Let p be the map: $A^1(\mathcal{O}) \to V(\mathcal{O})$ where $\Omega(p(\alpha), X) = i(X)\alpha$. Then it is easily checked that $p(i^*(d\psi))_{f_1} = \sigma(d_{f_1}\psi) f_1$.

The lift $\lambda_{\mathcal{O}}: \mathfrak{g} \to A(\mathcal{O})$ is given by $\lambda_{\mathcal{O}}(X) = \Psi^Y|_{\mathcal{O}}$.

Thus $\lambda_{\mathcal{O}}(X)(f) = \langle f, X\rangle$. We leave it to the reader to check that $\lambda_{\mathcal{O}}$ is a lift of $\sigma(\mathcal{O}, X)_f$ and that $\mathcal{L}(\sigma(\mathcal{O}, X))\Omega = 0$ and $d\Omega = 0$.

COROLLARY 7.1.7. For each X in $\mathfrak{g}$ the function $\lambda_X(f) = \langle f, X\rangle$ on $\mathfrak{g}^*$ satisfies $\langle d_X(f), Y\rangle = \Omega_f(\sigma(\mathcal{O}, X)_f, Y)$ – i.e. λ_X is the Hamiltonian function giving rise to the vector field $\sigma(\mathcal{O}, X)$.

To relate to functions on $\mathfrak{g}$ we have

THEOREM 7.1.8. If for 1-form φ on $\mathcal{O}$ we set $p(\varphi)$ to be the vector field with $\Omega(p(\varphi), X) = \langle \varphi, X\rangle$ for vector field X on $\mathcal{O}$, we have $p(i^*_{\mathcal{O}}(d\psi))_f = \sigma(d_f\psi))$ for function ψ on $\mathfrak{g}^*$.

Proof. $\Omega(\sigma(d_f\psi), \sigma(Y)_f) = \langle d_f\psi, Y.f\rangle$ and also the left-hand side is equal to $\langle f, [d_f\psi, Y]\rangle = \langle Y.f, d_f\psi\rangle$. The Poisson brackets on $\mathcal{O}$ are thus given by

$$\{\psi_1, \psi_2\} = \Omega(\mathrm{d}\psi_2)\psi_1 .$$

And under the Poisson bracket, $A(\mathcal{O})$ is a Lie algebra.

THEOREM 7.1.9. If ψ_1, ψ_2 are two functions on $\mathfrak{g}^*$ with restrictions $\psi_1|_{\mathcal{O}}$ and $\psi_2|_{\mathcal{O}}$ we have for f in $\mathcal{O}$

$$\langle f, \{\psi_1|_{\mathcal{O}}, \psi_2|_{\mathcal{O}}\}\rangle = \langle f, [\mathrm{d}_f\psi_1, \mathrm{d}_f\psi_2]\rangle .$$

Proof. $\langle\{\psi_1|_{\mathcal{O}}, \psi_2|_{\mathcal{O}}\}, f\rangle = p(\mathrm{d}\psi_2|_{\mathcal{O}})_f\psi_1|_{\mathcal{O}} = \sigma(\mathrm{d}_f\psi_2)_f\psi_1 =$
$= \langle \mathrm{d}_f\psi_2 . f, \mathrm{d}_f\psi_1\rangle = \langle f_1, [\mathrm{d}_f\psi_1, \mathrm{d}_f\psi_2]\rangle .$

COROLLARY 7.1.10. For X, Y in $\mathfrak{g}$ we have $[\lambda_{\mathcal{O}}(X), \lambda_{\mathcal{O}}(Y)] = \lambda_{\mathcal{O}}([X, Y])$.

EXAMPLE 7.1.11. Let $G = SU(2)$ and $\mathfrak{g} = su(2)$. The Killing form can be used to identify adjoint and coadjoint orbits. The coadjoint action for f defined by

$$f(Y) = -\operatorname{Tr}(XY) \quad \text{for} \quad X = \begin{pmatrix} i & 0 \\ 0 & i \end{pmatrix}$$

and Y in $\mathfrak{g}$ has isotropy subgroup $\mathfrak{g}(f) = U(1)$. Thus $\mathcal{O}_f = SU(2)/U(1) =$ $= CP(1)$.

EXAMPLE 7.1.12. Let

$$G = \left\{ \begin{pmatrix} a & b \\ 0 & 0 \end{pmatrix} \middle| \, a > 0, b \in R \right\}.$$

The Lie algebra of G is

$$\mathfrak{g} = \left\{ \begin{pmatrix} \alpha & \beta \\ 0 & \alpha \end{pmatrix} \middle| \, \alpha, \beta \text{ in } R \right\}.$$

The exponential map is

$$\exp\begin{pmatrix} \alpha & \beta \\ 0 & \alpha \end{pmatrix} = \begin{pmatrix} e^{\alpha} & \beta \sinh(\alpha) \\ 0 & e^{-\alpha} \end{pmatrix}.$$

f in $\mathfrak{g}^*$ has the form $f(X) = u\alpha + v\beta$, u, v in R. The adjoint action of G on $\mathfrak{g}$ is

$$\operatorname{Ad} g(X) = gXg^{-1} = \begin{pmatrix} \alpha & \alpha^2\beta - 2ab\alpha \\ 0 & -\alpha \end{pmatrix}.$$

Thus the coadjoint action is $g^{-1}f = f(\mathrm{Ad}g(X)) = u\alpha + v(a^2\beta - 2ab\alpha)$. Hence there are two types of orbits:

(i) the case $f = (u, 0)$; these orbits are just points.

(ii) $\mathcal{O}^{\pm} = \{f = (u, v) | v \gtrless 0\}$, the upper and lower half planes.

Kostant extended a result of Wang by dropping the assumption of compactness in his classification of homogeneous symplectic manifolds.

THEOREM (Kostant) 7.1.13. If $H^1(\mathfrak{g}, R) = 0 = H^2(\mathfrak{g}, R)$ (e.g. if $\mathfrak{g}$ is semisimple), then the most general G-symplectic homogeneous space (M, Ω) covers an orbit $\mathcal{O}_f = G/H$ for some f in $\mathfrak{g}^*$ and H is the centralizer of a torus. Furthermore, the covering map and orbit are unique. For a proof see Kostant K16 or Wallach W3.

As we introduced above, the method of Kirillov is a theory of parameterizing the space $\hat{G}$ by the orbits in $\mathfrak{g}^*$. We saw that in certain cases we are able to identify $\mathfrak{g}^*/G$ with $\hat{G}$. Basically the idea is to find a subalgebra $\mathfrak{h}$ of $\mathfrak{g}$ such that $f([\mathfrak{h}, \mathfrak{h}]) = 0$ for f in $\mathcal{O}$. Then the irreducible unitary representation $U(f, \mathfrak{h})$ of G is formed by inducing by the character $\mathfrak{X}_f$ of $H = \exp(\mathfrak{h})$ where $\mathrm{d}\mathfrak{X}_f = \mathrm{i}f$ on $\mathfrak{h}$. This construction is related to the integrality of the symplectic form on the orbit as follows:

THEOREM 7.1.14. Let (M, Ω, λ) be a G-homogeneous Hamiltonian G-space. Assume $H^1(M, R) = 0$. Let $\tilde{G}$ be the universal cover of G with covering homomorphism $c: \tilde{G} \to G$. For a point m in M let $\tilde{G}_m = \{g \text{ in } G | c(g)m = m\}$. Then Ω is integral iff there is a character $\mathfrak{X}: \tilde{G}_m \to T^1$ such that $\mathrm{d}\mathfrak{X}(X) = 2\pi i \lambda(X)(m)$ for X in the Lie algebra of $\tilde{G}_m$.

EXAMPLE 7.1.15. The Heisenberg Lie group will be studied in full generality in the next chapter. We consider here the simplest case. Let N denote the pairs (z, t) in $C \times R$ where $z = (p, q)$. The product structure is $(z, t)(z', t') = (z + z', t + t' + B(z, z'))$ where $B(z, z') = pq' - qp'$. The inverse is clearly $(-z, -t)$. The Lie algebra $\mathfrak{n}$ of N is the set of pairs $(\mathfrak{z}, t)$ under the bracket $[(\mathfrak{z}, t), (\mathfrak{z}', t')] = (0, B(\mathfrak{z}, \mathfrak{z}'))$. The center $\mathfrak{z}$ of $\mathfrak{n}$ is the one dimensional subspace generated by $(1, 0)$. Thus $\mathfrak{n} = \mathfrak{z} \oplus C$. Let Z denote the subgroup corresponding to $\mathfrak{z}$.

Checking that $(z, t)(\mathfrak{z}, s)(-z, -t) = (\mathfrak{z}, z + B(z, \mathfrak{z}))$ we see that the adjoint action of N on $\mathfrak{n}$ is given by $\mathrm{Ad}(z, t)(\mathfrak{z}, s) = (\mathfrak{z}, z + B(z, \mathfrak{z}))$. So the coadjoint action for f in $\mathfrak{n}^*$ is $g.f(\mathfrak{z}, s) = f(\mathfrak{z}, s) - B(z, \mathfrak{z}) f(0, 1)$ where $g = (z, t)$. Thus the orbits for which $f|_{\mathfrak{z}} \neq 0$ are hyperplanes in $\mathfrak{n}^*$ characterized by $f(0, 1) = \mu$. If $X = (\mathfrak{z}, 0)$, $X' = (\mathfrak{z}', 0)$ are in $\mathfrak{n}$ then $\Omega_f(\sigma(\mathcal{O}, X)_f, \sigma(\mathcal{O}, X')_f) =$

$= \langle f,[X,Y]\rangle = f(0,B(\mathfrak{z},\mathfrak{z}')) = \mu B(\mathfrak{z},\mathfrak{z})$. One can easily check that $\mathcal{O}_\mu = N/Z$. $\mathcal{O}_\mu$ can also be identified with R^2 under the map $\tau: R^2 \to \mathcal{O}_\mu$ where $\tau(z) = (z,0).f$. And under this map $\tau^*\Omega_\mu = \mu\ \mathrm{d}p \wedge \mathrm{d}q$. We leave it to the reader to show that $(R^2, \mathrm{d}p \wedge \mathrm{d}q)$ is a Hamiltonian N-space.

At this time we summarize what we have demonstrated for the Heisenberg group so far and include the Kirillov character formula for later reference.

THEOREM (Kirillov) 7.1.16. If U is an irreducible unitary representation of N with $U(0,t) = \exp(itl)I$, $l \neq 0$, then U is equivalent to $T_l(z=(x,y),t)\varphi(x) = \exp(il(t + \langle x, z - \frac{1}{2}y\rangle))\varphi(x-y)$ on $L^2(R)$. The orbits $\mathcal{O}_l$ in $\mathfrak{n}^*$ are symplectic manifolds and correspond to the representations $T_{2\pi l}$. If f is in $\mathfrak{n}^*$ there is a polarization at f and U_l is equivalent to the induced representation Ind $(\mathfrak{X}_f)$ where $\mathfrak{X}_f$ is the character of T whose differential is $2\pi if$ where $f(0,1) = l$. If we let $U_\mathcal{O}$ denote the representation corresponding to orbit $\mathcal{O}$, then $U_\mathcal{O}$ is of trace class with

$$\mathrm{Tr}\ U_\mathcal{O}(\varphi) = \int_\mathcal{O}(\varphi|_\mathcal{O})\,\mathrm{d}\beta = \frac{1}{|l|}\int_{-\infty}^{\infty} \varphi(0,t)\exp(2\pi ilt)\,\mathrm{d}t.$$

Thus $U_\mathcal{O}, U_{\mathcal{O}'}$ are equivalent iff $\mathcal{O} = \mathcal{O}'$. If $\lambda \in \hat{G}$, then $\lambda = [\mathrm{U}_\mathcal{O}]$ for some orbit $\mathcal{O}$. Thus we have an isomorphism (in fact Borel isomorphism) $\mathfrak{n}^*/N \to \hat{N}$. Finally the infinitesmal representation generator

$$\lim_{s\to 0} T_{-2\pi}\frac{(sz,st)\varphi - \varphi}{s} = -X_\psi + 2\pi i(y\,\partial\varphi/\partial y - \varphi),$$

where $z = (v,w)$ and ψ is a polynomial of degree at most one – so $\psi = vx + wy$; here

$$X_\varphi f = \frac{\partial f}{\partial x}\frac{\partial \varphi}{\partial y} - \frac{\partial \varphi}{\partial X}\frac{\partial f}{\partial y}$$

for f in $A(R) \subset L^2(R)$. This is just the van Hove quantum map.

We relate invariant polarizations on orbit spaces to Lie algebra polarizations by the next theorem.

THEOREM 7.1.17. Let F be a G-invariant polarization of $(\mathcal{O}_f, \Omega)$. Let $\sigma_f: \mathfrak{g}^C \to T_f^C$ be the map $\sigma_f(X)\psi^Y = \langle f,[X,Y]\rangle$ for X, Y in $\mathfrak{g}^C$. Let $\mathfrak{h}_f = \sigma_f^{-1}(F_f)$ where $F_f \subset T_f^C$. Then $F \to \mathfrak{h}_f$ is a bijection between the set of G-invariant polarizations on $\mathcal{O}_f$ and the set of subalgebras of $\mathfrak{g}^C$ which satisfy:

(i) $\mathfrak{g}^C(f) \subset \mathfrak{h}_f \subset \mathfrak{g}^C$;
(ii) $\mathfrak{h}_f$ is $\mathrm{Ad}(G(f))$-stable
(iii) $\dim_C(\mathfrak{h}_f/\mathfrak{g}^C(f)) = \dim_C(\mathfrak{g}^C/\mathfrak{h}_f)$;
(iv) $\langle f, [\mathfrak{h}_f, \mathfrak{h}_f] \rangle = 0$.

Finally F is an admissable polarization iff $\mathfrak{h}_f$ also satisfies

(v) $\mathfrak{h}_f + \bar{\mathfrak{h}}_f$ is a subalgebra of $\mathfrak{g}^C$.

7.2. Complete integrability

Let (M, Ω) be a symplectic manifold and let $f_1, \ldots, f_k \in A(M)$. Then we say that the set $f_1, \ldots, f_k$ is in *involution* iff

(i) $\{df_i(m)\}$ are linearly independent in T^*M for all m in M; and
(ii) $\{f_i, f_j\} = 0$ for all $i, j = 1, \ldots, k$.

If $\{f_1, \ldots, f_n\}$ are in involution and $c = (c_1, \ldots, c_n) \in R^n$ we set $\Lambda(c) = \{m \in M \mid f_i(m) = c_i,\ i = 1, \ldots, n\}$. By (ii) if $\Lambda(c)$ is nonempty it is a submanifold of M of dimension n. Clearly $\Lambda(c)$ is a Lagrangian submanifold of M. Thus a system in involution determines a foliation of M whose leaves $\Lambda(c)$ are Lagrangian submanifolds of M. By (ii) we have $X_{f_i}(f_j) = 0$ for all j so X_{f_i} is tangent to $\Lambda(c)$ for all i and by (i) $X_{f_1}, \ldots, X_{f_n}$ gives a family of commuting vector fields on $\Lambda(c)$ which are linearly independent and so span the tangent space to $\Lambda(c)$.

Theorem 7.2.1. $\Lambda(c)$ has a unique flat affine connection such that X_{f_i} are parallel.

We recall that to say that X_{f_i} is tangent to $\Lambda(c)$ means that the integral curves of X_{f_i} remain in $\Lambda(c)$.

Definition 7.2.2. A function f in $A(\Lambda(c))$ is called linear if $X_{f_i}(X_{f_j}(f)) = \{f_i, \{f_j, f\}\} = 0$ for all i, j.

Given a dynamical system with n degrees of freedom and with n first integrals in involution, then Liouville showed that the dynamical system is integrable by quadratures. Thus these dynamical system are called *completely integrable*. For a discussion of completely integrable systems see Arnol'd A7, Abraham–Marsden A2 or Vinogradov-Kupershmidt V7.

A class of completely integrable systems arises from Hamiltonian systems on Lie groups. We consider a Hamiltonian H on the symplectic manifold T^*G. It is a simple exercise to check that H is left invariant, then X_H is also left invariant. Thus X_H defines a vector field on $\mathfrak{g}^*$ which is specified by the *Euler equations* $\dot{f} = \{f, \mathrm{d}H(f)\}$.

THEOREM 7.2.3. If ψ is a function on $\mathfrak{g}^*$ which is constant on orbits of the coadjoint representation, then ψ is a first integral of the Euler system.

THEOREM 7.2.4. The Euler equations form a Hamiltonian vector field on the orbit $(\mathcal{O}, \Omega)$ with respect to the Kirillov symplectic structure where the Hamiltonian is just the restriction of $H(f)$ to $\mathcal{O}$.

The proof follows from the results of Section 7.1.

The case most frequently studied is the case of geodesic flow for the left invariant metric on group G. In this case the Hamiltonian function is a nondegenerate quadratic form on $\mathfrak{g}^*$ and $\mathrm{d}H\colon \mathfrak{g}^* \to \mathfrak{g}$ is a linear map.

THEOREM (Miscenko–Fomenko) 7.2.5. Let $\mathfrak{g}$ be a complex semisimple Lie algebra and assume H belongs to the space $\mathfrak{m}$ defined below; (this inclusion can be expressed in Lie algebra terms, but we omit it). Then the Euler equations are completely integrable á la Liouville on the orbits of the adjoint representation which are in general position.

Proof. The basic idea of the proof is to examine functions on $\mathfrak{g}$ which are functionally generated by functions of the form $\bar{f}_\lambda(x) = f(x + \lambda a)$; here f are functions which are constant on orbits of the adjoint representation and a in $\mathfrak{g}$ is in general position. It is easy to check that all functions in $\mathfrak{m}$ are in involution and that $H(f)$ is in $\mathfrak{m}$. This means that all functions from $\mathfrak{m}$ are integrals of the Euler equation. One must then check that the number of functionally independent functions on $\mathfrak{m}$ is not less than half the dimension of an orbit. For this we refer the reader to M32.

EXAMPLE 7.2.6. The classic example is the case of the n dimensional top. In this case see D9 and M7 for details.

Rather than developing the general theory we specialize the discussion of this topic to the example of the Toda lattice.

EXAMPLE (Nonperiodic Toda lattice.) 7.2.7. Let G be the identity component of the group of invertible lower triangular real $n \times n$ matrices. Then its Lie algebra $\mathfrak{g}$ is the set of all lower triangular matrices and $\mathfrak{g}^*$ is the set of all upper triangular matrices. We pair f in $\mathfrak{g}^*$ and X in $\mathfrak{g}$ by $f(X) = \mathrm{Tr}(fX)$

where fX is matrix multiplication. The adjoint action of G on $\mathfrak{g}$ is $\mathrm{Ad}(g)X = gXg^{-1}$. Let $A\hat{}$ denote the upper triangular part of matrix A. Then the adjoint action is given by $g.f(X) = f(\mathrm{Ad}(g)^{-1}X) = f(g^{-1}Xg) = = \mathrm{Tr}(fg^{-1}Xg) = \mathrm{Tr}(gfg^{-1}X) = \mathrm{Tr}((gfg)\hat{}X) = (g^{-1}fg)\hat{}(X)$. Thus we have $g.f = (g^{-1}fg)\hat{}$. If we take f of the form

$$f = \begin{pmatrix} c & e_1 & 0 & & 0 \\ 0 & c & e_2 & & 0 \\ \cdots & \cdots & & & \\ & & & c & e_{n-1} \\ & & & & c \end{pmatrix}$$

then one can check that the coadjoint orbit is

$$\mathcal{O}_f = G.f = \begin{pmatrix} b_1 & a_1 & & & 0 \\ 0 & b_2 & a_2 & & 0 \\ \cdots & \cdot\cdot & & & \\ & & & b_{n-1} & a_{n-1} \\ & & & & b_n \end{pmatrix},$$

where a_k, b_k are in R, $a_k > 0$, $\sum b_k = nc$.

The tangent space to f in $\mathcal{O}_f$ is $T_f(\mathcal{O}_f) = \{\langle f, X\rangle\hat{}\,|\,X \text{ in } \mathfrak{g}\}$. Thus the symplectic form Ω_f on $\mathcal{O}_f$ is given by $\Omega_f(f)(\langle f, X_1\rangle\hat{},\ \langle f, X_2\rangle\hat{}) = \mathrm{Tr}(f([X_2, X_1]))$. The orbit $\mathcal{O}_f$ is diffeomorphic to R^{2n-2} by using $(a_1, \ldots, b_{n-1})$. In this chart

$$\Omega_f = \sum_{j=1}^{n-1}\left(\sum_{i=j}^{n-1}\frac{\mathrm{d}a_i}{a_i}\right)\wedge \mathrm{d}b_j$$

The Hamiltonian vector field for F in $A(\mathcal{O}_f)$ is given by

$$X_f(F) = \sum_{i=1}^{n-1}\left[a_i\left(\frac{\partial f}{\partial b_{i+1}} - \frac{\partial f}{\partial b_i}\right)\frac{\partial}{\partial a_i} + \left(a_i\frac{\partial f}{\partial a_i} - a_{i-1}\frac{\partial f}{\partial a_{i-1}}\right)\frac{\partial f}{\partial b_2}\right]F,$$

where $a_0 = 0$. That is, X_f in $\mathfrak{g}$ has diagonal entries $(\partial f/\partial b_1, \ldots, \partial f/\partial b_n)$ and subdiagonal entries $(\partial f/\partial a_1, \ldots, \partial f/\partial a_{n-1})$.

EXAMPLE (Nonperiodic Toda lattice, cont.) 7.2.8. The Toda lattice has the Hamiltonian $H = \frac{1}{2}\sum p_i^2 + \exp(q_1 - q_2) + \cdots + \exp(q_{n-1} - q_n)$. One integral of this system is $M = \sum p_i$; i.e. $X_H(M) = 0$. The flow of M is given by $q_i \to q_i + t, p_i \to p_i$, $i = 1,\ldots,n$. Thus the functions $p_1,\ldots,p_n$, $q_1 - q_2,\ldots,$ $q_{n-1} - q_n$ are constant under the flow of M. If B is the set of integral curves of M lying in $M^{-1}(0)$, then B is a symplectic manifold on which the coordinates $p_1, \ldots, p_n$, $q_1 - q_2, \ldots, q_{n-1} - q_n$ are defined and satisfy $\sum p_i = 0$. The induced symplectic structure is

$$\Omega = \sum dp_i \wedge dq_i = \sum_{i=1}^{n-1} dp_i \wedge dq_i - \sum_{i=1}^{n-1} p_i \wedge dq_n =$$
$$= \sum_{i=1}^{n-1} dp_i \wedge (dq_i - dq_n) = \sum_{j=1}^{n-1} \sum_{i=1}^{j} dp_i \wedge (dq_j - dq_{j+1}).$$

Setting

$$a_j = \exp(q_j - q_{j+1}) \quad \text{and} \quad b_i = p_i,$$

then

$$\Omega = \sum_{j=1}^{n-1} \left(\sum_{i=1}^{j} db_i \right) \wedge \frac{da_j}{a_j} \quad \text{and} \quad H = \tfrac{1}{2} \sum_{i=1}^{n} b_i + a_1 + \cdots + a_{n-1}.$$

The approach to show that the Toda lattice is completely integrable centers on rewriting the Hamiltonian in the form $H = \frac{1}{2}\mathrm{Tr}(L^2)$ for an appropriate L. Then the Hamiltonian system of equations is shown to be equivalent to the Lax isospectral system $\dot{L} = [B, L]$ for an appropriate B. Setting $F_k = (1/k)\,\mathrm{Tr}\,(L^k)$ the F_k are found to be preserved by the flow–i.e. $\{F_2, F_i\} = 0$ for $i \geqq 2$. And $F_2, \ldots, F_n$ form an independent set of integrals. Thus Moser was able to show that the nonperiodic Toda lattice is a completely integrable dynamical system.

EXAMPLE 7.2.9. We introduced the Lie algebra description of symmetric spaces in Section 0.4. E.g. $CP(1) = SL(2, C)/P = G/P$ where the Lie algebra of G is of the form $\mathfrak{g} = \mathfrak{n}^- + \mathfrak{p}$ where $\mathfrak{p} = CH_\alpha + CE_\alpha = \mathfrak{h} + \mathfrak{n}^+$, where

$$H_\alpha = \tfrac{1}{4}\begin{pmatrix} 1 & 0 \\ 0 & -1 \end{pmatrix}, \qquad E_\alpha = \tfrac{1}{2}\begin{pmatrix} 0 & 1 \\ 0 & 0 \end{pmatrix}$$

and $\mathfrak{n}^- = CE_{-\alpha}$ where

$$E_{-\alpha} = \tfrac{1}{2}\begin{pmatrix} 0 & 0 \\ 1 & 0 \end{pmatrix}.$$

Let N and H denote the groups corresponding to $\mathfrak{n}^+$ and $\mathfrak{h}$. Any point m in $M = G/P$ can be written as $m = n(m)h(m)n^+(m)$ where $n^+(m)$ is the complex conjugate of the matrix transpose and $n(m) \in N$, $h(m) \in H$. There is a natural imbedding of $M \to G$ as a completely geodesic submanifold. Thus a geodesic

$m(t)$ can be described by vector fields on G, viz. $X_+(t) = m(t)^{-1}\dot{m}(t)$ and $X_-(t) = \dot{m}(t)m^{-1}(t)$. We leave it to the reader to show that if $m(t)$ is a geodesic on M, then $(d/dt)(X_+(t) + X_-(t)) = 0$, and to show that the solution to this equation is $m(t) = g\exp(2Y)g^+$ where $g\in G$ and $Y\in\mathfrak{p}$. Rewriting this decomposition, using the identification stated above, as $m(t) = u(t)\exp(2aq(t))u(t)^{-1}$ where $u(t)\in K$, $q(t)\in\mathfrak{h}$, then the geodesic flow on M is equivalent to $L = [L, M]$ i.e. a Lax system. To see this take $M(t) = u(t)^{-1}\dot{u}(t)$ and so $X_\pm(t) = a\,\mathrm{Ad}(u(t))(2p(t) \mp \frac{1}{2}\,\mathrm{Ad}(\exp(\mp 2aq(t))M \mp (1/a)M)$ where $p(t) = \dot{q}(t)$. Thus $X_+ + X_- = 4a\mathrm{Ad}(u(t))(L)$ where $L = p - (1/4a)[\mathrm{Ad}(\exp(2aq(t)) - \mathrm{Ad}(\exp(-2aq(t))]M(t)$. In particular, we leave it to the reader to check that the Toda lattice in the case $n = 2$ is given by the Lax system

$$M = 2\begin{pmatrix} 0 & \exp_2(q_1 - q_2) \\ 0 & 0 \end{pmatrix} \quad \text{and} \quad L = \begin{pmatrix} p_1 & \exp_2(q_1 - \bar{q}_2) \\ 1 & p_2 \end{pmatrix}.$$

Thus we see that the Toda lattice is equivalent to the geodesic flow on CP(1) with this Lax pair.

7.3. Morse Theory of Orbit Spaces

In the introduction we treated examples of orbits spaces given by adjoint action of groups on Lie algebras. The Morse theory of these orbit spaces is outlined in this section. We view the orbit space M of dimension n as differentiably imbedded in a real Euclidean space $R = R^{n+k}$ (viz. the adjoint orbits in $\mathfrak{g}$ are viewed as imbedded in the Euclidean space $\mathfrak{g}$). The Morse function of interest is the function on M given by the square of the distance of points on M from a fixed point $p\in R\backslash M$. Let $L_p(x)$, $x\in M$, denote this function. Taking p as the origin of R we let $L_p(x) = (\mathbf{x}, \mathbf{x})$. Thus $dL_p(x) = (d\mathbf{x}, \mathbf{x})$ and so $dL_p(x) = 0$ iff $d\mathbf{x}$ is perpendicular to $\mathbf{x}$; and so q is a critical point for L iff the vector from p to q is normal to M at q.

The Hessian quadratic form is

$$d^2 L_p(x) = 2(\mathbf{x}, d\mathbf{x}) + 2(d\mathbf{x}, d\mathbf{x}).$$

At a critical point $q, \mathbf{x} = |\mathbf{x}|N$ where N is the unit normal vector; thus $d^2L_p(x)/2 = (d\mathbf{x}, d\mathbf{x}) + (|\mathbf{x}|N, d^2\mathbf{x})$, the first fundamental form and the second fundamental form for N.

If we choose the local coordinates for M near $q = (0,\ldots,0)$ as $x_1,\ldots,x_n$; then in R, M is given by $g_1(x_1,\ldots,x_n) = x_{n+1}$, $l = 1,\ldots,k$ where g are smooth functions; since pq is perpendicular to M we take $p = (0,\ldots,0,$

$p_1, \ldots, p_k$). Let $t_p = (0, \ldots, 0, tp_1, \ldots, tp_k)$. Then the Hessian of L_{tp} at 0 is

$$HL_{tp}(0) = I_n - \sum_{l=1}^{k} tp_l \frac{\partial^2 g}{\partial x_\tau \partial x_j}(0).$$

By results on quadratic forms we can find a basis form TM_q such that the matrix

$$\left(- \sum_{l=1}^{k} tp_l \frac{\partial^2 g}{\partial x_i \partial x_j}(0) \right)$$

is reduced to diagonal form where we have

$$HL_{tp}(0) = \begin{bmatrix} ta_{11} + 1 & & 0 \\ & \ddots & \\ 0 & & ta_{nn} + 1 \end{bmatrix}.$$

Thus q is a nondegenerate critical point iff $t \neq -1/a_{ii}$ for all i. Thus for only finitely many values of t, q is degenerate. The values $a_{11}, \ldots, a_{nn}$ are called the principal curvatures of M at q corresponding to N. The values $1/a_{11}, \ldots, 1/a_{nn}$ are the principal radii of curvatures. These need not be all distinct. We let $t_1 = 1/a_{11}, \ldots, t_m = 1/a_{mm}$ denote the distinct values.

As t goes from 0 to 1 we get the segment qp. If $t = 0$, then $HL_{tp}(0)$ is positive definite and the index of $HL_{tp}(0)$ is zero. The index of $HL_{tp}(0)$ is seen to be an increasing function of t and the entires of $HL_{tp}(0)$ change sign at $-t_1 p, \ldots, -t_m p$. The number of changes in sign at $t_i p$ is denoted $\nu(HL_{t_i p}(0))$ = dimension of the nullity of $HL_{t_i p}$. Thus we have

THEOREM 7.3.1. The index of the Hessian $HL_p(q) = \sum_{0 < t < 1} \nu(HL_{(1-t)p+tq})$.

Consider now the Morse theory of orbit spaces as considered in Section 0.9. Recall that a point p in the maximal torus $\mathfrak{z}$ is called a *general point* if $\mathfrak{g}_p = \{X \mid [X, p] = 0\} = \mathfrak{z}$.

THEOREM 7.3.2. If p is a general point then $N_p = \mathfrak{z}$.

We leave to the reader to check that orbits of general points are of maximal dimension. We denote the orbit through X in this section by M_X.

THEOREM 7.3.3. If a line is perpendicular to an orbit, then it is perpendicular to all orbits it intersects.

Proof. Let $B + At$ be perpendicular to the orbit M_B through B. Thus $([X, B], A) = 0$ for all X in $\mathfrak{g}$. Since $([X, A], A) = (X, [A, A]) = 0$ for all X in $\mathfrak{g}$ we have $([X, B + tA], A) = 0$ for these X.

THEOREM 7.3.4. Let M_X be any orbit and p a general point. Then the critical points of L_p in M_X are $M_X \cap \mathfrak{z}$; thus these points are independent of p.

Proof. Let $A \in M_X$ be a critical point for L_p. Then by Morse theory pA is perpendicular to M_X at A. By Theorem 7.3.3 pA is perpendicular to M_X at p. By Theorem 7.3.2 pA is in $\mathfrak{z}$ since $\mathfrak{p}$ is a general point. Hence $A \in \mathfrak{z}$. The converse is obvious.

Consider now the orbit spaces of Ad G. Let M be an orbit of any point of $\mathfrak{g}$ under Ad G. Let p be a regular point of $\mathfrak{g}\backslash M$ on an orbit of maximal dimesnsion. Let q be a nondegenerate critical point of $L_p(M)$. Then by using Theorem 7.3.1 and the theory of Jacobi fields one can show that the index of q is given by $\mathrm{Ind}(q) = \sum v(F_i)$ where F_i are points where the segment pq intersects orbits of lower dimension and $v(F_i)$, the multiplicity of F_i, is given by $v(F_i) = \dim M_p - \dim M_{F_i}$. Here M_p and M_{F_i} represent the orbits through p and F_i respectively.

As we noted in Section 0.9 if we have the decomposition $\mathfrak{g} = \mathfrak{p} + \mathfrak{k}$ we can also study orbits arising from the adjoint action of K on $\mathfrak{p}$. In this case $A \in \mathfrak{z}$ is a general point if $\mathfrak{p}_A = \{Z \in \mathfrak{p} \mid [Z, A] = 0\} = \mathfrak{z}$.

The reader can check that for the orbits M_X for the action of Ad K on $\mathfrak{p}$, we have the tangent space $T_X = \mathrm{ad}\, X(\mathfrak{t})$ while the normal space is $N_X = \{Z \in \mathfrak{p} \mid [Z, X] = 0\} = \mathfrak{p}_X$.

If we view the orbit $M \subset \mathfrak{p}$ and if $\langle , \rangle$ is the Killing form of $\mathfrak{g}$, the length function L_X on M from the point X of $\mathfrak{p}$ is defined by $L_X(Y) = \langle Y - X, Y - X \rangle$ for Y in $M \subset \mathfrak{p}$. The Morse theory on these spaces is conveniently parametrized by another function which is related to L_X: viz. expanding L_X we have

$$L_X(Y) = -2\langle X, Y \rangle + \langle Y, Y \rangle + \langle X, X \rangle.$$

It appears one also studies the function f_X on M given by $f_X(Y) = \langle Y, X \rangle$. We will find this plays a significant role in Section 10.3.

PROBLEMS

EXERCISE 7.1. The Lie algebra of $G = SL(2, R)$ has a basis

$$E_+ = \begin{pmatrix} 0 & 1 \\ 0 & 0 \end{pmatrix}, \quad E_- = \begin{pmatrix} 0 & 0 \\ 1 & 0 \end{pmatrix}, \quad \text{and} \quad H = \begin{pmatrix} -1 & 0 \\ 0 & 1 \end{pmatrix}.$$

H defines the Cartan subalgebra $\mathfrak{h} = RH$ and $E_\pm$ define the subalgebra $\mathfrak{n}^\pm = RE_\pm$. The subalgebra $\mathfrak{l} = \mathfrak{h} + \mathfrak{n}^\pm$ is a maximal solvable subalgebra of

$\mathfrak{g}$. Set $\bar{\mathfrak{l}} = \mathfrak{h} + \mathfrak{n}^-$. Let B, $\bar{B}$, N, $\bar{N}$ be the groups corresponding to $\mathfrak{l}, \bar{\mathfrak{l}}, \mathfrak{n}^+, \mathfrak{n}^-$. Show that the dual $\bar{\mathfrak{l}}^*$ may be identified with the upper triangular matrices, i.e. $\mathfrak{l}$; and show that the coadjoint action of $\bar{B}$ on $\mathfrak{l}$ is given by conjugation followed by replacing all entries below the main diagonal by zero. Show that the coadjoint orbit $\mathcal{O}$ corresponding to E_+ in $\mathfrak{l}$ under $\bar{B}$ is the manifold $Z = \{cH + aE_+ | a > 0\} = \bar{B}/[\bar{N}, \bar{N}]$. Then the manifold $\tilde{Z} = f + Z$ is of the form

$$\tilde{Z} = \left\{ \begin{pmatrix} b_1 & a \\ 1 & b_2 \end{pmatrix} \middle| a > 0,\, b_1 + b_2 = 0 \right\}.$$

Show that the Kirillov form on Z is $\Omega_Z = db_1 \wedge (da/a)$. Show that $\frac{1}{2}B(f + y, f + y) = \frac{1}{2}(b_1^2 + b_2^2) + a$ i.e. the Toda–Hamiltonian for the case $n = 2$. Show that if $M = \bar{B}/\bar{N}$, then $\mathcal{O} = T^*M$ and the orbit $N.x$ for x in $\mathcal{O}$ is the Lagrangian submanifold corresponding to a cotangent leaf.

Exercise 7.2. Let $G = GL(n, F)$ where $F = R, C$. Show that every nontrivial coadjoint orbit of G has dimension $\geqq 2\dim_R(F) \times (n-1)$. Show that equality holds only for those orbits $\mathcal{O}(\lambda, r) = G.f_{X(\lambda,r)}$ where $X \in \mathfrak{g} = gl(n, F)$ is of the form

$$X(\lambda, r) = \lambda I + \begin{pmatrix} r & 0 & & \\ & \cdot & & \\ & & \cdot & \\ & & & 0 \end{pmatrix} r \neq 0 \quad \text{and}$$

$$X(\lambda, 0) = \lambda I + \begin{pmatrix} 0 \ldots 0 & 1 \\ & 0 \\ & 0 \end{pmatrix}$$

(where r is central in F). Show that these minimal dimension orbits all have invariant real polarization given by maximal parabolic subalgebras $\mathfrak{p}$ of $\mathfrak{g}$ where

$$\mathfrak{p} = \left\{ \left(\begin{array}{c|c} a & b \\ \hline 0 & A \end{array} \right) a \text{ in } F^t,\, b \text{ in } F^{n-1} \text{ and } A \text{ in } gl(n-1, F) \right\}.$$

Show that $G/P = FP(n-1)$. Show that if $f_{X(\lambda,r)}$ exponentiates to a unitary character on the identity component of P for all (λ, r) in case $F = R$ and for all (λ, r) with Im λ, Im r in Z in the case $F = C$.

Exercise 7.3. If B is a nondegenerate antisymmetric bilinear form on R^{2n} show that the minimal dimensional coadjoint orbits M of $G = \mathrm{SP}(n, R)$ are

$G.f_{u,u} = R^{2u} - \{0\}$ $(*)$, where $f_{u,v} \in \mathfrak{g}^*$ is given by $f_{u,v}(A) = B(u, Av) + B(v, Au)$. Show that the lift

$$\begin{array}{ccccccc} 0 \rightarrow R \rightarrow A(M) & \rightarrow & \mathrm{ham}\,(M) \rightarrow 0 \\ & \nwarrow & \uparrow \\ & \lambda & \mathfrak{g} \end{array}$$

is given by $\lambda(X)v = \frac{1}{2}B(Xv, v)$. Show that the moment map $\Phi\colon M \rightarrow \mathfrak{g}^*$ is given by $\Phi(v)(X) = \frac{1}{2}B(Xv, v)$ and that Φ gives rise to the ismorphism $(*)$.

EXERCISE 7.4. Let $G = U(n)$ and consider the action on $\mathfrak{u}(n)$. Show that the ith Betti number of the complex flag manifold is equal to the number of critical points of index i.

EXERCISE 7.5. Let $V_{p+q,p} = SO(p+q)/SO(q)$. Take $G = SO(2p+q)$. Consider the involution $X \rightarrow I(p+q, p)\,X I(p+q, p)$ of G where $I(p+q, p)$ is the diagonal matrix with the first $p+q$ entries equal to 1 and the next p equal to -1. Show that the full fixed set of the involution is $\{0(p+q) \times 0(q)\} \cap SO(2p+q)$. If the Cartan decomposition is $\mathfrak{g} = \mathfrak{k} + \mathfrak{p}$ then $\mathfrak{p}$ has the form

$$\begin{array}{cc} \begin{pmatrix} 0 & * \\ * & 0 \end{pmatrix} & \begin{array}{l} p+q \\ p \end{array} \\ \begin{array}{cc} p+q & p \end{array} & \end{array}$$

and the Cartan subalgebra $\mathfrak{h}$ in $\mathfrak{p}$ is of the form

$$\begin{array}{cc} \left[\begin{array}{c|c} 0 & \begin{matrix} * & & & \\ & \cdot & & \\ & & \cdot & \\ & & & \cdot \\ & & & & * \end{matrix} \\ \hline \begin{matrix} * & & & & \\ & \cdot & & & \\ & & \cdot & & \\ & & & \cdot & \\ & & & & * \end{matrix} & 0 \end{array}\right] & \begin{array}{l} p+q \\ \\ \\ \\ \\ p \end{array} \\ \begin{array}{cc} p+q & \qquad p \end{array} & \end{array}$$

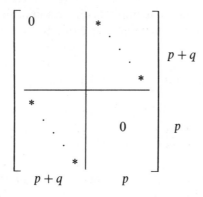

Taking

$$P=\left[\begin{array}{c|c} 0 & \begin{matrix} 1 & & & \\ & 2 & & \\ & & \ddots & \\ & & & p \end{matrix} \\ \hline \begin{matrix} -1 & & & 0 \\ & -2 & & \\ & & \ddots & \\ 0 & & & -p \end{matrix} & 0 \end{array}\right] \in \mathfrak{h}$$

show that P is a general point; describe the orbit of P. Show that there are 2^p critical points on $V_{p+q,p}$ for the length function $L_P(x)$.

Chapter 8

Fock Space

8.1. Fock space and cohomology

The last example of Chapter 2 motivates the following generalization of Fock space. Let V be a real vector space and let V^C be the vector space over C obtained by scalar extension. Let $T = R/Z$ be the 1-torus. The canonical homomorphism e: $R \to T$ is $\mathbf{e}(r) = \exp(2\pi i r)$ for r in R. Assume V has an alternating bilinear form A. Then $\mathrm{e}(\frac{1}{2}A(u, v))$ for u, v in V is a 2-cocycle; so it defines a central extension $\tilde{V}$ of V by T. $\tilde{V}$ is a group when endowed with the product

$$(x, u)(x', u') = (xx'\mathrm{e}(\tfrac{1}{2}A(u, u')),\ u + u').$$

If $\mathfrak{z}$ denotes the subspace spanned by $(1, 0)$ then clearly $\tilde{\mathfrak{v}} = \mathfrak{z} + V$ is the Lie algebra of $\tilde{V}$. And since A is nondegenerate $\mathfrak{z}$ is the center of $\tilde{\mathfrak{v}}$.

Under the present assumptions we have a complex structure J on V such that $A(x, Jy)$ is symmetric and positive definite. In this case the signature of $A(x, Jy)$ is of the form $(2r, 2n - 2r)$ where $2n = \dim V$.

Each element of V^C can be written as $w = u + iv$, u, v, in V. $\tilde{V}$ has a natural complexification $\tilde{V}^C$ gotten by taking pairs in $V^C \times C^*$ while the Lie algebra $\tilde{\mathfrak{v}}_C$ consists of pairs (z, u) z in C and u in V^C under the same bracket. Let $\mathfrak{t}$ denote the Lie algebra of T with complexification $\mathfrak{t}^C$.

Let $W_\pm$ denote the $\pm$ i-eigenspace of J in V^C. Then $W_\pm$ is totally isotropic for A – i.e. $A|W_\pm \times W_\pm = 0$. Each u in V^C is written the as $u = u_+ + u_-$ with $u_\pm$ in $W_\pm$. The map $u \to u_+$ gives a C-linear isomorphism $(V, J) \to W_+$. Let $\tilde{\mathfrak{m}}_+$ denote the subspace $\{(0, w) | w \in W_+\}$ of $\tilde{\mathfrak{v}}_C$ and let $\mathfrak{m} = \{(0, u) | u \text{ in } V\}$. The following properties are easily checked:

THEOREM 8.1.1.

(i) $\tilde{\mathfrak{m}}_+ W_- \subset W_+$,

$\tilde{\mathfrak{m}}_+ W_+ = \{0\}$;

(ii) $[\mathfrak{t}_C, \mathfrak{m}_+] \subset \mathfrak{m}_+$;

(iii) $\mathfrak{m} \cap \mathfrak{t} = \{0\}$ and $\operatorname{ad}(t)\, \mathfrak{m} \subset \mathfrak{m}$ for all t in T;

(iv) if $M_{\pm}$ and T are the subgroups corresponding to $\mathfrak{m}_{\pm}$ and $\mathfrak{t}_C$ then $G \subset M_+ T M_-$ and $G \cap T_C M_- = T$.

The exponential map $\exp: \mathfrak{m}_+ \to M_+$ is bijective – viz. $\exp(0, w) = (1, w)$ for w in W_+. Let $D = \tilde{V}/T$ and let $\pi: \tilde{V} \to D$ denote the canonical projection. Then D can be identified with W_+ by the map $\pi(t, u) \to u_+$. The element $g = (t, u)$ in $\tilde{V}$ acts on W_+ by $g(w) = w + u_+$. Define now

$$J(g, w) = t\mathrm{e}(A(u_-, w + \tfrac{1}{2}u_+)$$
$$K(w', w) = \mathrm{e}(A(\bar{w}, w'))$$

for w, w' in W_+.

THEOREM 8.1.2. For g, g' in V and w, w' in W_+ we have:

(i) $J(gg', w) = J(g, g'(w)) J(g', w)$;

(ii) $K(g(w'), g(w)) = J(g, w') K(w', w)/J(g, w)$.

DEFINITION 8.1.3. The map $J: \tilde{V} \times D \to T^C$ with the properties that it is smooth in V and holomorphic in D and satisfies (α): $J(t, \pi(e)) = t$ is called a *canonical automorphy factor*.

DEFINITION 8.1.4. Any map $K: D \times D \to T^C$ which satisfies property (ii), is holomorphic in D and satisfies (β): $K(z, z') = \overline{K(z', z)}^{-1}$ and (γ): $K(\pi(e), \pi(e)) = e$ is called a *normalized kernel function* associated with J.

DEFINITION 8.1.5. Two T^C-valued factors of automorphy J, J' of $(\tilde{V}, D)$ are *equivalent* if there is a holomorphic map φ: $D \to T^C$ such that $J'(g, z) = \varphi(g(z))\, J(g, z)\, \varphi(z)$.

THEOREM (Murakami) 8.1.6. For $(\tilde{V}, D)$ any two T^C-valued automorphy factors which satisfy (α) are equivalent.

THEOREM (Murakami) 8.1.7. For any T^C-valued factor of automorphy satisfying (α) a kernel $K: D \times D \to T^C$ satisfying ((ii) & (β)) is unique up to multiplication by an element in the center of T. Thus, for the canonical automorphy factor, K is uniquely characterized by conditions ((ii), (β), (γ)).

Under the natural injection $D = \tilde{V}/T \to M_+ \simeq \mathfrak{m}_+$ we see that there is an invariant complex structure on D. Now define a vector bundle over D as follows. Let τ be an irreducible unitary representation of T. Thus τ is of the form $t \to t^l$ for t in T and l in Z. Consider the action of T on $\tilde{V} \times C$ by

$t(g, z) = (gt, \tau(t)^{-1}z)$ where g is in $\tilde{V}$ and z in C. Then $E_\tau = (\tilde{V} \times C)/T$ is a vector bundle over D with one dimensional fiber C. Let $\tilde{\pi}: \tilde{V} \times C \to E_\tau$ be the canonical projection. Since E_τ can be identified with an open subbundle of $(\tilde{V}_C \times C)/T_C M_-$, E_τ has a natural structure of a holomorphic line bundle. Since C has a T-invariant Hermitian metric, E_τ is a Hermitian vector bundle.

Let $C^q(E_\tau)$ denote the space of all E-valued smooth forms of type $(0, q)$ on D. And let $C^q_c(E_\tau)$ denote the subspace of all $(0, q)$ forms with compact support. There is an L^2-norm on $C^q_c(E_\tau)$ given by $\|\varphi\|^2 = \int_D \varphi^t \Lambda^* \varphi \#$. This gives rise to the Hilbert spaces $L^q_2(E_\tau)$. Using the $\bar{\partial}$ operator we say φ in $L^q_2(E_\tau)$ is harmonic if both $\bar{\partial}\varphi$ and $\vartheta\varphi$ are defined and equal zero. $H^q(E_\tau)$ is set equal to the closed subspace of $L^q_2(E_\tau)$ of all harmonic forms.

The theorem of Kobayashi cited above states in this context that

THEOREM 8.1.8. *$H^0(E_\tau)$ is irreducible.*

However, $H^0(E_\tau)$ may be vacuous.

Let $C^q(\tilde{V}, T, \tau)$ denote the space of C-valued smooth forms of degree q on $\tilde{V}$ which satisfy $i(X)\varphi = 0$ for X in $\mathfrak{t}_C + \mathfrak{m}_+$ and $\varphi(gt, X) = \tau(t)^{-1}\varphi(g, \mathrm{ad}(t)X)$ for g in V and t in T.

THEOREM 8.1.9. *$C^q(\tilde{V}, T, \tau)$ is isomorphic to $C^q(E_\tau)$ under the map $\varphi \to \tilde{\varphi}(\pi(g), \pi(X_g)) = \tilde{\pi}(g, \varphi(g, X))$.*

Thus transferring the norm from $C^q(E_\tau)$ by this map we can complete $C^q(\tilde{V}, T, \tau)$ to give $L^q_2(\tilde{V}, T, \tau)$. And $H^q(\tilde{V}, T, \tau)$ is the closed subspace of harmonic forms.

THEOREM 8.1.10. *J is a canonical automorphy factor and K is a kernel function for $(\tilde{V}, W_+)$.*

COROLLARY 8.1.11. *$J(\tilde{u}, w)^l$ holomorphically triviallizes E_{τ_l} over W_+ – i.e. $E_{\tau_l} \simeq W_+ \times C$ by the mapping $\tilde{\pi}((t, u), z) \to (u_+, J(u, 0)^l z)$.*

THEOREM 8.1.12. *$L^0_2(V, T, l) \simeq L_2(W_+)^{(l)}$ by the map $\varphi \to f(u_+) = J(u, 0)^l \varphi(\tilde{w})$ where $L_2(W_+)^{(l)}$ is the space of measurable functions f on W_+ with $\|f\|^2 = \int_{W^+} |f(w)|^2 K(w, w)^{-1}\, đw$ where $đw = 2^{-n}\prod_i |dw_i d\bar{w}_i|$ where w_i is the orthonormal coordinate of w in W_+ with respect to the Hermitial form $2iA(w, w)$.*

COROLLARY 8.1.13. *$H^0(V, T, 1) \simeq \mathscr{F}(W_+)^{(l)}$ = holomorphic functions in $L_2(W_+)^{(l)}$.*

We note that $\tilde{V}$ acts on $\mathscr{F}(W_+)^{(l)}$ by the representation

$$(T(g)f)(w) = J(g^{-1}, w)^{-l}\, f(g^{-1}(w)).$$

THEOREM 8.1.14. The unitary representation $(T, \mathscr{F}(W_+)^{(l)}$ is irreducible.
Proof. Every element h of $\mathscr{F}(W_+)^{(l)}$ satisfies

$$h(z) = \int_{W_+} h(z)k(w', w)K(w, w)^{-1}\,\mathrm{d}w$$

for an appropriate k. But taking $T(g)f$ for h, we find that k is a kernel function. By uniqueness of the kernel one finds that $k = K$. In particular if $\mathscr{F}^0$ is a closed invariant subspace of $\mathscr{F}(W_+)^{(l)}$ and k' is a kernel function for $\mathscr{F}^0$, then k' is a scalar multiple of K. Thus $\mathscr{F}^0 = \mathscr{F}(W_+)^{(l)}$ or $\{0\}$.

Assume that $lA(x, Jx)$ is positive definite. Then we have

THEOREM 8.1.15. $L_2^q(W_+) \simeq \mathscr{F}^q(W_+)^{(l)} \otimes \overline{\mathscr{F}^q(W_+)^{(l)}}$.

THEOREM 8.1.16. The $\bar{\partial}$-cohomology is equivalent to the d-cohomology of $\mathscr{F}^q(W_+)^{(l)}$.

Setting $\Delta = d\delta + \delta d$ we have $\Delta\varphi = 2\pi l(\Sigma w_k(\partial/\partial w_k) + q)\varphi$ for φ in $\mathscr{F}^q(W_+)^{(l)}$. Thus, if φ is harmonic one has $\varphi = 0$ except for the $q = 0$ case when φ is a constant. Thus

$$H^q(\tilde{V}, T, l) = \begin{cases} \{0\} & \text{for} \quad q > 0 \\ \mathscr{F}(W_+)^{(l)} & \text{for} \quad q = 0. \end{cases}$$

By Serre duality $H^q(V, T, l)$ is dual to $H^{n-q}(V, T, -l)$. So $H^q(V, T, -l) = 0$ except when $q = n$ and $H^n(V, T, -l)$ is dual to $\mathscr{F}(W_+)^{(l)}$.

Dropping the positive definiteness the general case proceeds as follows. One decomposes V into positive and negative definite components: $V = V' + V''$. Then one finds that $H^q(\tilde{V}, T, l) \simeq \sum_{q'+q''=q} H^{q'}(V', T, l) \otimes H^{q''}(V'', T, l)$. Thus we have

THEOREM (Sataka) 8.1.17. Let $\tau_i \in \hat{T}$ for $l \neq 0$ and let the signature $lA(x, Jx) = (2r', 2n - 2r')$. Then $H^q(E_{\tau_l})$ is $\neq 0$ and is irreducible for $q = n - r'$ and $= \{0\}$ otherwise.

Define the intertwining operator $U_{z',z}$ by

$$(U_{z',z}f)(w') = \gamma_{z',z} \int_{W^{\mathbb{C}}} K(\tilde{z}', \tilde{z})\, K(\tilde{z}, \tilde{z})^{-1}\, f(w) \frac{dw}{z},$$

where

$$z = (w, z), z' = (w', z')$$

and

$$\gamma_{z',z} = \det\left(\frac{1}{2i}(z'-z)\right)^{-1/2} \det y^{1/4} \det y'^{(-1/4)}.$$

THEOREM 8.1.18. *$U_{g(z'),g(z)} T(\tilde{u}) = T(\tilde{u})\, U_{z',z}$ for $\tilde{u}$ in $\tilde{V}$.*

COROLLARY (Stone–von Neumann) 8.1.19. *$U_{z',z}$ gives a unitary equivalence of all Fock representations.*

8.2. NILPOTENT LIE GROUPS

The results for the Heisenberg group generalize to the class of connected simply connected nilpotent Lie groups. This was a major discovery of Kirillov. His results are beautifully summarized in Wallach W3. Thus we merely state the following:

THEOREM (Kirillov) 8.2.1. Let N be a nilpotent Lie group as described with Lie algebra $\mathfrak{n}$ and $\exp: \mathfrak{n} \to N$ being the identity. For f in $\mathfrak{n}^*$, then there is a Lie algebra polarization $\mathfrak{h}$ at f such that if $\mathfrak{X}_f$ is the character of $\exp(\mathfrak{h})$ given by $\mathfrak{X}_f(\exp X) = \exp(if(X))$ X in $\mathfrak{h}$ then:

(i) the induced representation $U_{f,\mathfrak{h}}$ from $\mathfrak{X}_f$ is irreducible;

(ii) $U_{f,\mathfrak{h}}$ and $U_{f',\mathfrak{h}'}$ are equivalent iff f and f' are on the same orbit $\mathcal{O}$ in $\mathfrak{n}^*$ and $\mathfrak{h}'$ is a polarization for f;

(iii) let $\pi: \mathcal{O}_f \to U(\mathcal{O}_f) = [U_{f,\mathfrak{h}}]$ denote the map that associates the orbit of f to the equivalence class $[U_{f,\mathfrak{h}}]$; thus

$$\pi: \mathfrak{n}^*/N \to \hat{N};$$

(iv) $U(\mathcal{O}_f)$ is of trace class and by the Kirillov character formula $U(\mathcal{O})$ is equivalent to $U(\mathcal{O}')$ iff $\mathcal{O} = \mathcal{O}'$;

(v) if $\lambda \in \hat{N}$, then $\lambda = [U(\mathcal{O})]$ for some orbit $\mathcal{O}$ in $\mathfrak{n}^*$.

Kostant and Auxlander and Pukansky have extended this philosophy to a wider class of groups.

DEFINITION 8.2.2. $\mathfrak{g}$ is called *exponential solvable* if exp: $\mathfrak{g} \to G$ is a diffeomorphism.

EXAMPLE 8.2.3. The group in Example 7.1.12 is exponential solvable.

For solvable groups we need another condition.

DEFINITION 8.2.4. The *Pukansky condition* is that the orbit of f in $\mathfrak{g}^*$ contains the affine space $f+\mathfrak{h}^\perp$ where $\mathfrak{h}^\perp=\{f_1\in\mathfrak{g}^* \mid f_1(\mathfrak{h})=0\}$ or equivalently that $\exp(\mathfrak{h}).f=f+\mathfrak{h}^\perp$.

This condition holds automatically in the nilpotent case.

THEOREM (Bernat–Pukansky) 8.2.5. Let G be an exponential solvable Lie group. For each f in $\mathfrak{g}^*$ there is a subordinate subalgebra $\mathfrak{h}$ in $\mathfrak{g}$ such that $U(f,\mathfrak{h})=\mathrm{Ind}^G_{\exp(\mathfrak{h})}(\mathfrak{X}_f)$, where $\mathfrak{X}_f(\exp X)=\exp(if(X))$, is irreducible iff $\mathfrak{h}$ is maximal and satisfies the Pukansky condition. $U(f,\mathfrak{h})$ is independent of $\mathfrak{h}$ and depends only on the G-orbit of f in $\mathfrak{g}^*$. Thus we have a bijection $G.f\to U(f)$ from $\mathfrak{g}^*/G$ onto $\hat{G}$.

Let U be a representation of $\mathfrak{g}$ in a complex vector space V. One can choose a base of V so that $U(\mathfrak{g})$ is triangular. The elements along the diagonal of $U(X)$ for X in $\mathfrak{g}$ are the values of complex linear forms on $\mathfrak{g}$, viz. the roots of the representation. If $\lambda\in\mathfrak{g}^*\otimes C$ we set $S_\lambda(X)=$ $=\sinh(\lambda(X)/2)/\lambda(X)/2$ for every X in $\mathfrak{g}$.

Kirillov and Pukansky showed that Kirillov's character formula holds for exponentially solvable groups. Viz. we have

THEOREM (Kirillov Pukansky) 8.2.6. Let G be an exponentially solvable group and let T be an irreducible unitary representation of G. Then there exists an orbit $\mathcal{O}$ of G in $\mathfrak{g}^*$ and a set $\lambda_1,\ldots,\lambda_n$ of roots of the adjoint representation such that if λ is a root, then $\bar{\lambda}$ is also a root; the function $\prod_{i=1}^n S_{\lambda_i}$ is positive. Setting $p_{\mathcal{O}}(X)=\{\prod_{i=1}^n S_{\lambda_i}(X)\}^{1/2}$ and taking $\alpha\in A_0(\mathfrak{g})$ such that $T(\alpha)$ has a trace, then

$$\mathrm{Tr}\,T(\alpha)=\int_{\mathcal{O}}\{\int_{\mathfrak{g}} p_{\mathcal{O}}^{-1}(X)\alpha(X)\exp(i\langle f,X\rangle)\,\mathrm{d}X]\,\mathrm{d}\beta(f),$$

where $f\in\mathfrak{g}^*$. If $\alpha\in A_0(\mathfrak{g})$ is such that $T(\alpha)$ is positive and the integral just cited converges, then $T(\alpha)$ has a trace.

Thus Bernat showed that Kirillov's method held true for exponential solvable groups. However, Bernat showed that for nonexponential solvable groups it may fail that $U(f,\mathfrak{h})$ is irreducible and two real polarizations associated to f may exist such that $U(f,\mathfrak{h})$ and $U(f,\mathfrak{h}')$ are not equivalent. The classic example of a nonexponential solvable group arises in physics – viz. the oscillator group which we introduced in Example 5.2.4.

EXAMPLE 8.2.7. Let D denote the oscillator Lie group. D is the semidirect product of the Heisenberg group and R. If (H, P, Q, I) denotes the basis of the Lie algebra $\mathscr{D}$ of D, we let H^*, P^*, Q^*, E^* denote the canonical dual basis in $\mathscr{D}^*$. Then if $f = aE^* + bP^* + cQ^* + dH^*$ is an arbitrary point in $\mathscr{D}^*$ we set $f = (a, b, c, d)$. Then one can show that the coadjoint orbit associated to $f_0 = (a_0, b_0, c_0, d_0)$ are cylinders if $b_0^2 + c_0^2 \neq 0$ or the line given by $d = d_0$ for the case $a_0 = 0$. If $a_0 \neq 0$ the orbits are paraboloids of revolution $d = d_0^{\pm}[(b_0^2 + c_0^2)/2a_0]$.

$\mathscr{D}$ has no real polarization so one must consider complex polarizations. $\mathfrak{h} = (H, E, P + iQ)$ and $\bar{\mathfrak{h}} = (H, E, P - iQ)$ are complex polarizations. The induced representation from $\mathfrak{h}$ at f_0 for $a_0 > 0$ is unitarily equivalent to

$$W(n, t)\, f(y) = \exp(it(h - \tfrac{1}{2}))\, U_a(n)\, V(t),$$

where $V(t) = \exp[-(it/2a)(P^2 + Q^2)]$ where $P = -d/\mathrm{d}x$, $Q = -iax$, and $U_a(n)$ is an irreducible representation of N such that on the center $U_a = \exp(ia)$. To prove this one notes that if $U(n = (x, y, a))$ is the Schrödinger representation, then upon time translation $n \to t(n)$, $U_t(n) = U(t(n))$ is an irreducible unitary representation of N. Thus by Stone–von Neumann there is a unitary operator $V(t)$ such that $U_t(n) = V(t)\, U(n)\, V(t)^{-1}$. It is left to the reader to check that $V(t) = \exp(it(P^2 + Q^2 + h - \tfrac{1}{2}))$.

For the general solvable Lie group we must turn to the results of Auslander and Kostant:

THEOREM (Auslander–Kostant) 8.2.8. Let G be a connected simply connected solvable Lie group with Lie algebra $\mathfrak{g}$. If $f \in \mathfrak{g}^*$ then:

(i) there is a strongly admissable Pukansky positive polarization $\mathfrak{h}$ at f;

(ii) the induced representation defined by $\mathfrak{h}$ is independent of the choice of $\mathfrak{h}$;

(iii) If G is of type I, then every irreducible unitary representation is equivalent to one of this form;

(iv) let $L_c(\mathcal{O})$ denote the set of all equivalence classes of line bundles with connection on $\mathcal{O}$ with the Kirillov 2-form $\Omega_{\mathcal{O}}$ as curvature class. Then if G is of type I, we have

$$\bigcup_{\mathcal{O} \in \mathfrak{g}^*/G} L_c(\mathcal{O}) \to \hat{G}.$$

Thus if G is of exponential type $\mathfrak{g}^*/G \to \hat{G}$ is a bijection.

PROBLEMS

EXERCISE 8.1. Show that the representation T^1 of $\tilde{V}$ on $\mathscr{F}(W_+)^{(1)}$ is equivalent to $\mathrm{e}(-t)\, V(c)$ where $V(c)$ is defined in Exercise 2.1.

Chapter 9

Borel–Weil Theory

9.1. Representation theory for compact semisimple Lie groups

In this chapter we want to develop enough of the representation theory for compact semisimple Lie groups to show that Kirillov's method applies to this case. This development requires somewhat of a review of the classical theory of representations for Lie groups and algebras. One major result is the Borel–Weil theory for geometric realization of the representations. This is important, for as we show in the following chapters the geometric quantization of certain mechanical systems is embodied in the Borel–Weil Theorem, which allows us to calculate the multiplicities of the eigenvalues given by quantization.

Let M be a compact Lie algebra $\mathfrak{m}_0$.

Definition 9.1.1. The *rank* of M is the dimension of a maximal abelian subalgebra of $\mathfrak{m}_0$.

If M is connected, the rank of M is then the dimension of a maximal torus (i.e. a connected abelian subgroup) H in M.

Example 9.1.2. If $G = SO(n)$ then a maximal torus is given by

$$H = \begin{pmatrix} \cos t_1 & \sin t_1 & & 0 \\ -\sin t_1 & \cos t_1 & & \\ & & \ddots & \\ & 0 & & \end{pmatrix}.$$

Thus the rank $(SO(n)) = [n/2]$.

The irreducible representations of H are all one dimensional and form a multiplicative group, $\hat{H} = \mathrm{Hom}(H, C^*)$, the character group of H.

Assume now that M is a connected simply connected compact Lie group with maximal torus H in M. Let $\mathfrak{m}_0$ resp. $\mathfrak{h}_0$ denote the real Lie algebras of M, H respectively. Let $\langle , \rangle$ be a M-invariant bilinear form on $\mathfrak{m}_0$ say given by the Killing form $\langle X, Y\rangle = \mathrm{Tr}(\mathrm{ad}\, X\ \mathrm{ad}\, Y)$. Then $\langle \mathrm{Ad}(g)X, \mathrm{Ad}(g)Y\rangle = \langle X, Y\rangle$ for g in M and X, Y in $\mathfrak{m}_0$. Thus we can identify $\mathfrak{m}_0$ with $\mathfrak{m}_0^*$ by $X \to \hat{X}$ where $\hat{X}(Y) = \langle X, Y\rangle$. Similarly we can identify the adjoint and coadjoint actions.

DEFINITION 9.1.2. If $N(H)$ is the normalizer of H in M, the finite group $W(H) = N(H)/H$ is called the *Weyl group*.

THEOREM 9.1.3. The orbit space of $\mathfrak{m}_0$ under the adjoint action of M is given by $\mathfrak{m}_0/\mathrm{Ad}(M) = \mathfrak{h}_0/W(H)$.

The complexification $\mathfrak{m} = \mathfrak{m}_0 \otimes C$ has complex conjugation induced from this representation by identifying $\mathfrak{m}_0$ with $\mathfrak{m}_0 \otimes 1$. To any representation T of M on a complex vector space E, $T \in \mathrm{Hom}(H, GL(E))$, there is a representation $\dot{T} \in \mathrm{Hom}(M, \mathrm{End}(E))$ given by the differential of T at the identity where

$$\dot{T}(X) = \lim_{t \to 0} \frac{T(\exp(tX)) - T(e)}{t},$$

where $X \in \mathfrak{m}_0$. Here $\dot{T}$ commutes with $i = \sqrt{-1}$; i.e. $\dot{T}(iX) = i\dot{T}(X)$ for X in $\mathfrak{m}_0$.

Restricting a representation $T: G \to GL(E)$ to H gives a decomposition $E = \bigoplus_{\mu} E_\mu$ where $E_\mu = \{v \text{ in } E \mid T(h)v = e^{\mu}(h)v \text{ for } h \text{ in } H\}$.

DEFINITION 9.1.4. μ is called a *weight* of T if $E_\mu \neq 0$. The *multiplicity* n_μ is defined to be $\dim E_\mu$.

If T is the adjoint representation Ad of M on $\mathfrak{m}$, then Ad restricted to H gives the decomposition

$$\mathfrak{m} = \mathfrak{h} \oplus \bigoplus_{\alpha \neq 0} \mathfrak{m}_\alpha,$$

where $\mathfrak{m}_\alpha = \{X \in \mathfrak{m} \mid \mathrm{ad}(H)X = [H, X] = \alpha(H)X \text{ for } H \text{ in } \mathfrak{h}\}$.

DEFINITION 9.1.5. The weights of the adjoint representation are called the *roots* of M with respect to H. Let Δ denote the set of roots.

We note that $\mathfrak{m}_\alpha = \mathfrak{m}_{-\alpha}$. Thus $\Delta = -\Delta$. And $\dim \mathfrak{m}_\alpha = 1$. T acting on $\mathfrak{m}_\alpha$ defines a character $\mathfrak{X}_\alpha$ of H where $\alpha \in \mathfrak{h}^*$; viz. $\mathfrak{X}_\alpha(\exp X) = \exp(\alpha(X))$ for X in $\mathfrak{h}$. Since $\mathrm{Ad}(h)X = \mathfrak{X}_\alpha(h)X$ for X in $\mathfrak{m}_\alpha$ and h in H we see that $\Delta \subset i\mathfrak{h}_0^*$.

We extend $\langle , \rangle$ to $\mathfrak{m}$ where it is also denoted $\langle , \rangle$.

DEFINITION 9.1.6. $\mathfrak{m}$ is called *semisimple* if $\langle , \rangle$ is nondegenerate. If $\mathfrak{m}$ is semisimple, we will say that M is semisimple.

One may choose an ordering in Δ which defines the positive root, Δ^+, e.g. so that $\alpha_1, \ldots, \alpha_m$ in $\mathfrak{h}_0^*$ are the positive roots. There is a minimal set Π of generators over Z in Δ^+. We write $\Pi = \{\alpha_1, \ldots, \alpha_l\}$ where $l = \dim \mathfrak{h} = \text{rank}$ $\mathfrak{m}$. Π is called the *system of simple roots*.

DEFINITION 9.1.7. For α in Δ we define

$$h'_\alpha \text{ in } i\mathfrak{h}_0 \text{ by } \langle h'_\alpha, Y\rangle = \alpha(Y) \qquad \text{for } Y \text{ in } \mathfrak{h}.$$

For α, β in Δ we set $(\alpha, \beta) = \langle h'_\alpha, h'_\beta\rangle$

Thus $(,)$ is a nondegenerate bilinear form on Δ.

DEFINITION 9.1.8. An element α in $\mathfrak{h}^*$ is called *integral* if $2(\alpha, \beta)/(\beta, \beta)$ belongs to Z for all β in Δ.

Select a point p in $\mathfrak{h}_0$; set $f_p(X) = \langle p, X\rangle$. Then $2\pi i f_p \in i\mathfrak{h}_0^* \subset \mathfrak{h}^*$. And if $2\pi i f_p$ is integral then $2\pi i \langle h'_\alpha, p\rangle / \langle h'_\alpha, h'_\alpha\rangle \in Z$.

Given p in $\mathfrak{h}_0$ let $M_p = \{g \text{ in } M \mid \text{Ad}(g)p = p\}$. The Lie algebra of M_p is $m_p = \{X \in \mathfrak{m}_0 \mid [X, p] = 0\}$. Under the present assumptions it can be shown that M_p is connected and, hence, M/M_p is simply connected. By Kostant's theorem M/M_p is thus realizable as an orbit $\mathcal{O}_p = M.p$ in $\mathfrak{m}_0^*$. And the Kirillov 2-form Ω_p on $\mathcal{O}_p$ is integral iff $X \to 2\pi i f_p(X)$, $X \in \mathfrak{m}_p$, is the differential of a character of M_p. Clearly $\mathfrak{X}(\exp X) = \exp(2\pi i f_p(X))$ defines a character of $H \subset M_p$. Combining these results with the representation theory of compact groups we have:

THEOREM 9.1.9. $X \to 2\pi i f_p(X)$ is the differential of the character $\mathfrak{X}_f(\exp X) = \exp(2\pi i f_p(X))$ iff $2\pi i f_p$ is integral iff the Kirillov form Ω_p is integral.

Thus we identify $\hat{H}$ with a lattice Λ in $i\mathfrak{h}_0^*$. Assume that Λ is nonempty.

The cases of interest are when $G_p = H$ is a maximal toral subgroup. (See Exercise 9.2 for the condition.) In this case H acts on $M \times C$ by $h(m, z) = (mh^{-1}, \mathfrak{X}_f(h)z)$. Let $L = M \times_H C$. Then L is a line bundle over M/H.

THEOREM 9.1.10. L has a natural connection ∇ and ∇-invariant Hermitian structure: $(L, \nabla) \in L_c(M/H, \Omega)$.

Proof. Let $[g, z]$ denote an element of L and set $\tilde{p}([g, z]) = gG_p$. The Hermitian structure h is defined by

$$h_{gG_p}([g, z], [g, w]) = z\bar{w}.$$

Clearly h is a Hermitian structure on L and $h(gy_1, gy_2) = h(y_1, y_2)$ for y_1, y_2 in L. The rest is left to the reader.

As we have just noted by Kostant's theorem there is a naturally defined connection ∇ on L such that h is ∇-invariant and such that $\text{curv}(L, \nabla) = 2\pi i \Omega_p$.

THEOREM 9.1.11. Assume $(M/H, \Omega, \lambda)$ is a Hamiltonian G-space. Then Ω is integral iff Λ is nonempty. $\Lambda \to L_c(M/H, \Omega)$: $\mathfrak{X} \to L_{\mathfrak{X}}$ is a bijection.

COROLLARY 9.1.12. Λ is the inverse image of Ω in $H^2(M/H, Z)$. The proof of the corollary is deferred until the next section.

DEFINITION 9.1.13. The *index* of λ in $\hat{H}$ is the number of α in Δ^+ such that $\dot{\lambda}(h'_\alpha) < 0$.

DEFINITION 9.1.14. λ in $\hat{H}$ is called singular if $\dot{\lambda}(h'_\alpha) = 0$ for some α in Δ^+. If λ is nonsingular, it is called *regular*.

DEFINITION 9.1.15. The *Weyl chamber* is $D = \{\lambda \in \hat{H} \mid \text{Ind}(\lambda) = 0\}$.

Thus D is given by $D(\mathfrak{m}) = \{\dot{\lambda} \in \mathfrak{h}^* \mid \langle \dot{\lambda}, h'_\alpha \rangle \geqq 0$, for all α in $\Delta^+\}$. D is a fundamental domain for the action of W on $\hat{H}$ where $w\lambda(h) = \lambda(w^{-1}(h))$ for w in W, h in H. We let D^0 denote the interior of D.

DEFINITION 9.1.16. Let $\delta = \frac{1}{2} \sum_{\alpha \in \Delta^+} \alpha$.

THEOREM 9.1.17. (i) δ is the minimal element in D such that $\delta + D = D^0$.
(ii) $\delta(h'_\alpha) = (\delta, \alpha) = (\alpha, \alpha)$.

As in Theorem 9.1.11 we can view the Weyl chamber D as the set of λ in $H^2(M/H, R)$ for which $(\psi^{-1}\lambda, \alpha) \geqq 0$ for α in Δ^+ where ψ is the transgression (v. B17). Thus the set $\Lambda \cap D$ is identified to be $H^2(M/H, Z) \cap D$. In the next section we will explicitly show that Λ lies in the inverse image of the Kirillov 2-form on M/H. We also note here that Λ is invariant under the coadjoint action of M on $\mathfrak{m}^*$.

We now connect the ideas of roots and rank by

THEOREM 9.1.18. Let M be a compact connected Lie group of dimension n and rank l. Then G has $2m$ roots $\pm\alpha_j$, $j = 1, \ldots 2m$, and $n = l + 2m$. M is semisimple iff it has l linearly independent roots.

Clearly $2m = \dim(M/H)$. We can relate the roots and the cohomology of M/H by

THEOREM 9.1.19. Let M be a compact connected semisimple Lie group. Then

$$\mathfrak{h}_0^* \simeq H^1(H, R) \xrightarrow{\psi} H^2(M/H, R).$$

The proof involves the study of transgression ψ of the fiber bundle $H \to \to M \to M/H$. For details v. B17.

We also note here that the space M/H has 2^m invariant almost complex structures determined by the roots $(\varepsilon_1\alpha_1, \ldots, \varepsilon_m\alpha_m)$ where $2m = \dim(M/H)$ and $\varepsilon_j = \pm 1$, $1 \leqq j \leqq m$. In this case we have.

THEOREM 9.1.20. The first Chern class of M/H is $c_1 = c_1(M/H) = \varepsilon_1\alpha_1 + + \cdots + \varepsilon_m\alpha_m$.

Since δ is a weight it follows that $c_1/2$ is a weight – i.e. $c_1(M/H) \in H^2(M/H, Z)$ vanishes when reduced to coefficients mod 2. In other words $w_2(M/H) \in H^2(M/H, Z_2)$ vanishes.

THEOREM 9.1.21. c_1 is nonsingular iff $\varepsilon_1\alpha_1, \ldots, \varepsilon_m\alpha_m$ is a positive system of roots of M iff the almost complex structure on M/H is integrable.

THEOREM 9.1.22. The number of invariant complex structures on M/H is equal to $|W(H)|$.

Thus we pick our system of positive roots $\Delta^+ = \{\alpha_1, \ldots, \alpha_m\}$ of M/H and let $2\delta = \sum \alpha_j$. We select the invariant complex structure on M/H which has Δ^+ as its system of roots.

EXAMPLE 9.1.23. Let $M = SU(2)$. Then $\mathfrak{m} = sl(2, C)$ and $\mathfrak{h}$ is the set of diagonal matrices in $sl(2, C)$. The nonzero roots are defined by

$$\alpha_{ij}\left[\begin{pmatrix} c_1 & 0 \\ 0 & c_2 \end{pmatrix}\right] = c_i - c_j \quad \text{where} \quad 1 \leqq i, j \leqq 2, i \neq j.$$

But $\bar{c}_k = -c_k$, so set $c_k = i\varphi_k$ and identify $\Delta = \{i\varphi_k - i\varphi_l\} 1 \leqq k \neq l \leqq 2$. In this case Π and Δ^+ can be taken as $\{\alpha_{12}\}$. Set $\alpha = \alpha_{12}$. Λ is the lattice in $i\mathfrak{h}_0^* \simeq R$ spanned by $i\varphi_1, i\varphi_2$. The root space decomposition is $sl(2, C) = CH_\alpha \oplus CE_\alpha \oplus CE_{-\alpha}$ where

$$H_\alpha = \frac{1}{4}\begin{pmatrix} 1 & 1 \\ 0 & -1 \end{pmatrix} \quad \text{and} \quad E_\alpha = \frac{1}{2}\begin{pmatrix} 0 & 1 \\ 0 & 0 \end{pmatrix}.$$

Clearly $[H_\alpha, E_\alpha] = \frac{1}{2}E_\alpha$, $\alpha(H_\alpha) = \frac{1}{2}$ and $[E_\alpha, E_{-\alpha}] = H_\alpha$. We take the Killing form to be $\langle\ ,\ \rangle = 4\mathrm{Tr}(XY)$; thus $\alpha(H_\alpha) = \langle H_\alpha, H_\alpha \rangle = \frac{1}{2}$, – i.e. the usual normalization.

The Weyl group of $SU(2)$ is just the group of permutations of $i\varphi_1$ and $i\varphi_2$ – i.e. $W = S_2$. Since $|W| = 2$ we know that M/H has two invariant complex structures which are complex conjugates of each other. They are realized as follows: Let $\mathfrak{l} = CH_\alpha \oplus CE_\alpha$. Then the associated Lie subgroup B of $SL(2, C)$ is

$$B = \left\{ \begin{pmatrix} a & b \\ 0 & c \end{pmatrix} \middle| ac = 1 \right\}.$$

And $M/H = SL(2, C)/B = CP(1)$. $\mathfrak{l}$ is a Lie algebra polarization giving rise to the Kāhler structure on $CP(1)$.

The Chern class of $CP(1)$ is calculated as follows. Let $\mathfrak{X}_0$ be the character of B defined by

$$\mathfrak{X}_0 \begin{pmatrix} a & b \\ 0 & c \end{pmatrix} = a.$$

Let $L_{\mathfrak{X}_0}$ be the line bundle over $CP(1)$ defined by $\mathfrak{X}_0$. Let $\mathfrak{X}$ be the character of B defined by $\mathfrak{X}(H_\alpha) = -\sum_{\beta \in \Delta^+} \beta(H_\alpha)$ and $\mathfrak{X}(E_\alpha) = 0$. The tangent bundle is the vector bundle associated to the linear isotropy representation $\rho: B \to GL(s1(2, C)/\mathfrak{l})$. It is easily checked that this is given by $\rho(Y)E_\alpha = -\alpha(Y)E_{-\alpha}$ for α in Δ^+, Y in $\mathfrak{h}$. The representation $\sigma = \det(\rho) = = \Lambda^2 \rho$ defines the line bundle det (T). From the last comment we see that $\sigma(Y) = \mathrm{Tr}(\rho(Y)) = -\sum_{\alpha \in \Delta^+} \alpha(Y)$. Thus we have det $(T) = L_{\mathfrak{X}}$ since $\mathfrak{X}_0^{-2} = \mathfrak{X}$ we have det $(T) = L_{\mathfrak{X}_0}^{-2}$. Thus, if $\mathfrak{g}$ is the positive generator of the infinite cyclic group $H^2(CP(1), Z)$. Then $c_1(CP(1)) = 2\mathfrak{g}$. So upon integrating over $CP(1)$ we have $c_1(CP(1))[CP(1)] = 2$.

This result is just a special case of the following theorem:

THEOREM 9.1.24. If G/U is of the form where U is the centralizer of a 1-dimensional torus S defined by $(\alpha_1 = \cdots = \alpha_{l-1} = 0)$ where $\{\alpha_1, \ldots, \alpha_l\}$ are simple roots of G and G is semisimple, then $c_1(G/U) = 2[(2\delta, \alpha_l)/(\alpha_l, \alpha_l)]\,\mathfrak{g}$, where $\mathfrak{g}$ is the generator of $H^2(G/U, Z)$.

As we saw in Chapter 1 the characters of irreducible unitary representations parametrize $\hat{G}$. In the present context this is specified by Weyl's character formula. For w in W we let $\varepsilon(w) = \det\{w: i\mathfrak{h}_0^* \to i\mathfrak{h}_0^*\}$. Thus ε maps W onto $\{\pm 1\}$.

THEOREM 9.1.25. Every character of an irreducible representation of a compact simply connected semisimple Lie group is of the form $\mathfrak{X}_\lambda$ for λ in

$\Lambda \cap D^0$ where

$$\mathfrak{X}_\lambda|_H = \frac{\sum_{w \in W} \varepsilon(w) e^{w\lambda}}{\prod_{\alpha \leftarrow \Delta\dagger} (e^{\alpha/2} - e^{-\alpha/2})}.$$

COROLLARY 9.1.26. $\hat{G} \simeq \Lambda \cap D^0$.

EXAMPLE 9.1.27 (9.1.23 cont.). $\lambda \in \Lambda \cap D^0$ is of the form $\lambda = n\alpha/2$, $n > 0$. Thus

$$\mathfrak{X}_\lambda|_H = \frac{\exp(-\lambda) - \exp(-\lambda)}{\exp(\alpha/2) - \exp(-\alpha/2)}$$

as noted in Chapter 1.

DEFINITION 9.1.28. An element μ in Λ is the *highest weight* of a representation T if μ is a weight of T and $\mu + \alpha$ is not a weight for every α in Δ^+.

THEOREM (Highest Weight) 9.1.29. Every irreducible representation U of M has a highest weight μ of multiplicity one where μ is in $\Lambda \cap D$. Conversely every μ in $\Lambda \cap D$ is the highest weight of some U in $\hat{M}$.

Neither the highest weight theorem nor the Weyl character formula gives a realization of the irreducible representations of M. The realization is given by the Borel–Weil theory to which we turn next.

9.2. BOREL–WEIL THEORY

We saw that M/H can be transformed uniquely into a homogeneous complex manifold; the tangent bundles of this manifold are determined by the appropriate Lie algebras as follows:

THEOREM 9.2.1.

(i) $T(M/H)_{x_0} = \mathfrak{m}_0/\mathfrak{h}_0$;

(ii) $T(M/H)^C_{x_0} = \mathfrak{m}/\mathfrak{h} = \bigoplus_{\alpha \in \Delta^+} \mathfrak{m}_\alpha \oplus \bigoplus_{\alpha \in \Delta^-} \mathfrak{m}_\alpha$;

(iii) the sheaf of germs of holomorphic functions over open U in M/H is given $\mathcal{O}(U) = \{f \in A(p^{-1}(U)) | f(gh) = f(g), Xf = 0$ for all X in $\oplus \mathfrak{m}_\alpha$. h in $H\}$ where $p: M \to M/H$.

Construct the line bundle $E(\lambda)$ for λ in Λ as $M \times_H C$ where $(m, z) \sim$

$\sim (mh, e^{\lambda}(h)z)$. Let $(m,z)_p$ be the element of $E(\lambda)$ which is the natural projection on $E(\lambda)$. M acts on $E(\lambda)$ by $m'(m,z)_p = (m'm, z)_p$. $E(\lambda)$ can be given the structure of a holomorphic line bundle; viz. $\mathcal{O}(E(\lambda))|_U =$ $= \{f \in A(p^{-1}(U)) | f(gh) = e^{\lambda}(h)f(g),\ Xf = 0\}$.

Thus $H^0(M/H, \mathcal{O}(E(\lambda))$, the space of global holomorphic sections of $E(\lambda)$ is an M-module. The sections of $E(\lambda) \to M/H$ over open U are maps $s: U \to$ $\to E(\lambda)$ of the form $s(mx) = [m, f_s(x)]$ where f_s satisfies $f_s(xh) = e^{-\lambda}(h) f_s(x)$ for all h in H, x in M/H.

This representation of M-module is realized locally as follows. Let $\varphi_\alpha : U_\alpha \times C \to E(\lambda)$ be the local trivialization given by the mapping $x \to g_x$ where $g_x.x_0 = x$ for g in M. Let 0 denote x_0. Then $(g,z) = (g_{g\cdot 0} g_{g\cdot 0}^{-1} g, z)$ $(g_{q\cdot 0}, \lambda(g_{g\cdot 0}^{-1} g)z)$ which we view as an element of $U_\alpha \times C$. That is

$$\begin{array}{ccc} (g.0, \lambda(g_{g.0}^{-1} g)z) & \xleftarrow{\ \phi_\alpha^{-1}\ } & (g,z) \\ \uparrow \phi_\alpha^{-1} & \swarrow \sim & \\ (gh, \lambda(h^{-1})z) & & \end{array}$$

commutes. Thus M acts on $E(\lambda)$ by

$$\begin{array}{l} (g'g, z) \sim (g_{g'g.0}, \lambda(g_{g'g.0}^{-1} g'g)z) \\ \big\downarrow \phi_\alpha^{-1} \\ g'(g.0, z) = (g'.g.0, \lambda(g_{g'g.0}^{-1} g' g_{g.0})z) \end{array}$$

or

$$g(x,z) = (g, x, \lambda(g_{gx}^{-1} g g_x)z).$$

Thus on local cross sections f of $E(\lambda)$ we have

$$(T(g)f)(x) = \lambda(g_x^{-1} g y_{g^{-1}x}) f(g^{-1}x).$$

Since $h_0(z,z') = \bar{z}z'$ is H invariant we can form a G invariant Hermitian metric by translating h_0 to give

$$\begin{aligned} h_{g.0}(z,z') &= |\lambda(g_{g.0}^{-1} g)|^{-2}\, \bar{z}z' \\ &= \exp(-F(g.0, \overline{g.0})\bar{z}z'. \end{aligned}$$

The G-invariant scalar product is then

$$\langle f_1 | f_2 \rangle = \int_M h(f_1, f_2)\Omega^n.$$

THEOREM (Borel–Weil) 9.2.2. Let λ in Λ be a character of H such that $\lambda + \delta$ is in D. Then:

(i) $H^q(M/H, \mathcal{O}(E(\lambda)) = 0$ for $q > 0$;
(ii) $H^0(M/H, \mathcal{O}(E(\lambda)) = 0$ if $\lambda + \delta$ is singular (i.e. $\lambda \notin D \cap \Lambda$).
(iii) $H^0(M/H, \mathcal{O}(E(\lambda))$ is the irreducible M-module with highest weight λ if $\lambda \in D \cap \Lambda$.

The proof will be outlined in the next section after the Kirillov geometry is presented.

EXAMPLE 9.2.3. (9.1.23 cont.). Let $\mathfrak{X}$ be the holomorphic character of B defined by

$$\mathfrak{X}\begin{pmatrix} a & b \\ 0 & c \end{pmatrix} = a^{2j}, j \leqq 0.$$

Then $\mathfrak{X}$ defines the line bundle $E(\mathfrak{X})$ given by $(g, z) \sim (gb, \mathfrak{X}^{-1}z)$ where g is in $SL(2, C)$, z in C. Define the map

$$z_1 \to g_{z^+} = \begin{pmatrix} z_1 & -1 \\ 1 & 0 \end{pmatrix}$$

on open

$$U_1 = \left\{\begin{pmatrix} a & b \\ c & d \end{pmatrix} c \neq 0, z_1 = a/c\right\}.$$

Thus the trivialization is

$$(g, z) \xrightarrow{\phi_1} \left(\frac{a}{d}, c^{2j}z\right).$$

Similarly let

$$z_2 \to g_{z_2} = \begin{pmatrix} 1 & 0 \\ z_2 & 0 \end{pmatrix}$$

be defined on

$$U_2 = \left\{\begin{pmatrix} a & b \\ c & d \end{pmatrix} a \neq 0 \quad z_2 = c/a\right\}.$$

Then

$$(g, z) \underset{\phi_2}{\to} \left(\frac{c}{a}, a^{2j}z\right).$$

Thus one has

$$(\phi_1 \circ \phi_2^{-1})(z_2, z) = \phi_1\left(\begin{pmatrix} 1 & 0 \\ z_2 & 1 \end{pmatrix}, z\right) = \left(\frac{1}{z_2}, z_2^{2j}z\right).$$

Thus the transition functions are $c_{12} = z_2^{ij}$.

A holomorphic section s of $E(\mathfrak{X})$ as noted above has the forms $s|U_i = f_i s_i$ where $f_1 = c_{12} f_2$ on $U_1 \cap U_2$ where $z_1 = z_2^{-1}$. Thus

$$f_1(z_1) = \sum c_k^1 z_1^k = c_{12} f_2(z_2) = z_1^{-2j} \sum c_k^2 z_1^{-k} = \sum c_k^2 z_1^{-2j-k}.$$

Thus f_1 and f_2 are polynomials of degree at most $2j$.

The local action of $SU(2)$ on sections over U_1 is given by

$$\begin{aligned}(T(g)f)(z_1) &= (g_{z_1}^{-1} g g_{g+z_1}) f(g^{-1} z) \\ &= \mathfrak{X}\left(\begin{pmatrix} 0 & 1 \\ -1 & z \end{pmatrix}\begin{pmatrix} a & b \\ c & d \end{pmatrix}\begin{pmatrix} \bar{a}z - y & -1 \\ \overline{b}z + a & 0 \end{pmatrix}\right) f\left(\frac{\bar{a}z_1 - b}{bz_1 + a}\right) \\ &= (\bar{b}z_1 + a)^{-2j} f_1(g^{-1} z_1).\end{aligned}$$

Thus $H^0(M/H, E(\mathfrak{X}))$ is just $D^{(j)}$, the spin $|j|$-representations.

The cohomology $H^0(M/H, \mathcal{O}(E(\lambda))$ can be computed in terms of the complex $\{A^k(E(\lambda)), \bar{\partial}\}$ where $A^k(E(\lambda))$ is the space of $E(\lambda)$-valued $(0, k)$-forms. By the Hodge isomorphism the cohomology groups of this complex are isomorphic to the spaces $\mathscr{H}^k(E(\lambda)) = \{\psi \in A^k(E(\lambda)) | \square \psi = 0\}$ of harmonic k-forms. Here $\square$ is the Laplace–Beltrami operator on M/H. Since the Hermitian metric used to define $\square$ is M-invariant, the subspaces $\mathscr{H}^k(E(\lambda))$ are preserved by M. Thus we have

THEOREM 9.2.4. $\mathscr{H}^k(E(\lambda)) \simeq H^k(M/H, \mathcal{O}(E(\lambda))$ is an isomorphism of M-modules.

Independence of polarization follows from the Riemann–Roch theorem which we present next.

THEOREM (Riemann–Roch–Hirzebruch) 9.2.5.

$$\mathfrak{X}(M/H, E(\lambda)) = T(M/H, E(\lambda)),$$

where T is the Todd index.

This implies that $\mathfrak{X}(M/H, E(\lambda))$ depends only on the $E(\lambda), c_1(M/H)$ and the Pontryagin classes of M/H which depend only on the underlying real symplectic manifold M/H.

The orbit space description of M/H is given as follows. Let $\mathcal{O}(f)$ in $\mathfrak{m}^*$ be given by the coadjoint action $\mathcal{O}(f) = M.f$. Thus $\mathcal{O}(f) = M/M(f)$. Using Kirillov's theorem the symplectic form is given by $\Omega_f(\sigma_f X, \sigma_f Y) = = \langle f, [X, Y] \rangle$.

Let e_α in $\mathfrak{m}_\alpha$ be the root vectors which satisfy $(e_\alpha, e_{-\alpha}) = \delta_{\alpha,\beta}$, $h_\alpha = [e_\alpha, e_{-\alpha}]$. Then we define $\omega^\alpha(e_\beta) = \delta_{\alpha\beta}$ and consider ω^α as 1-forms on M.

The exterior product $\omega^\alpha \wedge \bar{\omega}^\alpha$ is invariant under $\mathrm{Ad}(H^C)$. The restriction to M gives an M-invariant 2-form on M/H, which we denote by the same symbol.

Note that $(\omega^\alpha \wedge \bar{\omega}^\alpha)(e_\beta, e_\gamma) = 0$ unless $\beta = -\gamma = \pm\alpha$. Thus we have a left invariant 2-form on M given by

$$\Omega_f = \frac{i}{2\pi} \sum_{\alpha \in \Delta^+} (f, \alpha)\omega^\alpha \wedge \bar{\omega}^\alpha$$

Since

$$\Omega_f(e_\alpha, e_{-\alpha}) = \frac{i}{4\pi} f(h_\alpha)$$

is given by the derivative of the 1-form f by using $d\omega(X, Y) = -\omega([X, Y])$. Viz. viewing f in $\mathfrak{h}^*$ as a 1-form on M we have

$$\frac{i}{2} df(e_\alpha, e_\beta) = -\frac{if}{2}([e_\alpha, e_\beta]) = -\frac{if}{2}(h_\alpha).$$

And $\Omega_f(e_\beta, e_\gamma) = 0$ unless $\beta = -\gamma = \pm\alpha$.
Thus we have:

THEOREM 9.2.6. Ω_f is of type $(1, 1)$ and belongs to the Chern class of the $U(1)$ bundle $L_{\mathfrak{X}_f}$.

THEOREM 9.2.7.

$$\dim H^1(M/H, \Omega^1) = \dim H^2(M/H, C) = l = \text{rank } \mathfrak{m}.$$

Here the elements of Π are paired to $H^2(M/H, C)$ by

$$\alpha_j \in \Pi \to \Omega_{\alpha_j} = \sum_{\alpha \in \Delta^+} (\alpha_j, \alpha)\omega^\alpha \wedge \bar{\omega}^\alpha.$$

THEOREM 9.2.8. Ω_δ is in the Chern class of the line bundle $E(\delta)$ where $(\delta, h_\alpha) = (\alpha, \alpha)$.

THEOREM 9.2.9. $H^p(M/H, \Omega^q) = 0$ unless $p = q$.
(ii) $H^p(M/H, \Omega^p) = H^2(M/H, C)$.

COROLLARY 9.2.10. The Euler characteristic is $\mathfrak{X}(M/H) = |W(H)|$.

We now turn to an outline of the proof of the Borel–Weil Theorem. Let

$\lambda \in D \cap \Lambda$. By Theorem 9.2.6 the curvature form of $E(\lambda)$ is given by

$$\Omega_\lambda = \sum_{\alpha \in \Delta^+} \langle \lambda, h_\alpha \rangle \omega^\alpha A \wedge \bar{\omega}^\alpha$$

and the curvature of $E(-\lambda) \otimes K$ where $K = E(-2\delta)$ is

$$\Omega_{-\lambda-2\delta} = \sum_{\alpha \in \Delta^+} -\langle \lambda + 2\delta, h_\alpha \rangle \omega^\alpha \wedge \bar{\omega}^\alpha.$$

Since $\lambda \in D$ we have $\langle \lambda + 2\delta, h_\alpha \rangle > 0$ for all α in Δ^+. Thus by Griffith's vanishing theorem $H^{n-q}(M/H, \mathcal{O}(E(-\lambda) \otimes K)) = 0$ for $q = 1, \ldots, n = \dim M/H$. By the Serre duality $H^q(M/H, \mathcal{O}E(\lambda))$ and $H^{n-q}(M/H, \mathcal{O}(E(-\lambda) \otimes K))$ are dual M-modules. Thus we have $H^q(M/H, \mathcal{O}(E(\lambda)) = 0$ for $q > 0$. For the proof or irreducibility we wait until the next chapter.

Define the Liouville form β by $\beta = (1/n!)(\Omega)^n$; so $\int_{\mathcal{O}(f)} \beta > 0$. Here $\beta = \prod_{\alpha \in \Delta^+} i(\delta + \lambda, \alpha) \omega^\alpha \wedge \bar{\omega}^\alpha$.

THEOREM (Kirillov) 9.2.11. Let M be a compact simply connected Lie group $\delta = \frac{1}{2} \sum_{\alpha \in \Delta^+}$, T_λ a representation with highest weight λ.

Set

$$p(X) = \prod_{\alpha \in \Delta^+} \frac{\sin(\alpha, X)/2}{(\alpha, X)/2}.$$

Then

$$\operatorname{Tr} T_\lambda(\exp(iX)) = p(X) \int_{\mathcal{O}(f)} \exp(i\langle f, X \rangle) \beta,$$

where $f = i^{-1}(\lambda + \delta) \in \mathfrak{h}^* \subset \mathfrak{m}^*$. That is

$$\operatorname{Tr} T_\lambda(\exp iX)) = J(X) = p(X)^{-1} \int_{\mathcal{O}(\lambda+\delta)} \exp(i\langle f_j X \rangle) \beta$$

and

$$J(X) = \frac{\sum_w \det w \exp(iw(\lambda + \delta)X)}{\prod_{\alpha \in \Delta^+} 2i \sin(\alpha, X)/2}.$$

Taking $X = 0$ we have

COROLLARY (Borel–Hirzebruch) 9.2.12. Dim $T_\lambda = \int_{\mathcal{O}(\lambda+\delta)} \beta$.

Thus we see that if $\lambda \in \Lambda \cap D$ i.e. λ is a nonnegative integral root then if $\mathcal{O}(\lambda + \delta)$ is the orbit associated to this element the character $\mathfrak{x}_\lambda$ of the irreducible unitary representation is determined by $\mathcal{O}(\lambda + \delta)$

$$\operatorname{Tr}(T_\lambda(\varphi)) = \int_{\mathcal{O}(\lambda+\delta)} \tilde{\varphi}\, dv.$$

and T_{λ_1} is equivalent to T_{λ_2} iff $\mathcal{O}_1 = \mathcal{O}_2$.

THEOREM 9.2.13. The orbit space of $\Lambda \cap D$ under the coadjoint action of M is in bijective correspondence with $\hat{M}$:

$$(\Lambda \cap D)/\mathrm{M} \simeq \hat{M}.$$

9.3. Cocompact Nilradical Groups

R. Lipsman has generalized the Borel–Weil theory to include the class of groups with nilradicals N which are simply connected and such that G/N is compact. According to a standard structure theorem L12 such a group G is of the form $G = H Ⓢ N$ where N is the normal simply connected nilpotent group and H is a compact Lie group. This class includes then all compact Lie groups, all simply connected nilpotent Lie groups, all motion groups (i.e. N is abelian), generalized oscillator groups – i.e. N is a Heisenberg group H_n and H is a torus fixing the center of N or more generally $U(n) Ⓢ H_n$, and finally $G = M Ⓢ N$ where MAN is the minimal parabolic subgroup in semi-simple Lie groups.

Let $\mathfrak{g}$ be the real Lie algebra of such a group G. For $f \in \mathfrak{g}^*$ let $G^0(f)$ denote the identity component of $G(f)$.

DEFINITION 9.3.1. $\Lambda(G) = \{f \in \mathfrak{g}^* \mid$ there exists a unitary character $\mathfrak{X} = \mathfrak{X}_f$ of $G^0(f)$ such that $d\mathfrak{X} = if|\mathfrak{g}(f)\}$.

For f in $\Lambda(G)$ set $\vartheta = f|\mathfrak{n}$, $\xi = f|\mathfrak{h}(\vartheta)$, $\nu \in (H^0(\vartheta))\hat{}$ given by the Borel–Weil theorem, $\tau \in H(f)$ – i.e. an irreducible representation of $H(f)$ whose restriction to $H^0(f)$ is a multiple of $\mathfrak{x}_\xi$, ξ and τ define a representation $\sigma = \sigma_{\xi,\tau}$ of $\hat{H}(\vartheta)$ as we shall see. Let $\gamma = \gamma(\vartheta) \in \hat{N}$ have a canonical extension $\tilde{\gamma}$. It can be shown that $G(f) = H(f)N(f)$ and $G(\vartheta) = H(\vartheta)N(\vartheta)$. We identify $\check{G}(f)$ and $\check{H}(f)$ by the equation

$$\tau(hn) = \tau(h)\mathfrak{X}_\vartheta(n) \quad h \in H(f),$$
$$n \in N(f).$$

Set $T(f, \tau) = \operatorname{Ind}_{H(\vartheta)N}^{G} \sigma_{\xi,\tau} \otimes \tilde{\gamma}$ and $[T(f)] = \{T(f,\tau) \mid \tau \in \check{G}(f)\}$.

THEOREM (Lipsman) 9.3.2. The map $A(G) \to \hat{G}: f \to [\check{T}(f)]$ is surjective, G-equivalent multivalued map of the form

$$\check{G}(f) \to \hat{G} \to \Lambda(G)/G.$$

Lipsman's version of Plancherel becomes:

THEOREM (Plancherel) 9.3.3. If $\varphi \in A_0(G)$ for G as above

$$\varphi(e_G) = \int_{\Lambda(G)/G} \sum_{G(f)} \mathrm{Tr}(T(f,\tau)(\varphi)) \times \\ \times \dim \tau \, \mathrm{d}\bar{\mu}(G \cdot f).$$

Lipsman has conjectured that Kirillov's character formula holds for the present class of groups.

Finally Lipsman has shown that using the technique of holomorphic induction of Auslander and Kostant completely determines $\hat{G}$.

The basis of Lipsman's work is the following extension of the Borel–Weil theorem. Let H be a compact Lie group (*not* necessarily connected) with identity component H^0. Let $f \in \Lambda(H)$ and let $v = v_f$ be the irreducible unitary representation of H^0 given by the standard Borel–Weil theorem, $f \in \mathfrak{h}^*$. For h in H $h \cdot v_f = v_{h \cdot f}$. Since $H(v) = H^0 H(f)$, we have $H(v)/H(f) \approx H^0/H^0(f)$. $H^0/H^0(f)$ is a connected complex manifold as we have outlined. This permits a holomorphic structure to be induced in $H/H(f)$, since its connected components are of the form $hH(v)/H(f)$.

Since $H^0(f)$ is normal in $H(f)$ there is a finite dimensional irreducible unitary representation τ of $H(f)$ such that $\tau|(H^0(f)$ is a multiple of $\mathfrak{X}$. Let V be the space on which τ acts. A holomorphic vector bundle is defined by $E(\tau) = (Hx_\tau V)/H(f)$. Let $S = H^0(H/H(f), \mathcal{O}(E(\tau)))$ denote the space of holomorphic sections. H acts on S by $(\sigma_{f,\tau}(h)s)(x) = h \cdot f(h^{-1} \cdot x)$ for s in S.

THEOREM (Borel–Weil–Version II) 9.3.4. (i) S is nontrivial, $\sigma_{f,\tau}$ is irreducible and all irreducible representations of H are obtained this way;

(ii) Two points of $\Lambda(H)$ from the same H-orbit give rise to the same finite collection of equivalence classes of irreducible representations of H;

(iii) for f in $\Lambda(H) \tau \to \sigma_{f,\tau}$ is a bijective correspondence between irreducible representations of $H(f)$ whose restriction to $H^0(f)$ is a multiple of $\mathfrak{X}$ and the irreducible representations of H whose restriction to H^0 is a multiple of $\bigoplus_{h \in H/H(v_f)} h \cdot v_f$;

(iv) $\hat{H}$ can be viewed as a fiber space $H(f) \to \check{H} \to N(H)/H$.

Problems

Exercise 9.1. Let $\mathfrak{m}$ be the set of f in $\mathfrak{g}^*$ such that the orbit $\mathcal{O}(f)$ is of maximal dimension in $\mathfrak{g}^*$. Show that if $f \in \mathfrak{m}$, then $\mathfrak{g}(f)$ is commutative. Thus if $\mathfrak{g}(f)$ is of minimum dimension, then $\mathfrak{g}(f)$ is an abelian subalgebra of $\mathfrak{g}$.

Exercise 9.2. For f in Λ show that the parabolic subalgebra $\mathfrak{p} = \sum_{\alpha \in \Delta_f \cup \{0\}} \mathfrak{m}_\alpha$, where $\Delta_f^+ = \{\alpha \in \Delta^+ \mid \langle (1/i)\alpha, f \rangle < 0\}$ is the unique $M(f)$-invariant positive polarization for f.

Exercise 9.3. Show $\mathfrak{p}$ is Borel iff f in Exercise 9.1 is regular. In this case $\mathfrak{m}(f)$ is a maximal toral subalgebra; $\mathfrak{p} \cap \bar{\mathfrak{p}} = \mathfrak{m}(f)$. Set $\mathscr{D} = \mathfrak{m}(f)$, $\mathscr{E} = \mathfrak{m}$ with associated groups D and E; show that $E/D = M/M(f)$.

Chapter 10

Geometry of C-Spaces and R-Spaces

10.1. The geometry of C-manifolds

The basic examples of quantizable dynamical systems, viz. the harmonic oscillator, the Kepler problem or the hydrogen atom, the spinning particle, etc., are based on C-spaces.

Definition 10.1.1. A *C-space* is a compact simply connected homogeneous manifold. We write C-space $M = G/U$. M has a compact form gotten by taking the maximal compact subgroup G^u of $G.G^u$ acts transitively on M $M = G^u/K.K = U \cap G^u$. Here G^u is semisimple.

Example 10.1.2. The spaces M/H of the last chapter are C-spaces.

C-spaces were first classified by Wang and the subject of homogeneous vector bundles over C-spaces formed the basis of Bott's extension of the Borel–Weil theorem.

Let $B \to P \overset{\pi}{\to} M$ be a holomorphic principal bundle over manifold M. Let $G^0(P)$ denote the connected component of the group of all bundle automorphisms of P.

Definition 10.1.3. If $\pi(G^0(P))$ operates transitively on M, P is said to be a *homogeneous bundle.*

One way to construct homogeneous bundles is to take a coset bundle $U \to G \to M = G/U$ and a homomorphism $\rho \in \mathrm{Hom}(U, B)$. Let $P = = \{G \times B | (g, b) \sim (gh, \rho(h^{-1})b), g$ in Gb in B and h in $U\}$. G acts on the equivalence class $[g, b]$ by $g'[g, b] = [g'g, b]$. Clearly these actions are bundle automorphism; so $P = G \times_\rho B$ is homogeneous.

Theorem 10.1.4. Let $B \to P \to M$ be a holomorphic principal bundle over a C-manifold M. Then P is homogeneous iff it is of the form $P \times_\rho B$ for some ρ.

Example 10.1.5. The holomorphic tangent bundle $T(M)$ to $M = G/U$ at eU

is naturally isomorphic to the quotient of complex Lie algebras $\mathfrak{g}/\mathfrak{u}$. To see that it is a homogeneous bundle, we consider the map $gU \to ugU$ at eU. It corresponds to the automorphism of $\mathfrak{g}/\mathfrak{u}$ induced by $\mathrm{Ad}\,u$. Thus $T(M) \to M$ is the homogeneous vector bundle $T(M) = M \times_{\mathrm{Ad}} \mathfrak{g}/\mathfrak{u}$.

Let M be a C-manifold. Then M has the form $M = G/U$ where G is a connected complex Lie group. If M is not Kählerian, there is a subgroup $\tilde{U}$ of G such that $M = G/\tilde{U}$ is a Kählerian C-space. Here U is a closed normal subgroup of $\tilde{U}$ and $\tilde{U}/U$ is a complex toroidal group. Topologically we have the characterization

THEOREM 10.1.6. C-manifold $M = G/U$ is Kähler iff the Euler characteristic $\chi(M) \neq 0$.

For C-manifolds every linc bundle is homogeneous – i.e. we have

THEOREM 10.1.7. If $M = G/U$ is a C-manifold, the map $\mathrm{Hom}(U, C^*) \to$ $\to H^1(M, \mathcal{O}^*)$ given by $\lambda \to E(\lambda)$ is bijective.

If $M = G/U$ is a Kählerian C-space then there is a maximal solvable subgroup U_f of G contained in U.

DEFINITION 10.1.8. A *Borel subgroup* of G is a maximal complex solvable Lie subgroup. A *parabolic subgroup* of G is a complex Lie subgroup containing a Borel subgroup.

Let $\mathfrak{h}$ be a Cartan subalgebra of $\mathfrak{g}$. Let Π be a system of simple roots. Thus any root φ can be written as $\varphi = \sum_{\pi \in \Pi} n_\varphi(\pi)\pi$. We define for the system of roots

$$\Phi^r = \{\varphi \in \Phi \,|\, n_\varphi(\pi) = 0 \quad \text{for all} \quad \pi \in \Pi \backslash \Phi\}$$

$$\Phi^u = \{\varphi \in \Phi \,|\, n_\varphi(\pi) > 0 \quad \text{for some} \quad \pi \in \Pi \backslash \Phi\}.$$

Define the subalgebras

$$\mathfrak{p}^r_\Phi = \mathfrak{h} + \sum_{\varphi \in \Phi^r} \mathfrak{g}_\varphi \quad \text{and} \quad \mathfrak{p}^u_\Phi = \sum_{\varphi \in \Phi^u} \mathfrak{g}_\varphi.$$

Set $\mathfrak{p}_\Phi = \mathfrak{p}^r_\Phi + \mathfrak{p}^u_\Phi$.

DEFINITION 10.1.9. $\mathfrak{p}_\Phi$ is the *parabolic subalgebra* of $\mathfrak{p}$ defined $\mathfrak{h}$, Φ, and Π.

DEFINITION 10.1.10. If Φ is empty $\mathfrak{p}_\Phi = \mathfrak{h} + \sum_{\varphi > 0} \mathfrak{g}_\varphi$ is the *Borel subalgebra* of $\mathfrak{g}$ defined by $\mathfrak{h}$ and Π.

We summarize a few properties of parabolic algebras and groups.

THEOREM 10.1.11. A complex Lie subgroup P of G is a parabolic subgroup iff the Lie algebra $\mathfrak{p}$ of P is a parabolic subalgebra of $\mathfrak{g}$ iff G/P is compact.

To develop the Lie algebra set up we proceed as follows. Let $(X, Y) = \mathrm{Tr}(\mathrm{ad}(X)\mathrm{ad}(Y))$ be the Cartan Killing form on the Lie algebra $\mathfrak{g}$ of G. We are thus able to identify $\mathfrak{g}$ and $\mathfrak{g}^*$. Let K denote the maximal compact subgroup of G with Lie algebra $\mathfrak{k}$. If we define $\mathfrak{p} = \{X \in \mathfrak{g} \mid (X, Y) = 0$ for all Y in $\mathfrak{k}\}$, then we have $\mathfrak{g} = \mathfrak{k} \oplus \mathfrak{p}$ with the properties: $[\mathfrak{k}, \mathfrak{p}] \subset \mathfrak{p}, [\mathfrak{p}, \mathfrak{p}] \subset \mathfrak{k}$.

DEFINITION 10.1.12. A root α is called *compact* or *noncompact* as $\mathfrak{g}_\alpha \in \mathfrak{k}$ or $\mathfrak{g}_\alpha \in \mathfrak{p}$. Let $\Delta_\mathfrak{k}$(resp. $\Delta_\mathfrak{p}$) denote the set of all compact (resp. noncompact) roots.

Thus we have $\Delta = \Delta_\mathfrak{k} \cup \Delta_\mathfrak{p}$. As before we define h'_α in $\mathfrak{h}$ to be the element such that $(X, h'_\alpha) = \alpha(X)$ for all X in $\mathfrak{h}$.

Then we may choose a Weyl basis $e_\alpha \in \mathfrak{g}_\alpha (\alpha \in \Delta)$ which satisfies:

(i) $[e_\alpha, e_{-\alpha}] = h'_\alpha$;

(ii) $[e_\alpha, e_\beta] = 0$ if $\alpha + \beta \neq 0$ and $\alpha + \beta \notin \Delta$;

(iii) $[e_\alpha, e_\beta] = N_{\alpha,\beta}\, e_{\alpha+\beta}$ if $\alpha + \beta \in \Delta$;

(iv) $\bar{e}_\alpha = \varepsilon_\alpha\, e_{-\alpha}$,

where for X in $\mathfrak{g}$, $\bar{X}$ is the image of X under conjugation of with respect to the real form $\mathfrak{g}_0$ of $\mathfrak{g}$. Here $\varepsilon_\alpha = -1$ if $\alpha \in \Delta_\mathfrak{k}$ and $\varepsilon_\alpha = 1$ if $\alpha \in \Delta_\mathfrak{p}$. We denote by $\{\omega^\alpha \mid \alpha \in \Delta\}$ the left invariant 1-forms on G which are dual to $\{e_\alpha \mid \alpha \in \Delta\}$.

DEFINITION 10.1.13. X in $\mathfrak{g}_0$ is called *elliptic* if ad(X) is semisimple and the centralizer of X in G is compact.

We are interested in the orbits of G in $\mathfrak{g}_0$ corresponding to elliptic elements. First we characterize elliptic elements by

THEOREM 10.1.14. Let $\lambda \in \mathfrak{h}_0$ be an elliptic element. Set

$\Phi_\lambda = \{\alpha \in \Delta^+ \mid (\lambda, \alpha) = 0\}$. Then:

(i) $\Phi_\lambda \subset \Delta_\mathfrak{k}$.

(ii) $\exp(i\lambda)$ forms a k-dimensional total subgroup S_λ in H with rank $(G) - k = \mathrm{card}(\Phi_\lambda)$.

(iii) $G(\lambda)$ is the centralizer of the torus S_λ in G.

(iv) Let $\Delta_u = \{\alpha \in \Delta \mid \pm\alpha \in \Phi_\lambda\}$. Then the Lie algebra of $G(\lambda)$ is $\mathfrak{g}(\lambda) = \oplus_{\alpha \in \Delta_u} \mathfrak{g}_\alpha + \mathfrak{h}$.

(v) $\Delta^+ \backslash \Phi_\lambda$ is a closed set of roots–i.e. if.

$$\alpha, \beta \in \Delta^+ \backslash \Phi_\lambda \text{ and } \alpha + \beta \in \Delta^+, \text{ then } \alpha + \beta \in \Delta^+ \backslash \Phi_\lambda.$$

Let $\mathcal{O}(\lambda)$ denote the orbit corresponding to λ. Thus $\mathcal{O}(\lambda) = G/G(\lambda) = G/U$. We note that G-invariant tensor fields on $\mathcal{O}(\lambda)$ correspond to Ad($G(\lambda)$)-invariant tensor fields on $T_0 = \mathfrak{g}_0/\mathfrak{g}(\lambda)_0$, and we define a complex structure in T_0 by

$$T^{\pm} = \bigoplus_{\pm\alpha\in\Delta^+\setminus\Phi_\lambda} \mathfrak{g}_\alpha,$$

where $JX = \pm iX$ if $X \in T^{\pm}$. Since J commutes with conjugation $X \to \bar{X}$, we see that J defines a complex structure on T_0. Checking that J is Ad($G(\lambda)$)-invariant on T_0 we have a complex structure on the tangent bundle $T\mathcal{O}(\lambda)$. Generalizing the proof of Theorem 9.2.6 we have

THEOREM 10.1.15. The symplectic form on $\mathcal{O}(\lambda)$ is given by the 2-form Ω on $\mathfrak{g}_0/\mathfrak{g}(\lambda)_0$ where

$$\Omega = \frac{i}{2\pi} \sum_{\alpha\in\Delta\setminus\Phi_\lambda} (\lambda, \alpha)\omega^\alpha \wedge \bar{\omega}^\alpha - \frac{1}{2\pi} \sum_{\Delta^+\setminus\Delta} (\lambda, \alpha)\omega^\alpha \wedge \bar{\omega}^\alpha.$$

We now want to show that this symplectic form is the curvature form for the bundle $E(\lambda) \to \mathcal{O}(\lambda)$. By Theorem 11.1.2 there is a canonically associated connection and covariant differential D: $A^{p,q}(E(\lambda)) \to A^{p+1,q}(E(\lambda)) + A^{p,q+1}(E(\lambda))$ which is compatible with the Hermitian metric – i.e. $\mathrm{d}(f, f') = (\mathrm{D}f, f') + (f, \mathrm{D}f')$ where f, f' are sections of $E(\lambda)$.

The curvature Θ is a Hom$(E(\lambda), E(\lambda))$-valued (1, 1)-form which we shall see is related to the symplectic structure.

Consider the principal bundle $K \to G^u \to G^u/K$. There is an Ad(K)-invariant splitting $\mathfrak{g}_0 = \mathfrak{k}_0 + \mathfrak{p}_0$. If $a_1, \ldots, a_m$ is a basis for $\mathfrak{g}_0$ such that $\{a_1, \ldots, a_r\}$ is a basis for $\mathfrak{k}_0$ and $\{a_{r+1}, \ldots, a_m\}$ lie in $\mathfrak{p}_0$ and if $\varphi^1, \ldots, \varphi^m$ is a dual basis of left invariant forms on G, then $\vartheta = \sum_{j=1}^r a_j \otimes \varphi^j$ defines a G-invariant, $\mathfrak{k}_0$-valued differential form on G. Thus ϑ defines a connection with curvature form $\Theta = \mathrm{d}\vartheta + \frac{1}{2}[\vartheta, \vartheta]$.

THEOREM 10.1.16. $\Theta = -\frac{1}{2}\sum_{i,j=r+1} [a_i, a_j]_{\mathfrak{p}} \otimes \varphi^i \wedge \varphi^j$.

Proof. Since $\mathrm{d}\vartheta = \sum a_j \otimes \mathrm{d}\varphi^j$ and $[\vartheta, \vartheta] = \sum_{i,j=1} [a_i, a_j] \otimes \varphi^i \wedge \varphi^j$ we have $\mathrm{d}\vartheta + [\vartheta, \vartheta] = -\frac{1}{2}\sum c^i_{kl} a_i \otimes \varphi^l \wedge \varphi^k$ since $c^k_{ij} = 0$. Here c^k_{ij} are given by $[a_j, a_k] = \sum c^i_{jk} a_i$. Thus if we let $[a]_{\mathfrak{p}_0}$ denote the projection of $a \in \mathfrak{g}_0$ with respect to Cartan decomposition we have the theorem.

Extending this result to the complexification we have $\mathfrak{k} = \mathfrak{h} + \sum_{\pm\alpha\in\Phi} \mathfrak{g}_\alpha$ and $\mathfrak{p} = \sum_{\pm\beta\in\Delta^+\setminus\Phi} \mathfrak{g}_\beta$, where Φ is the set of positive roots for $\mathfrak{k}$. Since $[e_\alpha, e_\beta]_{\mathfrak{k}} = 0$ for α, β in Δ^+/Φ we have

THEOREM 10.1.17. Let $M = G/B$ be a Kähler C-space. Then the natural connection in $B \to G \to M$ is

$$\Theta_M = - \sum_{\alpha,\beta \in \Delta^+ \setminus \Phi} [e_\alpha, e_{-\beta}]_{\mathfrak{p}} \otimes \omega^\alpha \Lambda \bar{\omega}^\beta.$$

Checking that $\omega^{-\alpha} = -\bar{\omega}^\alpha$ for $\alpha \in \Delta_{\mathfrak{k}}$ and $\omega^{-\beta} = \bar{\omega}^\beta$ for $\beta \in \Delta^+ \setminus \Delta_{\mathfrak{k}}$ we have using the facts that $[e_\alpha, e_{-\beta}]_{\mathfrak{p}} = 0$ for $\alpha \in \Delta^+ \setminus \Delta_{\mathfrak{p}}$ noncompact and $\beta \in \Delta_k \setminus \Phi$ positive compact:

$$\Theta = \sum_{\alpha,\beta \in \Delta_{\mathfrak{k}} \setminus \Phi} [e_\alpha, e_\beta]_{\mathfrak{p}} \otimes \omega^\alpha \wedge \bar{\omega}^\beta$$

$$- \sum_{\alpha,\beta \in \Delta^+ \setminus \Delta_{\mathfrak{k}}} [e_\alpha, e_{-\beta}] \otimes \omega^\alpha \Lambda \bar{\omega}^\beta.$$

The connection ϑ induces a connection $\lambda(\vartheta)$ in the associated bundle $E(\lambda)$ where if f is a section of $E(\lambda)$, the differential by this connection is

$$\mathrm{D}f = \textstyle\sum_{j=r+1}^m a_j f \otimes \varphi^f.$$

THEOREM 10.1.18. D is compatible with the Hermitian metric on $E(\lambda)$.

Proof. Noting that

$$\mathrm{D}f = \mathrm{d}f + \lambda(\vartheta) f = \mathrm{d}f + \textstyle\sum \lambda(a_j) f \otimes \varphi^j = \mathrm{d}f - \textstyle\sum a_j f \otimes \varphi^j;$$

and

$${}^t\lambda(\bar{\vartheta}) = -\lambda(\vartheta)$$

we see that

$$(\mathrm{D}f, f') + (f, \mathrm{D}f') = (\mathrm{d}f, f') + (f, \mathrm{d}f') + (f, \lambda(\vartheta) f') + (\lambda(\vartheta) f, f')$$
$$= \mathrm{d}(f, f').$$

THEOREM 10.1.19. The curvature of the induced connection is

$$\Theta_M(\lambda) = \sum_{\alpha,\beta \in \Delta_{\mathfrak{k}} \setminus \Phi} \lambda([e_\alpha, e_{-\beta}]_{\mathfrak{k}}) \otimes \omega^\alpha \Lambda \bar{\omega}^\beta$$

$$- \sum_{\alpha,\beta \in \Delta^+ \setminus \Phi} \lambda([e_\alpha \cdot e_{-\beta}]_{\mathfrak{p}} \otimes \omega^\alpha \Lambda \bar{\omega}^\beta.$$

THEOREM 10.1.20. The symplectic form on $\mathcal{O}(\lambda)$ is the curvature form of the connection given by the natural connection in $K \to G^u \to G^u/K$ and the irreducible unitary representation $\lambda \in \Lambda \subset \mathfrak{h}^*$.

COROLLARY 10.1.21. If $M = G^u/K$ is a C-space and $E(\lambda) \to M$ is a homogeneous bundle defined by a unitary representation λ of K in space E, then the connection in $E(\lambda)$ induced by the natural connection in $K \to G^u \to G^u/K$ is

$$\Theta_{G^u/K} = \sum_{\alpha,\beta} \lambda([e_\alpha, e_{-\beta}]) \otimes \omega^\alpha \Lambda \bar{\omega}^\beta.$$

THEOREM (Borel–Weil–Bott) 10.1.22. Let $M = G/U$ be a Kahlerian C-space and let (ρ, F) be an irreducible representation of U. Considering (ρ, E) as a representation of $V^c = V(S)H(S)$ with highest weight λ then:

(i) if $\lambda + \delta$ is singular $H^p(M, \mathcal{O}(E(\rho))) = 0$ for all $p \geqq 0$;

(ii) if $\lambda + \delta$ is regular and $\operatorname{ind}(\lambda + \delta) = p$ then $H^q(M, \mathcal{O}(E(\rho))) = 0$ for $q \neq p$ and $H^p(M, \mathcal{O}(E(\rho)))$ is an irreducible G-module with highest weight $\lambda^{(p)} = \sigma_p(\lambda + \delta) - \delta$ where σ_p is the unique element of W such that $\sigma_p(\lambda + \delta) \in D$.

Proof. We will prove only the Borel–Weil portion of the theorem. Let $M = G/U = G^u/V$ be a Kählerian C-space with irreducible representation $\rho: V \to GL(E(\rho))$. We want to first show

THEOREM 10.1.23. If $H^0(M, \mathcal{O}(E(\rho)) \neq 0$, then $\rho \in D$ and $H^0(.)$ is the irreducible G module with highest weight ρ.

Proof. Let $S = H^0(M, \mathcal{O}(E(\rho)))$. Since G acts holomorphically on $E(\rho)$, S is a finite dimensional G-module. Let F_{m_0} be the kernel of the restriction map $S \xrightarrow[r_{m_0}]{} E_{m_0}(\rho)$. Since $Um_0 = m_0$ we see that F_{m_0} is a U-submodule of S. Thus S/F_{m_0} is a U-submodule of $E_{m_0}(p)$. Since $S/F_{m_0} \neq 0$ and ρ is irreducible $r_{m_0}(S) = E_{m_0}(\rho)$. Now we need the Frobenius formula:

THEOREM 10.1.24. For an irreducible $\mathfrak{g}$-module W^λ, $\dim \operatorname{Hom}_{\mathfrak{u}}(W', E(\rho)) = \delta^\lambda_\rho$.

Proof. If nonzero $f \in \operatorname{Hom}_{\mathfrak{u}}(W^\lambda, E(\rho))$ then $f \in \operatorname{Hom}_{\mathfrak{v}}(W^\lambda, E(\rho))$ and $f(W^\lambda)$ is a $\mathfrak{v}$-submodule of $E(\rho)$. By irreducibility $f(W^\lambda) = E(\rho)$. The rest is a straight forward calculation.

THEOREM 10.1.25. If ρ is in D, then $H^0(M, \mathcal{O}(E(\rho)))$ is the irreducible G-module W^ρ with highest weight ρ and $H^q(M, \mathcal{O}(E(\rho))) = 0$ for $q > 0$.

Proof. By the Frobenius result there is a nonzero f in $\operatorname{Hom}_U(W^\rho, E(\rho))$. If $W(\rho) = W^\rho \times M$ and $\pi: W(\rho) \to E(\rho)$ is defined by $\pi(w, gU) = [g, (g^{-1} f(w))]$, then π is a bundle mapping and $\pi(S(W(\rho))$ is a nonzero subspace of $S(E(p))$ (where S denotes the space of sections). Thus by Theorem 10.1.23 the first part follows. The second part follows from Griffith's vanishing theorem.

10.2. Kirillov character formula

We now turn to the generalization of Kirillov's character formula for the general C-space. Define $\tilde{X}$ in $S(T\mathcal{O})$ by $(\tilde{X}\varphi)(f) = (\mathrm{d}/\mathrm{d}t)(\exp(-t\,\mathrm{ad}\,X)f) = \langle \mathrm{d}\varphi, [f, X]\rangle$ where φ is in $A(\mathcal{O})$. Thus we map $g_0 \to T_f\mathcal{O}$ by $X \to \langle f, X\rangle$. Define as before h_X in $A(\mathfrak{g}_0^*)$ by $h_X(f) = \langle f, X\rangle$. Then we have $(\tilde{Y}h_X)(f) = (X, [f, Y])$.

Theorem 10.2.1. The critical points of h_X are given by f in $\mathcal{O}$ such that $[f, X] = 0$.

Proof. If $[f, X] = 0$ for all X in $\mathfrak{g}_0$ then $(\tilde{Y}h_X)(f) = (X, [f, Y]) = ([f, X], Y) = 0$ for all Y in $T_f\mathcal{O}$. That is $\mathrm{d}h_X(f) = 0$.

Let $\mathfrak{g}'$ be the set of regular elements in $\mathfrak{g}$ and let $\mathfrak{h}_0' = \mathfrak{g}_0' \cap \mathfrak{h}_0$.

Theorem 10.2.2. For X in $\mathfrak{h}_0'$, h_X is a Morse function on $\mathcal{O}(\lambda)$. For X in $\mathfrak{h}_0$ the Morse index of $-h_X$ is

$$\begin{aligned}\mathrm{ind}_\lambda(-h_X) = {} & 2\,\mathrm{Card}\,\{\alpha \in \Delta_{\mathfrak{k}} \backslash \Phi_\lambda \mid (\lambda, \alpha) < 0\} + \\ & 2\,\mathrm{Card}\,\{\alpha \in \Delta^+ \backslash \Delta_{\mathfrak{k}} \mid (\lambda, \alpha) > 0\}.\end{aligned}$$

Let $\beta = (1/n!)(\Omega)^n$. Then *Kirillov's integral* is given by

$$J(X) = \int_{\mathcal{O}} \exp\left(\frac{i}{\hbar}(f, X)\right)\beta.$$

By the symplectic Morse Lemma the exponential is locally Gaussian, and we have

Theorem 10.2.3.

$$\begin{aligned}\prod_{\alpha \in \Delta + \backslash \Phi}^{(-1)} \frac{k(\lambda)e^{i/\hbar(\lambda, X)}}{(i/\hbar)(\alpha, X)} = {} & \pi^{-n} e^{(i/\hbar)(\lambda, X)} \prod_{\alpha \in \Delta + \backslash \Phi} |(\lambda, \alpha)| \frac{(2\pi\hbar)^n}{|\det_R \mathscr{H}_\lambda|^{\frac{1}{2}}} \times \\ & \times \exp\left\{\frac{-i\pi}{4}\mathrm{sgn}(-\mathscr{H}_\lambda)\right\},\end{aligned}$$

where

$$\begin{aligned}\mathscr{H}_\lambda(Z, Z) = {} & -2 \sum_{\alpha \in \Delta_{\mathfrak{k}} \backslash \Phi} (\lambda, \alpha)(X, \alpha) Z_\alpha \bar{Z}_\alpha + \\ & + 2 \sum_{\alpha \in \Delta^+ \backslash \Delta_{\mathfrak{k}}} (\lambda, \alpha)(X, \alpha) Z_\alpha \bar{Z}_\alpha.\end{aligned}$$

Proof. Since

$$\det_R \mathscr{H}_\lambda = 2^{2n} \prod_{\alpha \in \Delta^+ \setminus \Phi} |(\lambda, \alpha)|\,|(X, \alpha)|$$

and

$$\exp\left(-\frac{i}{4}\operatorname{sgn}(\mathscr{H}_\lambda)\right) = (-1)^n(-1)^{\frac{1}{2}\operatorname{ind}\lambda^{(-h_x)}}$$

$$= (-i)^n(-1)^{k(\lambda)} \prod_{\alpha \in \Delta^+ \setminus \Phi} \varepsilon(\alpha, X),$$

where

$$\varepsilon(\alpha, .) = \begin{cases} 1 & \\ -1 & \end{cases} (\alpha, .) \gtrless 0$$

we have the result.

To complete the evaluation of $J(X)$ we first examine the asymptotic limit of $J_\hbar(X)$ as $\hbar \to 0$.

THEOREM 10.2.4.

$$J_\hbar(X) = (-1)^{k(\lambda)} \sum \det w \exp(i(w\lambda, X)/ \prod_{\alpha \in \Delta^+} \frac{i}{\hbar}(\alpha, X).$$

$+$ rapidly decreasing terms as $\hbar \to 0$.

THEOREM 10.2.5.

$$J_\hbar(X) = (-1)^{k(\lambda)} \sum_{w \in W/W(U_\lambda)} \frac{e^{i/\hbar(w.\lambda, X)}}{\prod_{\alpha \in \Delta^+ \setminus \Phi} \frac{i}{\hbar}(w.\alpha, X)}$$

for λ in $\mathfrak{h}_0$ and $\lambda \perp \Phi_\lambda \subset \Delta_t$.

10.3. GEOMETRY OF *R*-SPACES

The study of *C*-spaces contains a great deal of physics. We now turn our attention to real spaces whose complexifications are *C*-spaces. These are called *R-spaces*. Examples of *R*-spaces are:

(1) Hermitian symmetric spaces of compact type;
(2) Grassmann manifolds $O(p+q)/Sp(p) \times Sp(q)$;
(3) $Sp(p+q)/Sp(p) \times Sp(q)$;
(5) $SO(m)$, $U(m)$, $Sp(m)$;

(5) $U(2m)/Sp(m)$ and $U(m)/O(m)$;
(6) Cayley projection plane;
(7) real quadrics $Q_{n,v}(R)$ (v. Exercise 10.6);
(8) Stiefel manifolds $V_{p+q,p}$.

In their simplest form R-spaces are examples of orbit spaces that we mentioned in the introduction. Viz. let G be real connected semisimple Lie group with finite center and let $G = \text{KAN}$ be its Iwasawa decomposition. As usual we let $\langle , \rangle$ denote the Killing form of $\mathfrak{g}$, the Lie algebra of G. Here A is a maximal torus in G and $\mathfrak{a}$ is the Lie algebra of A. The set of roots of $(\mathfrak{g}, \mathfrak{a})$ is denoted Δ. Δ^+ is the positive system of roots defining $\mathfrak{n}$, the Lie algebra of N. If $\mathfrak{m}$ is the centralizer of $\mathfrak{a}$ in $\mathfrak{k}$ we have a parabolic subalgebra $\mathfrak{p} = \mathfrak{m} \oplus \mathfrak{a} \oplus \mathfrak{n}$.

As in Section 0.9 we are interested in orbits given by the adjoint action of K on $\mathfrak{s}$, the orthogonal complement of $\mathfrak{k}$ in $\mathfrak{g}$ with respect to the Killing form. We let K_X denote the centralizer of X in K.

DEFINITION 10.3.1. An *R-space* is a space of the form K/K_H where $H \in \mathfrak{s}$.

It is easy to see that the map $k \to \text{Ad}(k).H$ defines a diffeomorphism of the R-space K/K_H to the Ad K-orbit through H in $\mathfrak{s}$. If we set $H^k = \text{Ad}k.H$, then an interesting function on K is $f_{X,H}(k) = \langle X, H^k \rangle$ where $X \in \mathfrak{s}$. (As a function on M we have $f_X(ko) = \langle \text{Ad}, Z, X \rangle$ (where o is the origin of M), a spherical function on M associated to the representation (Ad, $\mathfrak{p}$) of K). This function is just the height function on M with respect to the direction X in p – i.e. $f_X(Y) = \langle Y, X \rangle$ for $Y \in M$. As we noted in Section 7.3. f_X is linearly related to the length function.

Checking that

$$\frac{\mathrm{d}f}{\mathrm{d}t} f_{X,H}(k \exp tY) \Big|_{t=0} = \langle X, [Y, H]^k \rangle = -\langle [X, H^k], Y^k \rangle$$

for Y in $\mathfrak{k}$, we see that

THEOREM 10.3.2. k is a critical point of $f_{X,H}$ iff $[X, H^k] = 0$. Restricting X, H to $\mathfrak{a}$ one can show that the critical set of $f_{X,H}$ is $K_{X,H} = \bigcup_{w \in W} K_X w K_H$ where W is the Weyl group. The Hessian of $f_{X,H}$ at its stationary points is easily calculated. Viz. if $k = u x_w v$ for u in K_X, x_w a representative of w in W and v in K_H, then for each Y in $\mathfrak{k}$ we have

$$\frac{\mathrm{d}^2 f}{\mathrm{d}t^2} X, H(k \exp t\, Y) \Big|_{t=0} = - \sum_{\alpha \in \Delta^+} \alpha(H) w\alpha(X) \, \| F_\alpha(Y) \|^2,$$

where F_α is the orthogonal projection from $\mathfrak{k}$ onto $\mathfrak{k}\cap(\mathfrak{g}_\alpha\cap\mathfrak{g}_{-\alpha})$.

If S is a set of simple roots in Δ^+, then for any $\Phi\subset S$ we let $\Delta^+(\Phi)$ denote the set of roots which are nonnegative integral linear combinations of roots in Φ. For an H in $C1(\mathfrak{a}^+)$, the closure of the Weyl chamber $\mathfrak{a}^+$, we can choose this H so that Φ is precisely the subset of S vanishing at H. Then $\Delta^+\backslash\Delta(\Phi)$ is the set of roots α with $\alpha(H)>0$. Let W_H denote the group of W generated by relections corresponding to roots vanishing on H. In this case we set $\mathfrak{p}_\Phi=\mathfrak{g}(H)=\mathfrak{m}\oplus\mathfrak{a}\oplus\sum_{\alpha(H)\geq 0}\mathfrak{g}_\alpha$; we let $P_\Phi=G(H)$ denote the normalizer of $\mathfrak{p}_\Phi$ in G. $\mathfrak{p}_\Phi$ contains $\mathfrak{p}$ so is parabolic. $G(H)$ is a parabolic subgroup of G. We leave it to the reader to verify that the C-spaces can be considered as complexifications of the corresponding spaces $G/G(H)$. The relationship of the R-spaces and the spaces $G/G(H)$ is

THEOREM 10.3.3. K/K_H is diffeomorphic to $G/G(H)$.

We consider now geodesics on K/K_H of the form $\gamma(t)=\exp(tX)$ for X in $\mathfrak{g}$. The velocity vector of $\gamma(t)$ is denoted v_X. If we let a point of K/K_H be denoted by $\bar{k}=kK_H$ for k in K and if $E_\mathfrak{k}$ is the projection $\mathfrak{g}\to\mathfrak{k}$ along $\mathfrak{a}+\mathfrak{n}$, then one can show

THEOREM 10.3.4. $v(k)=E_\mathfrak{k}(X^{k-1})\operatorname{mod}\mathfrak{k}_X$.

The map $\zeta: Y+\vartheta Y\to Y-\vartheta Y$ for Y in $\mathfrak{n}$ is a linear isomorphism from $\mathfrak{k}\backslash\mathfrak{m}$ to $\mathfrak{s}\backslash\mathfrak{a}$. We extend it to a map $\mathfrak{k}\to\mathfrak{s}$ by defining ζ to be zero on $\mathfrak{m}$. We define the bilinear form b_H on $\mathfrak{k}\times\mathfrak{k}$ by

$$b_H(Z,Z')=\langle H,[Z,\zeta(Z')]\rangle \text{ for } Z,\ Z' \text{ in } \mathfrak{k}.$$

The reader can check that b_H is a symmetric positive semidefinite bilinear form. Furthermore

$$b_H(Z,Z')=\sum_{\alpha\in\Delta,\,\alpha(H)\geq 0}\alpha(H)((F_\alpha)(Z),F_\alpha(Z')).$$

The radical of b_H is $\mathfrak{k}_H$, so b_H determines an inner product b_H on $\mathfrak{k}/\mathfrak{k}_H$.

THEOREM 10.3.5. (1) b_H extends to a k invariant Riemannian metric β_H on K/K_H;

(2) for every X in $\mathfrak{a}$, v_X is equal to the gradient of $f_{X,H}$ considered as a function on K/K_H–i.e. $\mathrm{d}f_{X,H}=\beta_H(.,v_X)$.

The dynamics of v_X on the R-spaces is then summarized as follows:

THEOREM 10.3.6. Let $X\in\mathfrak{a}$, $H\in\varphi(\mathfrak{a}^+)$, then $f_{X,H}$ has isolated critical points on K/K_H iff $\alpha(X)=0$ implies $w^{-1}(\alpha(H))=0$ for α in Δ, w in $W(*)$. And in this case the set of critical points of f (or equivalently the zero set of v_X) is $WK_H = W/W_H$. For each w in W the flow $\gamma(t)$ of v_X on K/K_H is hyperbolic at wK_H. The stable manifolds are defined by

$$S_w^+ = \{x \in K/K_H \,|\, lim^{t\to\infty}\, \gamma_t(x) = wK_H\}.$$

If $(*)$ holds, K/K_H is equal to the disjoint union of stable manifolds S_w^+, $w \in W/W_H$.

Under condition $(*)$ $f_{X,H}$ is a Morse function on K/K_H and its number of critical points is equal to $|W/W_H|$. It can be shown that $|W/W_H|$ is equal to the sum of Betti numbers of the homology modulo 2 of K/K_H. However, the Morse inequality states that the number of critical points of any Morse function is $\geqq$ the sum of the Betti numbers of the homology with coefficients in any field. Thus $f_{X,H}$ has the minimum number of critical points. Thus we have characterized the embedding $k \to \mathrm{Ad}(k).H$ of K/K_H into $\mathfrak{s}$.

For an introduction to stable manifolds and hyperbolic flows, the reader is referred to Abraham and Marsden A2.

10.4. Schubert Cell Decompositions

Schubert cell decompositions have arisen recently in various aspects of engineering. We briefly note the relationship of Schubert cell decompositions and the theory of C-spaces and R-spaces. The physics of Schubert cell decompositions has not been examined fully.

We begin with a classic example. Let $G = SL(n+1)$; let B be the group given by upper triangular matrices; finally let T be the group of diagonal matrices. Thus we have a group G, a maximal torus T of G and a Borel subgroup G, $B \supset T$. We take the parabolic subgroup P of matrices (g_{ij}) $0 \leqq i, j \leqq n$ with $g_{i0} = 0$ for $i > 1$. Thus we can identify $P\backslash G$ and the projective space $P(V)$ where V is a vector space of dimension $n+1$, say with coordinates $(x_0, \ldots, x_n)$. The action of G on the right can be identified with matrix multiplication – i.e. $(x_0, \ldots, x_n)(g_{ij}) = (\ldots, \sum_0^n x_i g_{ij}, \ldots)$. The linear subspaces of $P(V)$ defined by $(0, ****), \ldots, (0, \ldots, 0, *)$ are stable under B. The subvarieties of $P(V)$ given by taking $x_0 = 0$; $x_0 = x_1 = 0; \ldots$; $x_0 = x_1 = \ldots x_{n-1} = 0$ are called the *Schubert varities*. This decomposition generalizes to Kähler C-spaces as follows.

Let G be a complex semisimple Lie group of rank n, T a maximal torus of G and $B \supset T$ a Borel subgroup of G. Let $W = N(T)/T$ be the Weyl group of

G. Let R^+ denote the set of positive roots relative to B. Set $R^- = -R^+$. We write $\alpha > 0$ (resp. $\alpha < 0$) as $\alpha \in R^+$ (resp. $\alpha \in R^-$). Let $S = \{\alpha_1, \ldots, \alpha_n\} \subset R^+$ be the simple system of roots. For each α in R let s_α denote the reflection with respect to α. For simplicity we set $s_i = s_{\alpha_i}$, $1 \leqq i \leqq n$

DEFINITION 10.4.1. The *length function on* W relative to $s_1, \ldots, s_n$ is given by

$$l(w) = \min_k \{k \mid w = s_{i_1}, \ldots, s_{i_k} 1 \leqq i_1, \ldots, i_k \leqq n\}.$$

If $w = s_{i_1} \ldots s_{i_k}$ with $k = l(w)$ this is called a *reduced expression* for w.

THEOREM 10.4.2. There is a unique element w_0 of largest length in W such that $l(w) \leqq l(w_0)$ for all w in W.

The element w_0 has the property that $w_0(\alpha) < 0$ for all $\alpha > 0$ – i.e. $w_0(R^+) = R^-$. The reader can check that $l(w_0 w) = l(w_0) - l(w)$ for all w in W.

As we have discussed, if P is a parabolic subgroup of G containing B, then P is associated to a unique subset S_P of S; and conversely every subset of S defines a parabolic subgroup containing B. Here $S_B = \phi$ and $S_G = S$. Let R_P^+ denote the set of all positive roots spanned by the simple roots in S_P and set $R_P^- = R_P^+$. The subgroup of W generated by s_α, α in S_P, is called the *Weyl group of* P and is denoted W_P. For a simple root α in S the parabolic subgroup associated to α is denoted P_α and is called the *minimal parabolic subgroup associated to* α. The parabolic subgroup associated to $S \backslash \{\alpha\}$ is denoted P_{α}.

DEFINITION 10.4.3. For w in W set $R_P(w) = \{\alpha > 0 \mid w^{-1}(\alpha) \in R^- \backslash R_P^-\}$ and set $N_P(w) = \text{card } R_P(w)$.

Let (G, T, B, P, W) be defined as above. Then for w in W let $n(w) \in N(T)$ be such that its residue mod T is w. Then the (B, P) doublecoset $Bn(w)P$ in G depends only on the coset wW_P in W (not on w or $n(w)$). We set $C_P(w) = Bn(w)P$ and call it the *open Bruhat cell* in G associated to wW_P. The Zariski closure of $C_P(w)$ in G is denoted $X_P(w)$ and is called the *closed Bruhat* cell. The *Bruhat decomposition* of G relative to P asserts that G is the disjoint union of open Bruhat cells $C_P(w)$ in G, $G = \bigcup_{w \in W/W_P} C_P(w)$.

Let $\pi: G \to G/P$ denote the natural projection. Then for w in W we set $\pi(C_P(w))$ and call it the *open Schubert cell* in G/P associated with $C_P(w)$. If we let $e_0 = \pi(P)$ then under the natural action of B on the left, $\pi(C_P(w))$ is just the B-orbit (in fact the B^u-orbit where B^u is the unipotent part of B) through

we_0. The Schubert cells provide a cellular decomposition of G/P – i.e. $G/P = \bigcup_{w \in W/W_P} \pi(C_P(w))$. We let $X_P(w)$ denote the Zariski closure of $\pi(C_P(w))$ in G/P.

THEOREM 10.4.4. $\dim X_P(w) = N_P(w)$.

COROLLARY 10.4.5. $\dim X_B(w) = N_B(w) = l(w)$.

The reader can check that for the largest length element w_0 we have $X_B(w_0) = G/B$ and $\dim G/B = l(w_0) = \text{card } R^+ = \frac{1}{2}$ the number of roots. The open Schuber cell $C(w_0)$ is called the *big cell*. For all w in W the Schubert variety $X_B(w_0 w)$ is of codimension $l(w)$ in G/B.

EXAMPLE 10.4.6. Let $G = SL(n+1, C)$. G has Dynkin diagram

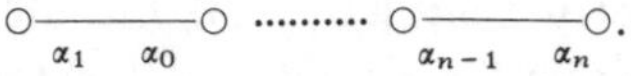

The number of roots in $n(n+1)$ and the order of the Weyl group is $(n+1)!$. A reduced expression for w in W is

$$w_0 = s_n(s_{n-1}s_n)\dots(s_i\dots s_n)\dots(s_1\dots s_n).$$

Here $l(w_0) = \frac{1}{2}n(n+1)$. The parabolic subgroup mentioned in the initial example in this section is the parabolic subgroup $P = P'_{\alpha_1}$. The semisimple part of this P is just $SL(n)$. Thus W_P has largest element $w_{0P} = s_n(s_{n-1}s_n)\dots(s_2\dots s_n)$; i.e. $w_0 = w_{0P}(s_1\dots s_n)$. The number of Schubert varieties in $P\backslash G$ is then $[W : W_P] = n+1$. These are given by the sequence $w_i = w_{0P}s_1\dots s_{n-i}$, $0 \leqq i \leqq n$.

PROBLEMS

EXERCISE 10.1 For a Kählerian C-space $M = G/U$ using Atiyah's exact sequence for principal bundle $U \to G \to G/U$ show that there is an imbedding $f_{\mathrm{Ad}}(gU) = \mathrm{Ad}(g)U$ of M into $G(n, m) = GL(n, C)/GL(n, m, C)$ where $G(n, m)$ is the Grassmanian viewed as $(n-m)$ dimensional subspaces of $\mathfrak{g}$. Here $n = \dim G$ and $m = \dim M$.

EXERCISE 10.2. Let $M = G/U$ be a C-space with homogeneous bundle $E(\lambda)$ with fiber F. Define the homomorphism $v : S = H^0(M, \mathcal{O}(E(\lambda)) \to F$ by $v(s) = s(e)$, e = identity of G. Let F' = kernel of v. Show that $0 \to F' \to S \to F \to 0$ is an exact sequence of U-modules. Take a basis $\{x_1, \dots, x_{n-m}, \dots, x_n\}$ of S

such that $\{x_1,\ldots,x_{n-m}\}$ belong to F'. Let T denote the action of G on $S:(T(g)x)(g') = s(g^{-1}g')$. Define the map $f_\lambda(gU) = T(g)GL(n, mC)$ mapping $M \to G(n, m)$. Show that if $E(\lambda)$ is the homogeneous bundle over a Kählerian C-space, then f_λ provides an imbedding of M into $CP(m)$. When is f biregular? When is the dimensional m a minimum? Work out the case $M = G(n, m)$.

EXERCISE 10.3. Let G be a connected semisimple Lie group with real Lie algebra $\mathfrak{g}_0$. Show that any semisimple element in $\mathfrak{g}_0$ has an admissible Lie algebra polarization. (Hint: embed the semisimple element X into a Cartan subalgebra $\mathfrak{h}_0$ of $\mathfrak{g}_0$. Decompose $X = X_1 + X_2$ where the eigenvalues of $\mathrm{ad}_{\mathfrak{g}_0}(X_1)$ resp. ad $\mathfrak{g}_0(X_2)$ are purely imaginary (resp. real). Let Δ be the set of roots of $(\mathfrak{g}, \mathfrak{h})$. Set

$$\mathfrak{p} = \mathfrak{h} + \sum_{\alpha(X)=0} \mathfrak{g}_\alpha + \sum_{\alpha(X_0)>0} \mathfrak{g}_\alpha + \sum_{\substack{\alpha(X_2)=0\\ i\alpha(X_1)>0}} \mathfrak{g}_\alpha$$

and show that $\mathfrak{p}$ is the Lie algebra polarization.)

EXERCISE 10.4. Let G be a reductive Lie group. Let f in $\mathfrak{g}$ be a nilpotent element (i.e. $\mathrm{ad}(f): \mathfrak{g} \to \mathfrak{g}$ is nilpotent as a linear transformation). Let $\mathfrak{q}$ be a complex polarization for f^* and let P be a parabolic subgroup of G corresponding $\mathfrak{p} = \mathfrak{q} \cap \mathfrak{g}$. Show that the orbit $\mathrm{Ad}(G).f$ is equivariantly diffeomorphic to an open G-orbit of the tangent bundle $T(G/P)$ iff $\mathfrak{q}$ is $\mathrm{Ad}(G(f))$-invariant.

EXERCISE 10.5. Using the last exercise in the case $\mathfrak{g} = \mathfrak{g}_l \otimes \mathfrak{c}$ where $\mathfrak{g}_l = [\mathfrak{q}, \mathfrak{g}] = so(n, 1)$, then if $f \in \mathfrak{g}$ is a regular nilpotent element $\mathrm{Ad}(P)$. $f = \mathfrak{p}^\perp - \{0\}$. But $T^*_f(G/P) = (\mathfrak{g}/\mathfrak{p})^* = \mathfrak{p}^\perp$. Thus show that the open G-orbit of F is $T^*(G/P) - \{0\}$.

EXERCISE 10.6. Show that in Example 10.4.6 we have

$$G = X_B(w_0) \supset X_B(w_1) \supset \cdots \supset X_B(w_n) = P;$$

show that each $X_B(w_i)$ is of codimension 1 in $X_B(w_{i-1})$; show that the $X_B(w_i)$ are the inverse images under π of the Schubert varieties in $P\backslash G$.

EXERCISE 10.7. Show that the complex manifold $V_w = \pi(C_P(w))$ is a complex $N_P(w)$-cell. Thus every Kähler C-space admits an analytic cell decomposition; show V_w is a CW-complex. Show that V_w is locally closed in

G/P in Zariski and in Hausdorff sense. Let $\bar{V}_w$ denote the cycle which the Schubert variety $X_P(w)$ represents. Show that $\{\bar{V}_w\}$ forms a basis of the integral homology of G/P, $\bar{V}_w \in H_{N_{P(w)}}(G/P, Z)$. Show that $\bar{V}_w \to \bar{V}_{w_0 w}$ gives Poincaré duality of G/P; i.e. show that for intersections of cycles we have $\bar{V}_w \cdot \bar{V}_{w_0 w} = 1$ and $\bar{V}_w \cdot \bar{V}_{w_1} = 0$ for any cycles $V_{w_1} \neq \bar{V}_{w_0 w}$ where $\dim \bar{V}_w + \dim \bar{V}_{w_1} = \dim G/P$.

EXERCISE 10.8. Let G be an irreducible real algebraic linear group with complexification G^C, a complex semisimple Lie group. Subgroup U of G is called parabolic if its complexification U^C is parabolic in G^C. $M = G/U$ is called an R-space Show that M admits a cellular decomposition $M = \bigcup V_m$ where V_m is homeomorphic to $R^{N_P(w)}$. Let $\bar{V}_w$ denote the closure of V_w in M. Show that $\bar{V}_w \to \bar{V}_{w_0 w}$ gives the Poincaré duality modulo 2 of M.

EXERCISE 10.9. The real quadric $M = Q_{n,v}(R)$ is defined to be all $(x) \in R^{P(n-1)}$ whose homogeneous coordinates $x_i (1 \leqq i \leqq n)$ satisfy

$$\sum_{i=1}^{v} x_i x_{n+1-i} + \sum_{k=1}^{v_0} x_{v+k}^2 = 0,$$

where $n = 2v + v_0$, $\dim M = n - 2$.

Let

$$A = \left(\begin{array}{c|c|c} & & \begin{smallmatrix} & & 1 \\ & \cdot^{\cdot^{\cdot}} & \\ 1 & & \end{smallmatrix} \\ \hline & \begin{smallmatrix} 1 & & \\ & \ddots & \\ & & 1 \end{smallmatrix} & \\ \hline \begin{smallmatrix} & & 1 \\ & \cdot^{\cdot^{\cdot}} & \\ 1 & & \end{smallmatrix} & & \end{array}\right) \begin{array}{l} v \\ \\ v_0 \\ \\ v \end{array}$$

$$G = \{g \in SL(n, R) \mid g^t A g = A\},$$
$$K = \{g \in SO(n) \mid Ag = gA\}$$

$$U = \left\{ g \in G \mid g = \left(\begin{array}{c|c|c} \lambda & * & * \\ \hline & * & * \\ \hline & & \lambda^{-1} \end{array}\right) \right\}$$

$$K^* = \left\{ k \in K \,\middle|\, k = \begin{pmatrix} \varepsilon & & \\ & * & \\ & & \varepsilon \end{pmatrix}, \varepsilon = \pm 1 \right\}.$$

Show $M = G/U = K/K^*$.

Define equivalence relation $\sim$ on $S^{v+v_0} \times S^{v-1}$ by $(x, y) \sim (-x, -y)$. Let $E_{n,v}$ denote the quotient space and let $[x, y]$ denote the equivalence class of (x, y). Define $p : E_{n,v} \to RP(v-1)$ by $[x, y] \to (y)$. Then $E_{n,v}$ is a S^{v+v_0-1}-bundle over $RP(v-1)$. Show that $E_{n,v}$ is diffeomorphic to M.

Show that the Schubert cell decomposition is given by $V_{x_i} = RP(i-1) - RP(i-2)$ for $1 \leqq i \leqq v$ for $v_0 \geqq 0$, where $RP(i) = \{(x) \in RP(n-1); x_{i+2} = \cdots = x_n = 0\}$ for $1 \leqq i \leqq n-1$. (Clearly $RP(i) \subset M$ in this case.)

Exercise 10.10. A smooth function f on a compact manifold M with nondegenerate critical points is said to be a *nice function* on M if $\operatorname{Ind}(f)(p) = f(p)$ for any critical point p of f. Show that if M is an R-space then $f = f_{-\delta} + \frac{1}{2}\dim M$ is a nice function on M, where

$$\delta = \tfrac{1}{2} \sum_{\alpha > 0} \alpha.$$

Exercise 10.11. Let $M = G/K$ be a Kähler C-space with almost complex structure J and symplectic form Ω. Let U be an irreducible unitary representation of G on Hilbert space $\mathscr{H}$. Assume there is an over complete basis of vectors $|z\rangle$ such that for every $|\Psi\rangle$ in $\mathscr{H}$ we have $\langle z|\Psi\rangle = \Psi(z)$ for every z in M. Let $H(z, \bar{z}) = \langle z|\hat{H}|z\rangle / \langle z|z\rangle$ and $\rho(z, \bar{z}) = N \langle z|\hat{\rho}|z\rangle / \langle z|z\rangle$ where $\hat{\rho} = \exp(-\beta\hat{H})$ is a density matrix in H. Show that $\operatorname{Tr}(\hat{\rho}\hat{H}) = \int_M \rho(z, \bar{z})(H(z, \bar{z}) + \frac{1}{2}\Delta H(z, \bar{z}))\Omega^n$. Show that $\int_M \rho(z, \bar{z}) H(z, \bar{z}) \Omega^n = \operatorname{Tr}(\hat{\rho}_{c1} \hat{H})$ where

$$\hat{\rho}_{c1} = \int_M \rho(z, \bar{z}) \frac{|z\rangle\langle z|}{\langle z|z\rangle} \Omega^n.$$

Show that $\rho(\beta, z, z) = \langle z|\exp(-\beta\hat{H})|z\rangle / \langle z|z\rangle$ satisfies $(\partial\rho/\partial\beta) = -H\rho + \frac{1}{2}J(X)\rho$ where $X = -\mathrm{d}H^\#$ (here $i_{\mathrm{d}H^\#}\,\Omega = \mathrm{d}H$; setting $g(X, Y) = \Omega(JX, Y)$ then $\mathrm{d}H = i_{\operatorname{grad} H}\, g$ and $\operatorname{grad} H = -J(\mathrm{d}H^\#)$).

Chapter 11

Geometric Quantization

11.1. Geometric quantization of complex manifolds

Definition 11.1.1. The quantum bundle of a C-space (M, Ω) is the line bundle $Q \to M$ defined as follows: (i) for the $(1,1)$ form Ω we have $\operatorname{sgn} \Omega(X, X) = \varepsilon$, for all nonzero holomorphic vector fields X; (ii) the class of Q is $q = [\Omega] + (\varepsilon/2)c_1(M) \in H^2(M, Z)$; (iii) $\operatorname{sgn} \Omega[a] \neq -\operatorname{sgn} q[a]$ for all $[a]$ in $H_2(M, R)$, $[a] \neq 0$.

Given a vector bundle $E \to M$, the connection on E is specified by a matrix valued one form $\vartheta = \{\vartheta^\rho_\sigma\}$. Writing $\vartheta = \vartheta' + \vartheta''$ in $(1,0)$ and $(0,1)$ components we say that ϑ is a *complex connection* if $\vartheta'' = 0$.

Theorem 11.1.2. If $E \to M$ is a holomorphic bundle with a Hermitian metric, then there is a unique complex connection ϑ which is compatible with the metric in E and the curvature Θ of ϑ is of type $(1,1)$.

Theorem 11.1.3. If $E \to M$ is a differentiable bundle with fiber C^r and ϑ is a connection for E, then the $(1,0)$ coframe $\{\omega^j\}$ and $\{\vartheta^\rho_\sigma\}$ define an almost complex structure J.

Theorem 11.1.4. If the curvature Θ of ϑ is type $(1,1)$ then J is integrable.

Let M be a compact Kãhler manifold and let $\gamma \in H^2(M, Z)$ be a cohomology class of type $(1,1)$. Then there is a smooth line bundle $E \to M$ with $c_1(E) = \gamma$. Let γ denote also the real $(1,1)$ form which represents $[\gamma]$ and let ϑ be a connection for E. Then $(i/2\pi)\,\mathrm{d}\vartheta = \gamma + \mathrm{d}\alpha$ where α is a global 1-form on M. Thus $(2\pi/i)\alpha$ is a connection on E whose curvature is type $(1,1)$. This is precisely the content of a theorem of Kodaira–Spencer. Viz. we have

Theorem 11.1.5. If Q is a quantum bundle over M, then $Q \to M$ is a holomorphic line bundle.

Let $M = G^u/V$ be a Kãhler C-space. If π: $V \to GL(E)$ is an irreducible unitary representation on a complex vector space E then the homogeneous vector bundle $E(\pi) = G^u \times_\pi E \to M$ has a G^u-invariant Hermitian metric.

$E(\pi)$ can be explicitly realized as a holomorphic bundle by taking the complex realization of $M = G/B$ and extending π to a holomorphic irreducible representation π: $B \to GL(E)$. Then $G \times_B E \to M$ is a holomorphic vector bundle over M which gives a complex structure to $E(\pi)$.

In summary, to construct quantum bundles over Kählerian C-spaces we proceed as follows. Let π: $V \to GL(E)$ be an irreducible unitary representation with highest weight λ. The square root of the canonical line bundle $K = E(2\delta)$ is the bundle associated with $(c_1(M)/2)$. The quantum bundle is then $Q = E(\pi) \otimes E(\delta)$ with $\lambda + \delta$ being nonnegative, for λ integral.

Let $H_2^0(M, \mathcal{O}(Q))$ denote the Hilbert space of square integrable sections of Q where the inner product is

$$(s_1, s_2) = \int (s_1 | s_2) \Omega^n,$$

where $\dim M = 2n$, with $(\ \ |\ \ )$ denoting the Hermitian structure on Q which is compatible with the connection $D(Q)$ on Q with curvature curv $(D(Q), Q)$ representing q.

11.2. Harmonic oscillator

The harmonic oscillator may be quantized as follows. The symplectic manifold (M,Ω) is $M = R^{2n}$ and $\Omega = (1/h) \sum dp_j \wedge dq_j$ where h = Planck's constant. The classical Hamiltonian is $H = (1/2m^2) \sum (p_j^2 + m^2 v^2 q_j^2)$. M has a natural complex structure with complex coordinates $z_j = p_j - imq_j$. The surface of constant energy E in C^n is the $(2n-1)$ sphere $\sum_{j=1}^n |z_j|^2 = 2mE$ and each orbit is a great circle $z(t) = \exp(-vt)z(0)$. The space of orbits may then be identified with $CP(n-1)$ by π: $(z) \to [z]$ where $[z]$ denotes as usual the set of all nonzero complex scalar multiples of z. On each subset $U = \{z \in CP(n-1) | z_n \neq 0\}$ we may introduce complex coordinates $w_1 = z_1/z_n, \ldots, w_{n-1} = z_{n-1}/z_n$. Ω is invariant along the classical orbits, which gives upon restriction to the constant energy surface and then to $CP(n-1)$ a symplectic form or Fubini-Study metric

$$\Omega_E | U = \frac{-iE}{\hbar} \sum_{k=1}^{n-1} \frac{dw_k \wedge d\bar{w}_k - \sum \bar{w}_k w_l \, dw_k \wedge d\bar{w}_l}{(1 + |w|^2)^2}$$

We write $\Omega_E = (-E/v\hbar)\gamma$. $CP(n-1)$ as we know from Theorem 9.1.22 has exactly two invariant complex structures, and these are complex conjugates of each other. Let F denote the usual holomorphic structure. Thus the base space of the quantum bundle is $(M = CP(n-1), \Omega, F)$. The first chern class of $CP(n-1)$ is $c_1(CP(n-1)) = ng$ where g is a positive generator of $H^2(CP(n-1), Z$. Clearly $\gamma \in g$.

From the definition (M, Ω, F) has a quantum bundle iff $q = [\Omega_E] + (ng/2)$ is a nonpositive element of $H^2(CP(n-1), Z)$. That is if $(-E/vh) + (n/2) = = 0, -1, -2, \ldots$ or $E = E_N = v\hbar((N+n)/2)$ for $N = 0, 1, 2, \ldots$

The connection ∇ in the quantum bundle Q_N, which admits a ∇-invariant Hermitian structure and is such that $\operatorname{curv}(Q, \nabla) = \eta = (Nvh/E)\Omega_E$, is given by the 1-form

$$\alpha_N | U \times (C - \{0\}) = \frac{d\mathfrak{z}}{2\pi i \mathfrak{z}} + \frac{N}{2\pi i} \sum_{k=1}^{n-1} w_k \, d\bar{w}_k,$$

where

$$|w|^2 = \sum_{j=1}^{n-1} |w_j|^2.$$

Let s be any holomorphic section of Q_N over open U. E.g.

$$w \to t(w) = \left(w, \left(\frac{1}{1 + |w|^2} \right)^N \right).$$

Thus each holomorphic section s U is of the form $s(w) = p(w)t(w)$ where p is a holomorphic function on U. p cannot have any poles; thus p must be a polynomial of degree $\leqq N$.

The ∇-invariant Hermitian structure on Q_N is

$$|(w, \mathfrak{z})|^2 = |\mathfrak{z}|^2 (1 + |w|^2)^N.$$

The Hilbert space inner product is given by

$$(s_1, s_2) = (-1)^{(n(n-1))/2} i^{(n-1)} \left(\frac{N + (n/2)}{2\pi} \right)^{(n-1)} \int_{C^{n-1}} p_1(w) \overline{p_2(w)} \, dw \wedge d\bar{w},$$

where

$$s_i | U = \frac{p_{i(w)}}{(1 + |w|^2)^N}.$$

In summary we have

THEOREM 11.2.1. Any section from $H_2^0(M, \mathcal{O}(Q))$ may be identified with $p(w)/(1 + |w|^2)^N$ where p is a polynomial of degree $\leqq N$.

COROLLARY 11.2.2. The multiplicity of energy eigenvalue E is given by

$$\operatorname{multip}(E_N) = \dim H_2^0(M, \mathcal{O}(Q_N)) = \binom{n + N - 1}{N}.$$

This last result is contained in Kodaira–Hirzebruch H23.

11.3. The Kepler Problem – Hydrogen Atom

The phase space associated to a free particle in $R^3 - \{0\}$ is $M = \{(p, q) | p, q \in R^3 q \neq 0\}$. M then has a natural 2-form

$$\Omega = \sum_{i-1}^{3} \mathrm{d}p_i \wedge \mathrm{d}q_i$$

and Poisson bracket

$$\{f_1, f_2\} = \sum_{i=1}^{3} \left(\frac{\partial f_i}{\partial p_i} \frac{\partial f_2}{\partial q_i} - \frac{\partial f_2}{\partial p_i} \frac{\partial f_1}{\partial q_i} \right).$$

The seven functions $h = (1/2m)p^2 - (K/q)$ (the energy) $l = q \times p$ (angular momentum), $a = l \times p + (mKq/|q|)$ (Runge–Lenz vector) satisfy the identities: $a.l = 0, |a|^2 - 2mh|l|^2 = m^2K^2$ and

$$\{h, l_i\} = \{h, a_i\} = \{l_i, a_i\} = 0$$
$$\{l_1, l_2\} = -l_3$$
$$\{l_1, a_2\} = -a_3$$
$$\{a_1, a_2\} = 2mhl_3.$$

These latter relationships are the Lie bracket relations for $O(4)$ for $h > 0$ and $O(3, 1)$ for $h < 0$ and Euclidean group for $h = 0$.

The solutions of Newton's equations

$$\ddot{q} = -Kq/|q|^3 \quad \text{in} \quad R^3 \backslash \{0\}$$

correspond to the solutions of Hamilton's equations

$$\dot{q} = \partial h/\partial p, \qquad p = -\partial h/\partial q$$

or the trajectories of the vector field $X_h = \{h, \ . \ \}$. These trajectories are called the Kepler orbits.

From the bracket relationships $\{h, l_1\} = 0, \{h, l^2\} = 0$ and $\{l_1, l^2\} = 0$ we see that $f_1 = h$, $f_2 = l^2$, and $f_3 = l_1$ is a complete collection of pairwise commuting symmetries of h. Thus the Kepler's problem is completely integrable.

Let $M(E)$ denote the set of Kepler orbits for which $h = -E, E > 0$. Let $\rho = \sqrt{2mE}$ and let x and y denote the functions on $M(E)$ induced by $\rho l \pm a$. Thus $|x|^2 = |y|^2 = m^2K^2$ and (x, y) provides the map which shows that $M(E)$ is diffeomorphic to $S^2(r) \times S^2(r)$ where $r = mK$.

Theorem 11.3.1. For the hydrogen atom (Kepler's problem) the space of classical orbits corresponding to energy $E < 0$ is $M = S^2 \times S^2 =$

$= \{(x, y) \in R^3 \times R^3 \,\big|\, |x|^2 = |y|^2 = m^2 K^2 \}$ where M has the symplectic form

$$\Omega_E = \frac{1}{\sqrt{8m|E|}} \left(\frac{dx_1 \wedge dx_2}{x_3} + \frac{dy_1 \wedge dy_2}{y_3} \right)$$

$x_3 \neq 0$ and $y \neq 0$. M has a natural complex structure when considered as $M = CP(1) \times CP(1)$ with

$$\Omega_E | U = K \sqrt{\frac{m}{2|E|}} \, i \left(\frac{dw \wedge d\bar{w}}{(1 + |w|^2)^2} + \frac{dw' \wedge d\bar{w}'}{(1 + |w'|^2)^2} \right)$$

$$= 2\pi K \sqrt{\frac{m}{2|E|}} \gamma,$$

where $c_1(M) = 2[\gamma]$.

The class of the quantum bundle is

$$q = (2\pi K \sqrt{\frac{m}{2|E|}} - 1)[\gamma]$$

and q is nonnegative in $H^2(M, Z)$ iff $q = (N - 1)[\gamma]$, $N = 1, 2, \ldots$. Thus we have Bohr's formula: $E_N = - m^2 e^4 / 2\hbar^2 N^2$.

The Riemann-Roch formula in this case states that

$$\mathfrak{x}(M, Q_N) = \tfrac{1}{2}(q_n^2 + q_n c_1(M))[M] + \tfrac{1}{4}(\mathfrak{x}(M) + \tau(M)),$$

where $\tau(M) =$ Hirzebruch signature $= 0$ since M is a product of two manifolds. $\mathfrak{x}(M) = 4$ and $q_N^2[M] = (N - 1) \int_M \gamma \wedge \gamma = 2(N - 1)$. Here $(c_1(M)q_N)[M] = (N - 1)\gamma^2[M] = 4(N - 1)^2$. Thus $\mathfrak{x}(M, Q_N) = N^2$. We need the following theorem of Kodaira

THEOREM 11.3.2. If any Kähler form is a representative of $c(B) + c_1(M)$ then $\mathfrak{x}(M, B) = \dim H^0(M, \mathcal{O}(B))$.

Since $q_N + 2[\gamma]$ contains a Kähler form $(N + 1)\gamma$ we have

COROLLARY 11.3.3. The multiplicity of the energy eigenvalue E_N is $\dim H^0(M, \mathcal{O}(Q)) = N^2$.

11.4. MASLOV QUANTIZATION

Czyz C17 has related Maslov quantization to geometric quantization as follows. Let $(M, \Omega) = (R^2, \sum dp_i \wedge dq_i)$ and for Hamiltonian $h: M \to R^1$ set

$M_E = h^{-1}(E)$. Assume M_E is connected, simply connected, compact (2n − 1)-dimensional submanifold of M. Assume all the orbits of the Hamilton equations on M_E are closed with orbit space $\mathcal{O}$, Let $\pi: M_E \to \mathcal{O}$. Then we have

THEOREM (Czyz) 11.4.1.

(i) $\mathcal{O}$ is simply connected;

(ii) there is a unique 2-form $\Omega_{\mathcal{O}}$ on $A^2(\mathcal{O})$ which is symplectic and such that $\pi^* \Omega_{\mathcal{O}} = \Omega | M_E$;

(iii) if $\omega = \sum p_i \, dq_i$ then $\int_{\mathcal{O}} \omega$ for orbit $\mathcal{O}$ in M_E does not depend on $\mathcal{O}$;

(iv) for each closed orientable simply connected 2-dimensional surface a in $\mathcal{O}$ there is a surface b in M_E such that $\pi(b) = a$ and $\int_a \Omega_{\mathcal{O}} = \int_b \Omega = \int_{\partial b} \omega$, where ∂b is either an orbit with some integer multiplicity or an empty set.

Let L be a Lagrange surface in M_E. If $\sum_1$ is a certain (v. C17) closed $(n-1)$-dimensional submanifold of L, then for a trajectory $d(t)$ $t \in [0,1]$ in L such that $d(0), d(1) \notin \sum_1$, then the *Maslov index* $\operatorname{ind}_L d$ is the number of points where $d(t)$ crosses $\sum_1$ going from the 'negative' side to the 'positive' side minus the number of crossings from 'positive' to 'negative' (for definition of sign v. M9). Thus for $\mathcal{O}_L$ in L we define the Maslov index $\operatorname{ind}_L O_L$. Since all orbits in L belong to the same homology class in $H_1(L, R)$ we have $\operatorname{ind}_L O_L =$ constant for all $\mathcal{O}_L$ in L. We call this value $\operatorname{ind}_L \mathcal{O}$, and set, following the notation of Theorem 11.4.1,

$$\tilde{q}(a) = \int_a \Omega_{\mathcal{O}} - \tfrac{1}{4} \operatorname{ind}_L(\partial b).$$

DEFINITION 11.4.2. The *Maslov quantization condition* states that for a Lagrangian surface L in M_E

$$\int_c \omega = k - \tfrac{1}{4} \operatorname{ind}_L c$$

for $k \in Z$ and c any closed curve in L.

THEOREM 11.4.3. If $n > 1$ and the Maslov quantization condition is satisfied, then $\tilde{q} \in H^2(\mathcal{O}, Z)$.

PROBLEMS

EXERCISE 11.1. Let G be the Poincaré group where $g = (a, \Lambda)$ $a \in R^4$ and $\Lambda \in O(3, 1)$. Let $\mathfrak{g}$ be the Lie algebra of the Poincaré group G with basis Ξ_α,

$\Lambda_{\alpha\beta} = -\Lambda_{\beta\alpha}$ ($\alpha, \beta = 0, 1, 2, 3$). The metric is $g_{oo} = -1$ and $g_{ii} = 1$, $i = 1, 2, 3$. The dual basis is given by P^α, $M^{\alpha\beta}$ where $(P^\alpha, \Xi_\beta) = \delta_{\alpha\beta}$ and $(M^{\alpha\beta}, \Lambda_{\gamma\delta}) = = \delta_{\alpha\gamma}\delta_{\beta\delta} - \delta_{\alpha\delta}\delta_{\beta\gamma}$. And $(P^\alpha, \Lambda_{\beta\gamma}) = (M^{\alpha\beta}, \Xi_\gamma) = 0$. A point x in $\mathfrak{g}^*$ has coordinates p_α, $m_{\alpha\beta}$ given by $x = p_\alpha P^\alpha + \frac{1}{2} m_{\alpha\beta} M^{\alpha\beta}$. Define the polynomials $p^2 = p_\alpha p^\alpha$ (where $p^\alpha = g^{\alpha\beta} p_\beta$) and $w^2 = w_\alpha w^\alpha$ (where $w^\alpha = \frac{1}{2}\varepsilon^{\alpha\beta\gamma\delta} m_{\beta\gamma} p_\delta$). Show that the space M defined by $-p^2 - m^2 > 0$, $w^2 = \sigma^2 m^2$ and $p^0 > 0$ is homomorphic to $R^6 \times S^2$. Let f in M be the point with coordinates $p^0 = m$, $p^i = 0$, $m_{12} = -\sigma$, $M_{31} = M_{23} = 0$, $M_{oi} = 0$. Write $M = G/G(f)$. Show that for $\sigma \neq 0$, $X \to 2\pi i \langle f, X \rangle$ lifts to a character iff $4\pi\sigma \in N$. Show that for $\sigma \neq 0$ the G-invariant polarizations on M are $\mathfrak{h}_f$ and $\bar{\mathfrak{h}}_f$ has basis $\{\Xi_\alpha, \Lambda_{12}, \Lambda_{23} + + i\Lambda_{31}\}$. Show that if $\sigma = 0$ the unique G-invariant polarization is real and has basis $\{\Xi_\alpha, \Lambda_{ij}\}$. Take standard coordinates of point gf in M as q in R^3, $p = mk$ in R^3, and $z \in C$. Show that $\{q_1, p^j\} = \delta_{ij}, \{p_i, p_j\} = 0$ but $\{q_i, q_j\} \neq 0$. Find coordinates x such that $\{x_i, p^j\} = \delta_{ij}$ and $\{x_i, x_j\} = 0$.

Show that M has the symplectic 2-form

$$\Omega_{\sigma,m} = (2\sigma + 1)\delta + m \sum_{r=1}^{3} \mathrm{d}p_r \wedge \mathrm{d}q_r$$

where

$$\delta | S^2 = \frac{i}{2\pi} \frac{\mathrm{d}w \wedge \mathrm{d}\bar{w}}{(1 + |w|^2)^2}.$$

Set $q = [\Omega_{\sigma,m}] - [\delta]$. Show that q determines a quantum bundle over M iff $\sigma = \frac{1}{2}$, $l = 0, 1, 2 \ldots$. Write down the connection ∇ for this bundle, the ∇-invariant Hermitian structure, and the covariant constant sections.

Show that the natural representation of $(SU(2)$ on the quantum Hilbert space over $S^2 = CP(1)$ is the Wigner representation of $SU(2)$ for spin σ.

Show that if $\sigma = 0$, $M \simeq R^6$.

Show that the orbit M corresponding to $p^2 = 0$, $w^2 = 0$, $w_\alpha = \lambda p_\alpha$, $p^0 > 0$ is of dimension 6 with integral 2-form iff $2\lambda \in Z$. λ is called *helicity*. If $\lambda = 0$, show $M \simeq R^6 \backslash \{p_i = 0\}$. Find Ω and the position operators. For $\lambda \neq 0$ show that there does not exist on M functions which can be interpreted as position variable, so there does not exist a quantum mechanical position operator.

EXERCISE 11.2. Let $G = SU(2)$ and $M = T^*G$. Define the bundle $L = M \times C$ and set $(m_1, z_1) \sim_i (m_2, z_2)$ iff $m_1 = \varepsilon m_2$ and $z_1 = \eta_i z_2$ where $\varepsilon \in$centre of G $\varepsilon \neq (e)$ and $\eta_1 = +1$, $\eta_2 = -1$. Let $L_i = L/\sim_i$ $i = 1, 2$ and set $\bar{M} = M/\sim$ where $m_1 \sim m_2$ iff $m_1 = \varepsilon m_2$ as above. Show $\bar{M} = T^*SO(3)$. Let $\bar{\Omega}$ be the canonical symplectic form on $\bar{M}$. L_i are bundles over $\bar{M}$. Show that $c_1(L_1) = 0$, but $c_1(L_2) \neq 0$. Define connection forms α_i such that $(L_i, \alpha_i) \in L_c(\Omega)$. Show that the canonical polarization of $(\bar{M}, \bar{\Omega})$ is $SO(3)$-invariant. Show that the angular momentum of the solid is able to assume all integer values in the case $i = 1$ and all half-integer values in the case $i = 2$.

Extend this example to study the free solid body where the kinematic group is the Galilean group and $M = T^*(R^2 \times SO(3))$.

Chapter 12

Principal Series Representations

12.1. Representation theory for noncompact semisimple Lie groups

Part I: Principal Series Representations

In the next few chapters we will need an understanding of elements of the representation theory of noncompact semisimple Lie groups – esp. those representations which occur in the Plancherel theory. These representations fall into two large classes – the discrete series and the principal series. We will study the principal series in this chapter.

The principal series arises as follows. Let G^0 be a connected semisimple Lie group. The Iwasawa decomposition is given by the following prescription. Let σ be the Cartan involution of G with fixed point set K. Let $\mathfrak{a}$ be a maximal abelian subspace of the (-1)-eigenspace of σ on G^0. Let $\Delta_\mathfrak{a}$ be the root space in $\mathfrak{a}^*$ where $\mathfrak{g}_0 = \mathfrak{z}_\mathfrak{a} + \sum_{\Delta\mathfrak{a}} \mathfrak{g}_0^\alpha$. Here $\mathfrak{z}_\mathfrak{a}$ = the central-of $\mathfrak{a}$ in $\mathfrak{g}_0$ and $\mathfrak{g}_0^\alpha = \{X \in \mathfrak{g}_0 | [Y, X] = \alpha(Y)X \text{ for } Y \text{ in } \mathfrak{g}_0\} \neq 0$. Let D denote a Weyl chamber of $\mathfrak{a}$. Then $\Delta_\mathfrak{a}^+ = \{\alpha \in \Delta_\mathfrak{a} | \alpha > 0 \text{ on } D\}$ where $\Delta_\mathfrak{a}^- = -\Delta_\mathfrak{a}^+$. Define the algebras $\mathfrak{n}^- = \sum_{\Delta_\mathfrak{a}^+} \mathfrak{g}_0^\alpha$ and $\mathfrak{n}^- = \sum_{\Delta_\mathfrak{a}^-} \mathfrak{g}_0^\alpha$, with subgroups N and A of G^0 corresponding to $\mathfrak{n}$ and $\mathfrak{a}$. Let M be the centralizer of A in K.

Definition 12.1.1. The composition $G^0 = \mathrm{KAN}$ is called the *Iwasawa decomposition*. The group $B = \mathrm{MAN}$ is the *minimal parabolic subgroup* of G.

Example 12.1.2. Let $G^0 = SL(2, R)$. Then

$$M = \begin{pmatrix} \pm 1 & 0 \\ 1 & \pm 1 \end{pmatrix}, \qquad A = \left\{ \begin{pmatrix} a & 0 \\ 0 & a^{-1} \end{pmatrix} a > 0 \right\}$$

and $K = SO(2)$ for an Iwasawa decomposition.

Let $\mathfrak{m}$ denote the Lie algebra of M and let $\mathfrak{t}$ be a Cartan subalgebra. $\mathfrak{t}$ determines a positive system of roots $\Delta_\mathfrak{t}^+$ which is consistent with $\Delta_\mathfrak{a}^+$. Viz. there is a positive system of roots for $\mathfrak{g}$ relative to its Cartan subalgebra $(\mathfrak{t} + \mathfrak{a})^C$ such that $\Delta_\mathfrak{a}^+ = \{\varphi|_\mathfrak{a} / \varphi \in \Delta^+, \Delta|_\mathfrak{a} \neq 0\}$ and $\Delta_\mathfrak{t}^+ = \{\varphi|_\mathfrak{t} / \varphi \in \Delta^+, \Delta|_\mathfrak{a} = 0\}$. We let δ_t denote half the sum of positive roots of $\Delta_\mathfrak{t}^+$. Set $T^0 = \exp(\mathfrak{t})$ and $L_\mathfrak{m}^+ = \{\nu \in i\mathfrak{t}^* | \exp(\nu - \delta_\mathfrak{t}) \in \hat{T}^0 \text{ and } \langle \nu, \varphi \rangle > 0 \text{ for every } \varphi \text{ in } \Delta_\mathfrak{t}^+\}$.

THEOREM 12.1.3. Let $\tilde{Z}$ denote the centralizer of M^0 and let $E = \tilde{Z} \cap M^0$. Then every $[\eta] \in \hat{M}$ is of the form $[\eta] = [\mathfrak{x} \otimes \eta^0]$ for $[\mathfrak{x}] \in \tilde{Z}$ and $\eta^0 \in \hat{M}^0$ and there is a bijection $L_{\mathrm{m}}^{+} \to \hat{M}$ given by $\nu \to [\eta_\nu^0]$ where $\nu - \delta_{\mathrm{t}}$ is the highest weight of η_ν^0.

Let Ψ be the simple $(\mathfrak{t} + \mathfrak{a})^C$-root system of $\mathfrak{g}$ and let $\Phi \subset (\Psi \cap \Delta_{\mathrm{t}}^{+})$ be an arbitrary set. Set $\mathfrak{z}_\Phi = \{X \in \mathfrak{t} / \Phi(X) = 0\}$ and let $Z_\Phi^0 = \exp \mathfrak{z}_\Phi$. If U_Φ is the M-centralizer of Z_Φ^0 then $S_\Phi = M/U_\Phi$ is a homogeneous Kähler manifold. Since $M = \tilde{Z} M^0$ we have $U_\Phi = \tilde{Z} U_\Phi^0$ and $E = \tilde{Z} \cap U_\Phi^0$. We can select $[\mu] \in \hat{U}_\Phi$ of the form $[\mu] = [\mathfrak{x} \otimes \mu^0]$ where $[\mathfrak{x}] \in \tilde{Z}$ with $\mathfrak{x}|_E$ is a multiple of $\xi \in \hat{E}$ and $[\mu^0] \in \hat{U}_\Phi^0$ with $\mu^0|_E$ a multiple of ξ.

For the minimal parabolic subgroup $B = \mathrm{MAN}$ the finite dimensional classes in $\hat{B}$ are exhausted by $\mu \otimes \tau$ where $\mu \in \hat{M}$ and $\tau \in \hat{A}$. Viz. let λ be an irreducible complex representation of MAN. By Lie's theorem $\lambda(N) = 1$. Thus $\lambda = \mu \otimes \tau$: man $\to \mu(m)\tau(a)$ where $\mu \in M$ and $\mathfrak{x}$ is a character of A. Let $\delta = \frac{1}{2}\sum \dim \mathfrak{g}_0^\alpha)\, \alpha \in \mathfrak{a}^*$. Set $\beta_{\mu,\eta}(\mathrm{man}) = \mu(m) \exp((\delta + i\eta)(\log a))\ \eta \in \mathfrak{a}^*$. This is an irreducible unitary representation of MAN on $V(\mu)$.

DEFINITION 12.1.3. The induced representation $T(\mu, \eta) = \mathrm{Ind}_B^G(\beta_{\mu,\eta})$ is called the *principal series* of G. Since μ is of the form $\mathfrak{x} \otimes \eta_\nu^0$ we sometimes write $T(\mathfrak{x}, \nu, \sigma)$.

We note that $T(\mu, \eta)$ is unitary iff η: $\mathfrak{a} \to R$ is real. Thus we have parametrized the principal series by $\hat{M} \times \mathfrak{a}^*$.

In general $T(\mu, \eta)$ is a finite direct sum of irreducible representations, although for certain class of groups (which class includes $SO_0(2n+1, 1)$) $T(\mu, \eta)$ is irreducible for all μ and η.

THEOREM 12.1.4.

(i) $T(\mu, \eta) = T(w\mu, w\eta)$ for w in $W(\mathfrak{a})$, the Weyl group associated to $\mathfrak{a}$, $W(\mathfrak{a}) = \{k \text{ in } K \mid \mathrm{ad}(k)\mathfrak{a} = \mathfrak{a}\}/M$;

(ii) if $\mu = 1_M$ then $T(1_M, \eta)$ is irreducible for all η in $\mathfrak{a}^*$.

DEFINITION 12.1.5. The principal series with $\mu = 1_M$ are called *spherical principal series.*

EXAMPLE 12.1.6 (12.1.2 cont.). If σ in $\hat{M}$ is given by

$$\sigma\begin{pmatrix} \pm 1 & 0 \\ 0 & \pm 1 \end{pmatrix} = \varepsilon = \pm 1$$

and if τ in $\hat{A}$ is given by

$$\tau\begin{pmatrix} a & 0 \\ 0 & a^{-1} \end{pmatrix} = a^{i\rho}.$$

$T(\sigma, \tau)$ is irreducible except when $\sigma \neq 1$ and $\tau = 1$.

The character of this representation is

$$\Theta(h) = \frac{|a|^{i\rho} + |a|^{-i\rho}\,\mathrm{sgn}^{\varepsilon}(a)}{|a - a^{-1}|}$$

for regular

$$h = \begin{pmatrix} a & 0 \\ 0 & a^{-1} \end{pmatrix}$$

in the Cartan subgroup

$$H = \left\{ \begin{pmatrix} a & 0 \\ 0 & a^{-1} \end{pmatrix} \middle| a \text{ in } R^* \right\}.$$

Example (The deSitter Group) 12.1.7. Let $\mathbb{Q}$ denote the field of quaternions whose elements are written $x = x_1 + x_2 i + x_3 j + x_4 k$, x_i in R and $i^2 = j^2 = k^2 = -1$ and $ij = -ji = k, jk = kj = i$ and $ki = -ik = j$. Let $\bar{x} = x_1 - x_2 i - x_3 j - x_4 k$ be the quaternionic conjugate. Let $\hat{x} = -k\bar{x}k = x_1 + x_2 i + x_3 j - x_4 k$. The norm is $|x| = (x.\bar{x})^{1/2}$.

Let $M_2(\mathbb{Q})$ denote the 2×2 matrices with elements in $\mathbb{Q}$. Then

$$G = \left\{ g = \begin{pmatrix} ab \\ cd \end{pmatrix} \middle| M_2(\mathbb{Q}) \middle| \bar{a}b = \bar{c}d, \quad |a|^2 - |c|^2 = 1, \quad |d|^2 - |b|^2 = 1 \right\}.$$

G is isomorphic to the universal covering group of the deSitter group $SO_0(4, 1)$. We define the following subgroups of G:

$$K = \left\{ g \text{ in } G \middle| g = \begin{pmatrix} u & 0 \\ 0 & v \end{pmatrix} |u| = |v| = 1 \right\}$$

$$A_t = \left\{ a_t \text{ in } G \middle| a_t = \begin{pmatrix} \mathrm{ch}t/2 & \mathrm{sh}t/2 \\ \mathrm{sh}t/2 & \mathrm{ch}t/2 \end{pmatrix} t \text{ in } R \right\}$$

$$N = \left\{ n \text{ in } G \middle| n = \begin{pmatrix} 1 - x & x \\ -x & 1 + x \end{pmatrix} \right.$$

$$\left. x = \tfrac{1}{2}(\xi_2 i + \xi_3 j + \xi_4 k) \quad \xi_\alpha \text{ in } R \right\}$$

$$M = \text{centralizer of } A_t \text{ in } K = \left\{ m \text{ in } G \middle| m = \begin{pmatrix} u & 0 \\ 0 & u \end{pmatrix} |u| = 1 \right\}.$$

The Iwasawa decomposition is just $G = \mathrm{KAN}$. Elements u in K with $|u| = 1$ are isomorphic to $SU(2)$, viz. by the mapping

$$u \to \begin{pmatrix} a & b \\ -\bar{b} & a \end{pmatrix} a = x_1 + x_2 i \quad \text{and} \quad b = x_3 + x_4 i.$$

Thus $\hat{M}$ is characterized by $\widehat{SU}(2)$. Let μ_n in $\widehat{SU}(2)$ act on $V(\mu)$. Let $\mathfrak{x}(\vartheta_t) = \exp(ist)$. Then the induced representation $T(n, s)$ from $\beta(ma_t n) = \mu_n(u)\exp(-ist)$ is irreducible iff n is a demiinteger ≥ 0 and $s = 3/2 + iv$ where v is real. $T(n, s)$ is irreducible except $n = \frac{1}{2} \bmod l$ and $v = 0$. Finally $T(n, s)$ is equivalent to $T(n', s')$ for $\mathrm{Re}(s') = \frac{3}{2}$ iff $n = n'$ and $\mathrm{Im}(s) = \pm \mathrm{Im}(s')$.

Wolf has developed a version of the Borel–Weil and Borel–Weil–Bott theorems for principal series. The analogue of the Borel–Weil Theorem is realized on a closed orbit in a complex flag manifold. Viz. let $G = G^{0C}$ and set $M = G/P$ where P is a parabolic subgroup of G. Let $x_0 \in M$ and set $P_{x_0} = \{g \text{ in } G | gx_0 = x_0\}$. Let $L = M \cap P_{x_0}$.

Consider the irreducible unitary representations v: $L \to GL(W_v), \mu: M \to GL(V_\mu)$ and $\kappa: K \to GL(U_\kappa)$. From the M-homogeneous holomorphic vector bundle $W(v) \to M/L$ over the flag manifold $M/L = S[x_0]$ such that $H^0(M/L, \mathcal{O}(W(v))) = V_\mu$.

Let $\sigma_{v,\eta}$ be an irreducible representation of LAN on W_v given by $\sigma_{v,\eta}(lan) = v(l)\exp((\delta + i\eta)(\log a))$. Let $W(v, \eta) \to G^0/\mathrm{LAN} = Y$ be the G^0-homogeneous complex vector bundle over Y. Thus $W(v, \eta)|_{M/N}$ is holomorphic since it is just $W(v)$. Wolf calls $W(v, \eta)$ in this case *partially holomorphic*.

Let $H_2^0(Y, \mathcal{O}_p(W(v, \eta))$ denote the Hilbert space of square integrable partially holomorphic sections – i.e. f: $G^0 \to W_v$ which satisfies:

(i) $f(glan) = \sigma_{v,\eta}(lan)^{-1} f(g)$;

(ii) $f | g$ MAN is a holomorphic section of $W(v, \eta) \,|\, S[gx_0]$;

(iii) $\int_K (f(k), f(k))_{W_v} dk < \infty$.

Let $\Pi(v, \eta)$ denote the representation of G^0 on $H_2^0(Y, \mathcal{O}_p(W(v, \eta))$.

THEOREM 12.1.6. The representation $\Pi(v, \eta)$ so defined is unitarily equivalent to $T(\gamma, \eta)$.

The characters of $T(v, \eta)$ are realized as follows.

THEOREM 12.1.7. Let $T \in \hat{P} = \mathrm{MAN}$. Then there is an element $f_0 \in \mathfrak{a}^*$ such that $T(m\exp(Y)n) = \exp(i\langle f_0, Y\rangle)\, T(m)$ for m in M, Y in $\mathfrak{a}$ and n in N. And $T|_{M_0}$ is irreducible.

Since M^0 is compact its universal covering space is the product of an abelian group and a compact semisimple Lie group. Thus by Kirillov's Theorem 9.2.11 there is a regular element f_1 in $\mathfrak{m}^*$, orbit $\mathcal{O}(f_1)$ in $\mathfrak{m}^*$ under M and measure $d\beta(f_1)$ such that

$$\operatorname{Tr} T(\exp X) = j_m(X)^{-1/2} \int_{\mathcal{O}(f_1)} \exp(i\langle f_1, X\rangle)\, d\beta(f_1).$$

Assume T and f_0 from Theorem 12.1.7 are such that f_0 is distinct from all transforms by $W(\mathfrak{a})$. Let V be the set $V = \{X \in \mathfrak{g} \mid \text{eigenvalues } \lambda \text{ of } X \text{ in the adjoint representation have } |\operatorname{Im} \lambda| < \pi\}$. Let $\mathcal{O}$ be the orbit of G in $\mathfrak{g}^*$ which contains $\mathcal{O}(f_1) + f_0$. Let $d\beta$ be the associated Kirillov measure on $\mathcal{O}$. Then for a distribution φ on V and $f = f_0 + f_1$ we have

THEOREM 12.1.8. $\operatorname{Tr} T(\mu, \nu)(\varphi) = \int_{\mathcal{O}} [\int_{\mathfrak{g}} j(X)^{1/2} \varphi(X) \exp(i\langle f, X\rangle)\, dX]\, d\beta(f)$

Wolf's version of the Bott–Borel–Weil theorem is given as follows. Similarly to the definition of $\mathcal{O}_p(W(\mu, \sigma))$ we define the space $A_p^{0,q}(W(\mu, \sigma))$ of smooth partially holomorphic $(0, q)$-forms on M/L with values in $W(\mu, \sigma)$. As usual the Laplace operator $\square$ is defined on $L_2^{0,q}(W(\mu, \sigma))$, the Hilbert space completion of $A_p^{0,q}(W(\mu, \sigma))$. Let $H_2^{0,q}(W(\mu, \sigma)) = \{w \in L_2^{0,q} \mid \square w = 0\}$ and let $\pi_{\mu,\sigma}^q$ denote the natural action of G on $H_2^{0,q}(W(\mu, \sigma))$.

THEOREM (Wolf) 12.1.9.

(i) $\pi_{\mu,\sigma}^q$ is a unitary representation of G;

(ii) if $[\mu] = [\mathfrak{x} \otimes \mu^0]$ is as above and β is the highest weight of μ^0, then

(a) if $\langle \beta + \delta_t, \varphi \rangle = 0$ for some φ in Δ_t^+, $H_2^{0,q} = 0$ for all q;

(b) if $\langle \beta + \delta_t, \varphi \rangle \neq 0$ for all φ in Δ_t^+ let q be the number of φ in Δ_t^+ such that $\langle \beta + \delta_t, \varphi \rangle < 0$ and let ν be the unique element of L^+ that is conjugate to $\beta + \delta_t$ by an element of the Weyl group $W(M^0, T^0)$. Then $\pi_{\mu,\sigma}^{q_0} = [\pi_{\mathfrak{x},\nu,\sigma}]$ is a principal series class and $H_2^{0,q} = 0$ for all $q \neq q_0$.

In particular given a principal series class we can realize it on $H^{0,0}(W(\mu, \sigma))\mu = [\mathfrak{x} \otimes \mu^0]$ with μ^0 having the highest weight $\nu - \delta_t$.

12.2. Application to the Toda Lattice

We conclude this section with an application of the theory of principal series representations to the quantization of the Toda lattice. If $S(\mathfrak{g})$ denotes the symmetric algebra of $\mathfrak{g}$, then by a theorem of Chevalley the algebra of symmetric invariants $S(\mathfrak{g})^G$ is a polynomial algebra with homogeneous generators $I_1,\ldots,I_l$, called primitive elements: $S(\mathfrak{g})^G = R[I_1,\ldots,I_l]$. We may take I_1 to be the Killing form.

The subspace $S(\mathfrak{g})$ of $A(\mathfrak{g}^*)$ inherits a Poisson structure by regarding its elements as left invariant functions on T^*G. And it can be checked that if $\varphi \in S(\mathfrak{g})^G$ then φ Poisson commutes with $S(\mathfrak{g})$.

THEOREM 12.2.1. $S(\mathfrak{g})^G \simeq S(\mathfrak{b})_f$ where $S(\bar{\mathfrak{b}})_f \subseteq S(\bar{\mathfrak{b}})$ is defined by $S(\bar{\mathfrak{b}})_f = R[I_1^f,\ldots,I_l^f]$ where $I_j^f = I_j(f+Z)$ with $f = \sum_{i=1}^{l} e_{-\alpha_i}$, $Z \in \mathfrak{b}$. Furthermore $S(\mathfrak{b})_f$ is Poisson commutative.

Thus we have l commuting elements $I_1,\ldots,I_l$ in $S(\mathfrak{b})$. The program then is to construct l elements $\tilde{I}_1,\ldots,\tilde{I}_l$ in $\mathfrak{u}(\mathfrak{b})$ and a simultaneous spectral resolution of $U_{\mathcal{O}}(\tilde{I}_j)$, $j = 1,\ldots,l$ where $U_{\mathcal{O}}$ is a unique representation of $\bar{B}$ where $\mathcal{O} = \bar{B}.e$.

Let $X \in \bar{\mathfrak{n}}$. Then $X = \sum_{\alpha \geq 0} d_\alpha e_\alpha$ and the character of $\bar{N}$ corresponding to e is defined by $\mathfrak{x}(\exp X) = \exp(2\pi i \sum d_i c_i)$ where $c_i = B(e_{\alpha_i}, e_{-\alpha_i})$. Then $U_{\mathcal{O}} = \mathrm{Ind}(\mathfrak{x})$.

DEFINITION 12.2.2. Let $v' \in H_{-\infty}$, the dual of the Gårding space. Then v' is called a *Whittaker vector with respect to* $\mathfrak{x}$ if $U(n)v' = \mathfrak{x}(n)v'$ for all n in $\bar{N}$.

Kostant has shown that if U is any member of the principal series then there exists up to a scalar multiple a unique Whittaker vector v' in $H_{-\infty}$.

THEOREM 12.2.3. Let λ in $\mathfrak{h}_{\mathbb{C}}^*$ determine a principal series representation U_λ on $H(\lambda)$. Let $v' \in H(\lambda)_{-\infty}$ be the Whittaker vector and $v \in H(\lambda)_\infty$ be a spherical vector (i.e. $U_\lambda(k)v = v$ for all k in K). Then $d_\lambda = d_{v,v'} = \langle v, U(\varphi)v' \rangle$ is an eigenfunction of $U_{\mathcal{O}}(\tilde{I}_j)$:

$$U_{\mathcal{O}}(\tilde{I}_j)d_\lambda = c_j(\lambda)d_j.$$

Thus the functions d_λ give joint eigenfunction for the commuting operators $U_{\mathcal{O}}(\tilde{I}_1),\ldots,U_{\mathcal{O}}(\tilde{I}_j)$ and the $c_1(\lambda),\ldots,c_1(\lambda)$ are the joint spectral functions as λ varies through $\mathfrak{h}^*$.

PROBLEMS

EXERCISE 12.1. Let $G = SU(1,1)$ with Lie algebra basis

$$J_1 = \tfrac{1}{2}\begin{pmatrix} i & 0 \\ 0 & i \end{pmatrix},\quad J_2 = \tfrac{1}{2}\begin{pmatrix} 0 & 1 \\ 1 & 0 \end{pmatrix} \quad \text{and} \quad J_3 = \tfrac{1}{2}\begin{pmatrix} 0 & i \\ -i & 0 \end{pmatrix}.$$

Let P_k be the dual basis $\langle P_k, J_j \rangle = \delta_{kj}$. Take $f = \pm\lambda P_0$ in $\mathfrak{g}^*$. Then show

$$G(f) = \left\{ \begin{pmatrix} \exp(i\varphi) & 0 \\ 0 & \exp(-i\varphi) \end{pmatrix} \middle| 0 \leqq \varphi \leqq 2\pi \right\}; \ \mathfrak{g}(f) = RJ_0.$$

Writing g in G as

$$g = \begin{pmatrix} ab \\ \overline{ba} \end{pmatrix} = (1 - z\bar{z})^{-1/2} \begin{pmatrix} 1z \\ \bar{z}1 \end{pmatrix} \begin{pmatrix} \exp(i\varphi) & 0 \\ 0 & \exp(-i\varphi) \end{pmatrix} \Bigg) \Bigg)$$

where $\varphi = \arg(a)$ and $z = b/\bar{a}$, show that the Kirillov form on orbit $\mathcal{O}_f = G/G(f)$ is

$$\Omega = 2i\lambda \frac{\mathrm{d}z \wedge \mathrm{d}\bar{z}}{(1 - z\bar{z})^2}.$$

Show there are two polarizations $\mathfrak{h}_{\lambda P_0} = CJ_3 + C(J_1 - iJ_2) = \bar{\mathfrak{h}}_{-\lambda P_0}$. F is then generated by $\{\partial/\partial\bar{z}\}$ and then (M, Ω, F) is a Kähler manifold. Show that the holomorphically induced representation from $G(f)$ is D_k^+ if $k = 4\pi\lambda > 1$.

Show that the orbit $\mathcal{O}_f$ for $f = \lambda P_1$ for $\lambda > 0$ has

$$G(f) = \left\{ \pm \begin{pmatrix} \mathrm{cha} & \mathrm{sha} \\ \mathrm{sha} & \mathrm{cha} \end{pmatrix} \middle| a \text{ in } R \right\} \quad \text{and} \quad \mathfrak{g}(f) = RJ_1.$$

Show that $\mathcal{O}(f)$ has two G-invariant real polarizations $\mathfrak{h}_f = CJ_1(J_0 \pm J_2)$.

Show that the F-induced representations $H^0(M, \mathbf{E}(\lambda)_F)$ are $\pi(\varepsilon, \rho)$ (i.e. $C_q^h = \frac{1}{4} + \sigma^2, \sigma = \pm 2\pi$).

Show that the cone $\mathcal{O}(f)$ for $f = P^0 + P^2$ has

$$G(f) = \left\{ \begin{pmatrix} 1 + ia & -ia \\ ia & 1 - ia \end{pmatrix} \right\} \quad \text{and} \quad \mathfrak{g}(f) = R(J_0 - J_2).$$

$\mathcal{O}(f)$ has a real polarization $\mathfrak{k}_f = C(J_0 - J_2) + CJ$.

Show that the F-induced representation is $\pi(\varepsilon, \rho)$ and $D_{1/2}^+ + D_{1/2}^-$.

Chapter 13

Geometry of De Sitter Spaces

13.1. De Sitter spaces

The de Sitter group Spin (4, 1) is a simply connected semisimple ten dimensional Lie group. The de Sitter group has been studied in cosmology (see Robertson–Noonan) and in the dynamical symmetry group studies of the hydrogen atom or Kepler problem (see Souriau S27).

The elementary classical de Sitter system is the Hamiltonian G-space where $G = \text{Spin}\,(4, 1)$. We show in this section that all the orbits in $\mathfrak{g}^*$ are simply connected; thus any deSitter system, since it is a covering space of an orbit by Kostant–Souriau theorem, must be isomorphic to one of these orbits.

Let C be a Clifford algebra with antiautomorphism τ. If $C = \sum_{r=0}^{5} V_r$ and $C_0 = \sum_{r=0}^{2} V_{2r}$, then Spin (4, 1) is the group G of units s of C_0 which satisfy $\tau(s)s = 1$ and $sV_1s^{-1} \subset V_1$. G has a representation ρ on V_1 given by $\rho(s)v = svs^{-1}$ for s in G and v in V_1. Then $\rho(G) = SO_0(4, 1)$ and ρ is the spin covering.

The Lie algebra $\mathfrak{g}$ of G may be identified with V_2 and the exponential map exp: $\mathfrak{g} \to G$ is just the ordinary exponential series in the algebra C. The adjoint representation of G on $\mathfrak{g}$ is of course given by $\text{Ad}(s)(X) = sXs^{-1}$ for s in G and X in $\mathfrak{g}$.

If p is the vector space projection from C to V_0 with respect to the decomposition $C = V_0 + \sum_{r=0}^{4} V_r$, then a symmetric nondegenerate bilinear form T on C is defined by $T(s, t)1 = P(st)$ for s, t in C. T allows us to identify the dual $\mathfrak{g}^*$ of $\mathfrak{g}$ with V_2 by making an element a in V_2 determine a linear form $\hat{a}$ on $\mathfrak{g}$ by $\langle \hat{a}, b \rangle = T(a, b)$ for all b in $\mathfrak{g} = V_2$. Checking that T is invariant under inner automorphisms of C we see that the coadjoint action of G on $\mathfrak{g}^*$ coincides with the adjoint action of G on $\mathfrak{g}$.

Two invariant polynomials are defined on $\mathfrak{g}$. Viz. $c(a) = T(a, a)$ and $w(a) = T(w_a, w_a)$ where $w_a = (a^2 - c(a).1)\gamma$ with $\gamma = e_0e_1e_2e_3e_4$. Here

$$e_0 = \begin{pmatrix} -1 & 0 \\ 0 & 1 \end{pmatrix}, \quad e_1 = \begin{pmatrix} 0 & i \\ i & 0 \end{pmatrix}, \quad e_2 = \begin{pmatrix} 0 & j \\ j & 0 \end{pmatrix},$$

$$e_3 = \begin{pmatrix} 0 & k \\ k & 0 \end{pmatrix} \quad \text{and} \quad e_4 = \begin{pmatrix} 0 & -1 \\ 1 & 1 \end{pmatrix}.$$

It is easily checked that $w_{\mathrm{Ad}s(a)} = \rho(s)w_a$ for a in $\mathfrak{g}$ and s in G. Thus the orbit of a in $\mathfrak{g}$ is completely determined by the $SO_0(4,1)$-orbit of w_a and the value of $c(a)$. This leads to

THEOREM (Rawnsley) 13.1.1. The following is a complete list of orbits of G on:

(i) $\{a \text{ in } \mathfrak{g} \mid c(a) = \lambda, \quad w(a) = k\} \quad k < 0;$

(ii) $\{a \text{ in } \mathfrak{g} \mid c(a) = \lambda, \quad w(a) = k, \quad T(w_a, e_0) \gtrless 0\}; \quad k > 0$

(iii) $\{a \text{ in } \mathfrak{g} \mid c(a) = \lambda, \quad w(a) = 0, \quad T(w_a, e_0) \gtrless 0\};$

(iv) $\{a \text{ in } \mathfrak{g} \mid c(a) = \lambda, \quad w_a = 0 \quad a \neq 0\};$

(v) $\{0\}.$

COROLLARY 13.1.2. Since $\mathfrak{g}$ is semisimple every G-symplectic homogeneous space is isomorphic to one of these orbits.

To proceed with the characterization of the orbits we need a few subgroups of G. First we recall the notation that if $x = x_0 + x_1 i + x_2 j + x_3 k \in Q$ then $\hat{x} = -k\bar{x}k$. Thus the image of Spin (4, 1) under $\tilde{\rho}: C \to M_2(Q)$ is $G = \{A \in M_2(Q) \mid A^* A = I\}$ where if

$$A = \begin{bmatrix} \alpha & \beta \\ \gamma & \delta \end{bmatrix} \quad \text{then} \quad A^* = \begin{bmatrix} \hat{\delta} & -\hat{\beta} \\ -\hat{\gamma} & \hat{\alpha} \end{bmatrix}.$$

The subgroups of G are

$$K = \left\{ \begin{bmatrix} a & b \\ -\bar{b} & \bar{a} \end{bmatrix} \in G \,\middle|\, |a|^2 + |b|^2 = 1, \quad a\hat{b} = \hat{b}a \right\}$$

$$A = \left\{ \begin{bmatrix} \lambda & 0 \\ 0 & \lambda^{-1} \end{bmatrix} \text{ in } G \,\middle|\, \lambda = \bar{\lambda}, \lambda > 0 \right\}$$

$$M = \left\{ \begin{bmatrix} \hat{u}^{-1} & 0 \\ 0 & u \end{bmatrix} \text{ in } G \,\middle|\, u\bar{u} = 1 \right\}$$

$$M_3 = \{g \text{ in } M \mid gk = kg\}$$

$$L = \{g \text{ in } G \mid |a|^2 - |b|^2 = 1, \quad a\hat{b} = b\hat{a}\}$$

$$N^+ = \left\{ \begin{bmatrix} 1 & z \\ 0 & 1 \end{bmatrix} \text{ in } G \,\middle|\, z = \hat{z} \right\}.$$

Then $G = KAN^+$ is the Iwasawa decomposition where K is the maximal-compact subgroup of G; K is isomorphic to Spin (4). N^+ is a nilpotent subgroup and A is an Abelian subgroup isomorphic of R and is interpreted as the group of time translations. L is the Lorentz subgroup of G, mapped by $\tilde{\rho}$ onto $SO_0(3,1)$ in $SO_0(4,1)$.

The homogeneous space G/L is sometimes called the de Sitter space. M is the subgroup of G isomorphic to Spin(3); M_3 is the group of rotations about the 3-axis; $E^{(3)} = N^+M$ is isomorphic to a simply connected cover of the Euclidean group. $P = MAN^+$ is the minimal parabolic subgroup of G.

THEOREM 13.1.3. $G/P = S^3$.

THEOREM 13.1.4. G/H can be identified with $T(S^3)$ or $T^*(S^3)$.

Proof. Let

$$g = \begin{pmatrix} \alpha & \beta \\ \gamma & \delta \end{pmatrix} = \begin{pmatrix} a & b \\ -\bar{b} & \bar{a} \end{pmatrix} \begin{pmatrix} \lambda & 0 \\ 0 & \lambda^{-1} \end{pmatrix} \begin{pmatrix} 1 & z \\ 0 & 1 \end{pmatrix} \in \mathrm{KAN}^+,$$

then $u(g) = (a + bk)(a - bk)$ is a unit quaternion such that $u(gh) = u(g)$ for all h in H. Ad $z(g) = \lambda^2(a + bk)z(a + bk)\hat{}$ is a quaternion which satisfies $z(gh) = z(g)$ and $\hat{z}(g)$. One realization of S^3 is the set of unit quaternions; and R^3 are just the quaternions which satisfy $\hat{z} = z$. Thus we have a smooth map $\Phi: G/H \to S^3 \times R^3$ which can be seen to be a diffeomorphism.

COROLLARY 13.1.5. G/H_3 is diffeomorphic to $S^3 \times R^3 \times S^2$.

We consider briefly the orbits. For orbits of type (i) with $a = \sigma e_1 e_2 + \mu e_0 e_4$ and $\lambda = \mu^2 - \sigma^2$, $k = -4\mu^2\sigma^2$, each orbit is obtained once for $\sigma > 0$ and $\mu > 0$. The isotropy subgroup $G(a)$ is $M_3 A$. The map $2\pi i\hat{a}$: $\mathfrak{g}(a) \to iR$ is given by $2\pi i\hat{a}\ (\alpha e_1 e_2 + \beta e_0 e_4) = 2\pi i(-\sigma\alpha + \mu\beta)$. It will integrate to a character $\mathfrak{x}_a$ of $G(a)$ iff $2\pi\sigma = n$, $n \in Z$ and β in R. There are four polarizations $\mathfrak{h}_{\varepsilon,\delta}$ given by $\{e_1 e_2, (e_1 - ie_2\varepsilon)e_3, (e_0 + \delta e_4)e_1 (e_0 + \delta e_4)e_1, (e_0 + \delta e_4)e_3\}$ for ε, $\delta = \pm 1$. The two with $\varepsilon = 1$ are positive. For $\mathfrak{h} = \mathfrak{h}_{1.1}$, $\mathscr{D} = \mathfrak{h} \cap \mathfrak{g}$ and $\mathscr{E} = (\mathfrak{h} + \mathfrak{h}) \cap \mathfrak{g}$ are just the Lie algebras of $D = M_3 AN^+ = G(a)N^+$ and $E = MAN^+ = P$. Thus $E/D = M/M_3 = S^2$.

Forming the induced representation from $\mathfrak{x}_a$ on $G(a)$ (extending it trivially on N^+ to D) we get a principal series representation, $U^{n,3/2 - 2\pi i\mu}$ of G. Finally one can check that $G/G(a) = S^3 \times S^2 \times R^3$.

Orbits (iia) and (iib) have no positive polarizations.

Orbits (iiia) and (iiib) each have one positive polarization. As manifolds, these orbits are $S^3 \times R^3 \times S^2$ with a different action of G from case (i). The representations induced to G from the character χ_a are members of the principal P-series representations not among those gained from orbit (i). (iiia) and (iiib) yield the *same* induced representations.

In case (iv) we must consider separately (a) $\lambda > 0$, (b) $\lambda = 0$, and (c) $\lambda < 0$. In case (a) there is a positive polarization yielding new principal P-series. The orbit is diffeomorphic to $S^3 \times R^3$. Case (b) has one positive polarization, gives the orbit $S^3 \times S^2 \times R^3$ and yields new principal P-series. Case (c) has no positive polarization.

We leave it to the reader to check these facts and to write down the Kirillov 2-form $\Omega(\mu, \sigma)$ for the orbits and show that its class is determined by the class of the 2-form on S^2 given by $(-2i\sigma\,(1 + z\bar{z})^{-2}\,dz \wedge d\bar{z})$. So $\Omega(\mu, \sigma)$ is integral iff $4\pi\sigma = 2n \in Z$.

This example illustrates the subtleties of geometric quantization. Here we have constructed the entire principal series of G (a fact to be checked by the reader), but occasionally different orbits gave the same representation; and orbits gave a member of the principal series only if they had a positive polarization. Finally we did not obtain the discrete series.

Chapter 14

Discrete Series Representations

14.1. Representations of noncompact semisimple Lie groups

Part II. Discrete Series

The Borel–Weil theory and the work of Bargmann, Harish–Chandra, Selberg, Bruhat and others led to a series of conjectures by Langlands on how to construct the discrete series representations of noncompact semisimple Lie groups.

Definition 14.1.1. Let $\pi \in \hat{G}$. π is *square integrable* if there is a nonzero vector ψ in $E(\pi)$ such that $g \to (\pi(g)\psi, \psi)$ is square integrable on G. If π has a square integrable matrix coefficients π is said to be a *discrete series representation.* Let $\hat{G}_d$ denote the classes of discrete series representations.

We review briefly the properties of square integrable representations in the following theorem:

Theorem 14.1.2. If π is in $\hat{G}_d$, then for every $\psi_1, \psi_2 \subset E(\pi)$, $f^{\pi}_{\psi_1\psi_2}(g) = (\pi(g)\psi_1, \psi_2)$ belongs to $L^2(G)$ and there is a constant (called the formal dimension of π) such that $\int f^{\pi}_{\psi_1,\psi_2}(g) f^{\pi}_{\psi'_1,\psi'_2}(g)\,\mathrm{d}g = d_\pi^{-1}(\psi_1, \psi'_1)(\psi_2, \psi'_2)$. And if $\pi_1, \pi_2 \in \hat{G}_d$ are inequivalent, then $\int_G f^{\pi_1}_{\psi_1\psi_2}(g)\, f^{\pi_2}_{\psi_1\psi_2}(g) = 0$.

Let G be a noncompact real semisimple Lie group which is the connected real form of a simply connected complex semisimple Lie group G^C. Let K be a maximal compact subgroup of G and let H be a compact Cartan subgroup of G; $H \subset K \subset G$. That H exists is due to Harish Chandra. Let $\mathfrak{g}^{\mathbb{C}}, \mathfrak{h}^{\mathbb{C}}$ be the complexifications of the Lie algebras of $\mathfrak{g}, \mathfrak{h}$ of G, H. Let Σ (resp. Δ^+) be a root system (resp. positive root system) for $(\mathfrak{g}^C, \mathfrak{h}^C)$. As before, since H is compact Δ lies in the vector space over R of all purely imaginary complex valued linear forms on $\mathfrak{h}$.

Let $\mathfrak{g} = \mathfrak{k} \oplus \mathfrak{p}$ denote the Cartan decomposition where $\mathfrak{p} = \{X \mid B(X, Y) = 0$ for all Y in $\mathfrak{k}\}$, where B is the Killing form on $\mathfrak{g}$. For a root α we let $\mathfrak{g}_\alpha$

denote the 1-dimensional eigenspace of α in $\mathfrak{g}^{\mathbb{C}}$. We set

$$\Delta_{\mathfrak{k}}^{+} = \{\alpha\in\Delta^{+} \mid \mathfrak{g}_{\alpha}\in\mathfrak{k}^{\mathbb{C}}\}$$

$$\Delta_{\mathfrak{p}}^{+} = \{\alpha\in\Delta^{+} \mid \mathfrak{g}_{\alpha}\in\mathfrak{p}^{\mathbb{C}}\}$$

to be the set of *positive compact*, resp *positive noncompact* roots. The half sums are denoted

$$\delta = \tfrac{1}{2}\sum_{\alpha\in\Delta^{+}} \alpha, \qquad \delta_{\mathfrak{k}} = \tfrac{1}{2}\sum_{\alpha\in\Delta_{\mathfrak{k}}^{+}} \alpha \quad \text{and} \quad \delta_{n} = \tfrac{1}{2}\sum_{\alpha\in\Delta_{\mathfrak{p}}^{+}} \alpha.$$

If $X_1,\ldots,X_m$, and $Y_1,\ldots,Y_n$ are orthonormal bases for $\mathfrak{k}$ and $\mathfrak{p}$ respectively, then the Casimir operator is $\mathfrak{c} = -\Sigma X_i^2 + \Sigma Y_j^2$.

As usual we define an inner product (,) on Hom $(i\mathfrak{h}, R)$. Then we set

$$\mathscr{F} = \left\{\mu\in\mathrm{Hom}(i\mathfrak{h}, R) \,\middle|\, 2\frac{(\mu,\alpha)}{(\alpha,\alpha)}\in Z, \quad \alpha\in\Sigma\right\}$$

which is isomorphic to $\hat{H}$ by $\mu\in\mathscr{F}\to e^{\mu}$. Let $\mathscr{F}' = \{\lambda\in\mathscr{F} \mid (\lambda,\alpha)\neq 0$ for all roots α in $\Sigma\}$. We set

$$\mathscr{F}_0 = \left\{\mu\in\mathrm{Hom}(i\mathfrak{h}, R) \,\middle|\, 2\frac{(\mu,\alpha)}{(\alpha,\alpha)}\in Z, \text{ is} \geqq 0 \text{ for all } \alpha\in\Delta_{\mathfrak{k}}^{+}\right\}.$$

Then associated to each μ in $\mathscr{F}_0$ is the irreducible K-module V_μ with highest weight μ.

We let $W(H) = N(H)/H$ denote the Weyl group of G and we let H' denote the *regular elements* of H; i.e. $H' = H - \exp(\mathfrak{h}_s)$ where $\mathfrak{h}_s = \{Y\in\mathfrak{h} \mid \alpha(Y) = 0$ for some $\alpha\in\Sigma\}$.

THEOREM (Harish Chandra) 14.1.3.

(i) G has a discrete series iff rank $(G) = \mathrm{rank}(K)$.

(ii) (Generalized Highest Weight Theorem) there is a natural surjective map $\omega: \mathscr{F}' \to \hat{G}_d$ and $\omega(\lambda') = \omega(\lambda)$ iff $w\lambda = \lambda'$ for w in $W(H)$.

(iii) (Generalized Weyl Character Formula) for λ in $\mathscr{F}'$ there is a unique tempered distribution Θ_λ such that

$$\Delta(\exp Y)\Theta_\lambda(\exp Y) = \sum_{w\in W(H)} \varepsilon(w)\exp(w\lambda(Y)),$$

where $\exp Y\in H'$ and $\Delta(\exp Y) = \prod_{\alpha\in\Delta^{+}}(\exp(\alpha(Y)/2 - \exp(-\alpha(Y)/2))$.

(iv) For λ in $\mathscr{F}'$ there is a unique discrete series class $\omega(\lambda)$ in $\hat{G}_d$ such that the character $\Theta_{\omega(\lambda)}$ of $\omega(\lambda)$ coincides with $(-1)^n\varepsilon(\lambda)\Theta_\lambda$ where $n = \frac{1}{2}\dim G/K$ and $\varepsilon(\lambda) = \mathrm{sign}\prod_{\alpha\in\Delta^{+}}(\lambda,\alpha)$.

The case $G = SO_0(n,1)$ has been studied extensively in this context. Noting that in this case rank $(G) =$ rank $(SO(n-1)) + 1 = [(n+1)/2] + 1$ we see that a discrete series occurs only when n is even.

EXAMPLE 14.1.4. The case $SL(2,R)$ is the classic example. Here

$M = \begin{pmatrix} \pm 1 & 0 \\ 0 & \pm 1 \end{pmatrix}$ and rank $(G) = \dim(A) =$ rank $(K) = 1$ and $B = K$.

The Lie algebra of B is

$$\mathfrak{b} = \mathfrak{k} = \left\{ \begin{pmatrix} 0 & b \\ -b & 0 \end{pmatrix} \middle| b \text{ in } R \right\}.$$

The discrete series is given on the Hilbert space $\mathscr{H}_n^{\pm}$ holomorphic, resp. conjugate holomorphic, functions on $\mathscr{P} = SL(2,R)/SO(2)$, the Poincaré upper half plane, with

$$(f_1, f_2) = \frac{1}{\Gamma(2n-1)} \int_{\mathscr{P}} f_1(x+iy) \bar{f}_2(x+iy) y^{-2+2n} \,\mathrm{d}x\,\mathrm{d}y$$

and the representation is given by

$$\pi_n^+(g) f(z) = (bz+d)^{-2n} f\left(\frac{az+c}{bz+d}\right), \quad f \in \mathscr{H}_n^+$$

and

$$\pi_n^-(g) f(z) = (b\bar{z}+d)^{-2n} f\left(\frac{az+c}{bz+d}\right), \quad f \in \mathscr{H}_n^-.$$

These are denoted as $D_n^{\pm}$ by Bargmann. G has two conjugacy classes of Cartan subgroups:

$$H = \left\{ \begin{pmatrix} h & 0 \\ 0 & h^{-1} \end{pmatrix} \middle| h \text{ in } R^* \right\}$$

and

$$B = \left\{ \begin{pmatrix} \cos\varphi & \sin\varphi \\ -\sin\varphi & \cos\varphi \end{pmatrix} \varphi \text{ in } R \right\}.$$

The minimal parabolic subgroup is

$$P = \left\{ \begin{pmatrix} a & b \\ 0 & a^{-1} \end{pmatrix} \middle| a \text{ in } R^* \text{ and } b \text{ in } R \right\}.$$

The characters of $\pi_n^{\pm}$, i.e. $\Theta^{\pm}|_B$ are given by

$$\Theta_n^{\pm}(b) = \frac{\exp(\pm i\varphi(-1+2n))}{\pm(\exp(i\varphi)-\exp(-i\varphi))}$$

for regular b in B.

The Plancherel formula is well known in this case

$$\int \|f\|^2 \, dg = \int_0^\infty \|\pi(1,\rho)(f)\|_2^2 \rho \tanh \pi\rho \, d\rho +$$

$$+ \int_0^\infty \|\pi(-1,\rho)(f)\|_2^2 \rho \coth \pi\rho \, d\rho +$$

$$+ \sum_{\substack{n \geq 1 \\ n \in \frac{1}{2}Z}} (n - \tfrac{1}{2})(\|\pi_n^+)f)\|_2^2 + \|\pi_n^-(f)\|_2^2$$

for f in $L^1(G) \cap L^2(G)$. This clearly shows that necessity of studying both the discrete series and the principal series.

Rather than giving a precise analogue of the Borel–Weil theorem we present an alternative version using the Casimir operator. Let $\mathfrak{u}(\mathfrak{g})$ denote the universal enveloping algebra of $\mathfrak{g}$. Every X in $\mathfrak{u}$ defines a left invariant differential operator $v(X)$ on $A(G)$. This is an isomorphism. It is also possible to identify $\mathfrak{u}(\mathfrak{g})$, as a linear space, to the space of polynomial functions on $\mathfrak{g}^*$. Let $\mathfrak{z}(\mathfrak{g}) = \text{Center}(\mathfrak{u}(\mathfrak{g}))$. Then under the identification just mentioned p in $\mathfrak{u}(\mathfrak{g})$ is in $\mathfrak{z}(\mathfrak{g})$ iff p is invariant under the coadjoint representation.

Extending T in $\hat{G}$ to $\mathfrak{u}$ then for z in $\mathfrak{z}(\mathfrak{g})$, $T(z) = \mathfrak{x}_T(z)$. $\mathfrak{x}_T : \mathfrak{z}(\mathfrak{g}) \to C$ is called the *infinitesimal character* of T.

If $\mathfrak{c}$ denotes the Casimir operator of G, then $\mathfrak{c}$ belongs to $\mathfrak{z}$ and so $T(\mathfrak{c})f = \mathfrak{x}_T(\mathfrak{c})f$.

If the trace of T in $\hat{G}$ is defined, then for f in $A_0(G)$ and z in $\mathfrak{z}(\mathfrak{g})$ we have $z\Theta_T(f) = \Theta_T(z * f) = \mathrm{Tr}(T(z * f) = \mathrm{Tr}\, T(z) * T(f) = \overline{\mathfrak{x}_T(z)}\Theta_T(f)$. In other words

THEOREM 14.1.5. The characters of irreducible representations are eigen distributions of $\mathfrak{z}(\mathfrak{g})$ with infinitesimal characters as eigenvalues.

Thus the Plancherel formula $\delta = \int_{\hat{G}} \Theta_\pi \, d\mu(\pi)$ can be interpreted as an eigenvalue expansion of δ. For more details see Maurin *M*21.

EXAMPLE (de Sitter Group) 14.1.6. Let $\rho^{n,0} \in \hat{K}$, where K is the maximal compact subgroup of the universal covering group G of the de Sitter group. Let $V(\rho)$ denote the K-module given by ρ. Let p be a demi-integer ≥ 1. Set

$H^{\rho,p}$ to be the Hilbert space of functions $f: G/K \to V(\rho)$ which are square integrable with respect to

$$(f_1, f_2) = \int_{G/K} (f_1(q), f_2(q))_{V(\rho)}(1 - |q|^2)^{2p-2}\, \mathrm{d}\mu(q), \quad q \in G/K.$$

Let G act on f by

$$T^{n,p}(g)\, f(q) = |cq + d|^{-2p-2}\, \rho(k(g^{-1}, q^{-1})\, f(aq + d) \times \\ \times (cq + d)^{-1})$$

where

$$g^{-1} = \begin{pmatrix} ab \\ cd \end{pmatrix},$$

f belongs to $H^{\rho,p}$, and

$$k(g, q) = \begin{pmatrix} (a + b\bar{q})/|cq + d| & 0 \\ 0 & (cq + d)/|cq + d| \end{pmatrix}.$$

Here n, p are demi integers $n \geqq p \geqq 1$ with $n - p$ integer.

The subspace of $H^{\rho,p}$ of solutions of the equation

$$[\tfrac{1}{4}(1 - |q|^2)\Delta - pD - \tfrac{1}{2}(D_1^{\nu} A_n + D_2^{\nu} B + D_3^{\nu} C_n) + \\ + (n(n + 1) - p(p + 1)]f = 0,$$

where Δ is the Laplacian and D_i^{ν} are first order differential operators (v. Takahashi T2 for explicit form and for coefficiences A_n, B_n, C_n), or equivalently of the equation

$$\nu(\mathfrak{c})f = [-n(n + 1) - (p + 1)(p - 2)]\, f \qquad (*)$$

is denoted $S^{\rho,p}$.

THEOREM (Takahashi) 14.1.7. The Casimir operator $\nu(\mathfrak{c})$ is elliptic and the subspace $S^{\rho,p}$ of smooth functions in $H^{\rho,p}$ which are solutions of the equation $(*)$ forms an irreducible G-module in $\hat{G}_d$.

The formal degree is given by $\mathrm{Tr}(T^{n,p}(f) * T^{n,p}(f)) = (2n + 1)(2p - 1) \times (n + p)(n - p + 1)/16\pi^2$. The Plancherel formula for f in $L^2(K/G/K)$ is given by

$$\int |f(g)|^2\, \mathrm{d}g = \frac{1}{8\pi^2} \sum (2n + 1) \int_0^{\infty} \mathrm{Tr}(U^{n,\frac{3}{2}+i\nu}(f) * U^{n,\frac{3}{2}+i\nu}(f))$$

$$\times \left[(n + \tfrac{1}{2})^2 + \nu^2(\nu \tanh(\nu + in)\mathrm{d}\nu + \frac{1}{16\pi^2} \sum_{n \geq 1} (2n + 1) \times \right.$$

$$\times \sum_{\substack{1 \leq p \leq n \\ n - p \text{ integer}}} (2p - 1)\,(n + p)(n - p + 1)\, \mathrm{Tr}[T^{n,p}(f) * T^{n,p}(f)].$$

For λ in $\mathscr{F}_0$ we associated a class $[\lambda] \in \hat{K}$. Viz. set $\varepsilon_k(\lambda) = \text{sign} \prod_{\alpha \in \Delta^+ \mathfrak{k}} (\lambda, \alpha)$ if the products is nonzero and 0 otherwise. If $\varepsilon_{\mathfrak{k}}(\lambda + \delta_{\mathfrak{k}}) \neq 0$ there is a unique element w in $W(H)$ such that $(w(\lambda + \delta_{\mathfrak{k}}), \alpha) > 0$ for all α in $\Delta_{\mathfrak{k}}^+$. Let $[\lambda]$ denote the equivalence class containing the irreducible K-module with highest weight $w(\lambda + \delta_{\mathfrak{k}}) - \delta_{\mathfrak{k}}$. Of course $[\lambda] \in \hat{K}$ can be realized by the Borel–Weil Theorem for (K, H).

Let V_λ be a finite dimensional unitary K-module. Let $V(\lambda) = G \times_\lambda V$ be the homogeneous vector bundle over G/K. Then the space of square integrable section of $V(\lambda)$ is isomorphic to $L^2(V(\lambda)) = \{f: G \to V | f \text{ in } L^2(G) \text{ and } f(gk) = k^{-1} f(g)\}$. The left regular representation of G on $L^2(G)$ induces a unitary representation of G on $L^2(V(\lambda))$. Let $S(V(\lambda))$ be the space of all V_λ-valued smooth functions f on G such that $f(gk) = k^{-1} f(g)$ for g in G and k in K. Let $\mathscr{H}(\lambda) = \{f \in S(V(\lambda)) \cap L^2(V(\lambda)) | v(\Omega) f = (\lambda + 2\delta_{\mathfrak{k}}, \lambda) f\}$.

The following is a generalization of earlier results due to Hotta, Schmid, and others.

THEOREM (Hotta–Parthasarathy) 14.1.8. Let $\Lambda \in \mathscr{F}'$ and choose a positive root system of $(\mathfrak{g}^C, \mathfrak{h}^C)$ such that $\Delta^+ = \{\alpha | (\Lambda, \alpha) > 0\}$. Let $\lambda = \Lambda - \delta$. Assume that (i) $(\lambda, \alpha) 0$ for every $\alpha \in_{\mathfrak{p}}^+$; and (ii) $(\lambda, \alpha) \geqq a = \max_{Q \subset \Delta_{\mathfrak{p}}^+} (\delta_n - \langle Q \rangle, \alpha)$ for every α in $\Delta_{\mathfrak{k}}^+$; here $\langle Q \rangle = \frac{1}{2} \sum_{\alpha \in Q} \alpha$. Then the discrete class $\omega(\Lambda) = \omega(\lambda + \delta) \in \hat{G}_d$ is realized by the left regular representation on the Hilbert space $\mathscr{H}(\lambda)$ of $V_{\lambda + 2\delta_n}$-valued square integrable functions f on G such $f(gk) = k^{-1} f(g)$ for g in G and k in K, which satisfy $v(\mathfrak{c}) f = (|\Lambda|^2 - |\delta|^2) f$ where $V_{\lambda + 2\delta_n}$ is the irreducible K-module with highest weight $\lambda + 2\delta_n$. The character of this representation is $\Theta_{\omega(\lambda + \delta)} = \Theta_{\omega(\Lambda)}$.

When G/K admits an invariant complex structure and all roots of $\Delta_{\mathfrak{p}}^+$ are totally positive, then Theorem 14.1.8 reduces to Narasimhan–Okamota's result N1. In this case the holomorphic cotangent space eK is identified with $\mathfrak{p}_+ = \sum_{\beta \in \Delta_{\mathfrak{p}}^+} C \mathfrak{g}_\beta$ and $V(\lambda)$ is a holomorphic bundle over G/K. The Laplace–Beltrami operator for the Dolbeault complex associated to $V(\lambda)$ is just

$$\square = -\tfrac{1}{2}(v(\mathfrak{c}) - (\lambda + 2\delta, \lambda)1).$$

Here we can consider $C^{0,q}(V(\lambda))$ resp. $L_2^{0,q}(V(\lambda))$ the space of smooth resp. square integrable differential forms of type $(0, q)$ with coefficients in $V(\lambda)$. As before we consider the space of harmonic forms $\mathscr{H}_2^{0,q}(V(\lambda))$. And $\mathscr{H}(\lambda)$ is equivalent to the square integrable cohomology space. In this case a may be chosen to be zero.

When G/K admits no invariant complex structure the Theorem 14.1.8 is just an extension of Takahashi's Theorem 14.1.7 on the covering group Spin(2*m*, 1) of the deSitter group $SO(2m, 1)$. In this case a may again be chosen to be zero and the Theorem 14.1.8 provides a realization of all discrete classes. In fact using Schmid's thesis it can be shown that the theorem holds without assumptions (i) or (ii) for all discrete classes for Spin (2*m*, 1).

PROBLEMS

EXERCISE 14.1. Consider the group $SO(n, 2)$ of linear transformations on R^{n+2} which leave invariant

$$(g_{ij}) = \begin{pmatrix} 1 & & & & 0 \\ & 1 & & & \\ & & -1 & & \\ & & & \ddots & \\ 0 & & & & -1 \end{pmatrix}.$$

Let $so(n, 2)$ denote its Lie algebra with basis $X_{ij} = -X_{ji}$, $[X_{ij}, X_{hk}] = g_{ih}X_{jk} + g_{jk}X_{ih} - g_{jh}X_{ih} - g_{jh}X_{ik} - g_{ik}X_{jh}$. Set $S = X_{12}$, $M_{\mu\nu} = X_{\mu+2,\nu+2}$, $Z_\mu = X_{1,\mu+2}$, $W_\mu = X_{2,\mu+2}$, $\mu, \nu = 1, \ldots, n$. $B(X, Y) = (1/2n)$ Tr(ad X ad Y) is nonsingular and we define the dual basis X^b by $\langle X^b, Y\rangle = B(X, Y)$ for Y in $so(n, 2)$. Show that $B(X_{ij}, Y_{hk}) = g_{ik}g_{jh} - g_{ih}g_{jk}$. A generic point of $so(n, 2)^*$ is then $\omega = sS^b + \sum m_{\mu\nu}M^b_{\mu\nu} + \sum(z_\mu Z^b_\mu + w_\mu W^b_\mu)$, which we denote by $(s, m_{\mu\nu}, z_\mu, w_\mu)$. Let $\mathcal{O}_{\omega_l} = \{\omega | s^2 + \sum m^2_{\mu\nu} - \sum(z^2_\mu + w^2_\mu) = l^2, sm_{\mu\nu} = z_\mu w_\nu - z_\nu w_\mu, l > 0\}$. Let $\zeta \in C^n$ and let $\sigma = (\sigma_1, \ldots, \sigma_n)$ be defined by $\sigma_\mu = z_\mu - iw_\mu$. Show that the map μ_l: $\mathcal{O}_{\omega_l} \to C^n$ given by

$$\zeta = \frac{i}{2s}\left(\sigma + \frac{\sigma\sigma^T}{2s(s+l) - \sigma\bar{\sigma}^T}\sigma\right)$$

is smooth nonsingular which identifies $\mathcal{O}_{\omega_1}$ and the homogeneous bounded domain of type IV

$$\mathcal{D} = \{\zeta \in C^n | \zeta\bar{\zeta}^T < 1, 1 - 2\zeta\bar{\zeta}^T + |\zeta\zeta|^2 > 0\}.$$

Define the representation of $SO(n, 2)$ on holomorphic functions on $\mathcal{D}$ by $(U_g\Psi)(\zeta) = \mu(g, \zeta)\Psi(g^{-1}\zeta)$ which is the holomorphically induced representation from $\exp(i(l + (n/2) - 1)S^b)$. Evaluate $\mu(g, \zeta)$. The Hilbert space H_l with inner product

$$\langle \Psi_1 | \Psi_2 \rangle = N \int_{\mathcal{D}} \overline{\Psi_1(\zeta)}\, \Psi_2(\zeta)(1 - 2\zeta\bar{\zeta}^T + |\zeta\zeta^T|^2)^{l-n/2-1} \prod_\mu \mathrm{d}\zeta_\mu \wedge \overline{\mathrm{d}\zeta_\mu}/2i,$$

where N is a normalization constant. Evaluate N. Show that H_l is nontrivial iff $l > \frac{1}{2}n$; and the representation then is irreducible and unitary.

Define the coherent states $|\zeta\rangle$ in H_l by

$$\langle\zeta|\Psi\rangle = N \int_{\mathscr{D}} \overline{\zeta(\xi)}\Psi(\xi)a(\xi)\mathrm{d}\xi$$

with

$$\zeta(\xi) = \langle\xi|\zeta\rangle = (1 - 2\xi\bar{\zeta}^T + \xi\xi^T\,\zeta\zeta^T)^{-l-n/2+1},$$

$$a(\xi) = (1 - 2\xi\bar{\xi}^T + |\xi\xi^T|^2)^{l-n/2+1}$$

and

$$\mathrm{d}\xi = \prod_{\mu} \mathrm{d}\xi_\mu \wedge \mathrm{d}\bar{\xi}_\mu/2i;$$

show $U(g)|\zeta\rangle = \overline{\mu(g^{-1}\zeta)}\,|g\zeta\rangle$. Differentiate this representation U_g to obtain the representation $X \to \hat{X}$ of the Lie algebra $so(n,2)$. In particular check that $\hat{S} = (l + (n/2) - 1) + \sum \zeta_\mu\partial_\mu, \hat{M}_{\mu\nu} = -i(\zeta_\mu\partial_\nu - \zeta_\nu\partial_\mu)$ where $\partial_\mu = (\partial/\partial\zeta_\mu)$. Evaluate $\langle\zeta|\hat{X}|\zeta\rangle/\langle\zeta|\zeta\rangle$.

Let p be the half sum of positive roots and p_c be the half sum of positive compact roots. For $n = 2r - 1$ show that $so(n,2) \simeq B_r$, $i\omega_l = le_1 = ilS^b$ Show $p - p_c = -(r - \frac{1}{2})e_1$, and $i\omega_l - p + p_c = (l + r - \frac{1}{2})e_1$. Show that the relation $\omega_l \to \lambda = i\omega_l$ where e^λ is a representation of the stability subgroup of ω_l, viz. $K = SO(2) \otimes SO(n)$, is not a W_K-invariant quantization rule whereas $\omega_l \to \lambda = i\omega_l - p + w^{-1}(p_c) + \lambda_0$ is a W_K-invariant rule if $W_K\lambda_0 = \lambda_0$ and if $w(i\omega_l + p_c)$ lies in the highest Weyl chamber.

Let

$$\mathscr{O}_{\omega_0} = \Big\{\omega | s^2 + \sum_{\mu<\nu} m_{\mu\nu} - \sum(z_\mu^2 + w_\mu^2) = 0,$$

$$sm_{\mu\nu} = z_\mu w_\nu - z_\nu w_\mu, \sum z_\mu^2 = \sum 2_\mu^2, \sum z_\mu w_\mu = 0\Big\}.$$

Show $\mathscr{O}_{\omega_0} \simeq R \times SO(n)/SO(n-2)$.

Relate $\omega = (s, m_{\mu\nu}, z_\mu, w_\mu)$ to the Kepler problem $H = (p^2/2m)(-k/r)$ on $T^*(R^{n-1}\backslash 0)$ with coordinates $(\mathbf{x}, \mathbf{p})$ by setting

$$p^2 = \mathbf{p}\cdot\mathbf{p}, r = (\mathbf{x}\cdot\mathbf{x})^{1/2}, \mathbf{z} = (z_1, \ldots, z_{n-1}), \mathbf{w} = (w_1, \ldots, w_{n-1}),$$
$$\rho = (m_{1n}, \ldots, m_{n-1,n}).$$

Take

$$s = mk/(-2mH)^{1/2}, \boldsymbol{\sigma} = \mathbf{z} - i\mathbf{w} = (rp + imk\,\mathbf{x}/r - (\mathbf{x}\cdot\mathbf{p})\mathbf{p})$$
$$\exp\{i(-2mH)^{1/2} + i\mathbf{x}\cdot\mathbf{p})\}(-2mH)^{1/2}$$

$$\sigma_n = z_n - iw_n = \left(r\frac{p^2 - mk}{(-2mH)^{1/2}} + i\mathbf{x}\cdot\mathbf{p} \right) \exp(i(-2mH)^{1/2} + i\mathbf{x}\cdot\mathbf{p})$$

$$m_{\mu\nu} = x_\mu p_\nu - x_\nu p_\mu, \mu, \nu = 1, \dots n-1,$$
$$\rho = (p^2\mathbf{x} - (\mathbf{x}\cdot\mathbf{p})\mathbf{p})/(-2mH)^{1/2} - mk\,\mathbf{x}/r.$$

Here ρ is the Runge–Lenz vector and z_μ, w_μ are the Bacry–Gyorgyi parameters.

Exercise 14.2. The harmonic oscillator Hamiltonian $H = (-\frac{1}{2}d^2/dx^2) + (x^2/2)$ can be written in the form $H = a^+a + \frac{1}{2}$ where $a = \frac{1}{2}(x + (d/dx)$ and $a^+ = \frac{1}{2}(x - (d/dx))$. Check that $[H, a^{+2}] = 2a^{+2}, [H, a^2] = -2a^2$ and $[a^{+2}, a^2] = 4H$. Thus a^2, a^{+2}, H from the Lie algebra of $SO(2,1)$ (matrices of determinant one which leave invariant $x_0^2 - x_1^2 - x_2^2$). Show that the Casimir operator

$$\mathfrak{c} = \frac{1}{4}\left(\frac{a^2a^{+2} + a^{+2}a^2}{2} - H^2\right)$$

reduces to $\mathfrak{c} = \frac{3}{16}$. Thus the set of all wave functions

$$\Psi_n = \sqrt{2}\sqrt{\frac{n!}{\Gamma(n+3/2)}} \times L_n^{1/2}(x^2)e^{-x^2/2}$$

(where $L_n^{1/2}$ are Laguerre polynomials) with energy levels $E_n = 2n + \frac{3}{2}$ transform according to the irreducible representation $D_{3/4}^+$ of the discrete series of $SO(2,1)$; (note the shift in the ground state energy level.) Show that an identical analysis applies to the case $\tilde{H} = H + V$ where $V = g/x^2$, with the replacements $a^{+2} \to B_2^+ - a^{+2} \quad g/x^2, a^2 \to B_2 = a^2 - (g/x^2)$, and $D_{3/4}^+ \to D_k^+$ where $k = (\alpha + \frac{1}{2})/2$ with $\alpha = \frac{1}{2} + \sqrt{(\frac{1}{4} + 2g)}$.

Chapter 15

Representations and Automorphic Forms

15.1. Geometric quantization and automorphic forms

From Definition 11.1.1 we see that

THEOREM 15.1.1. Let M be a compact complex manifold and assume the first Chern class contains a Kähler form. Then any cohomology class $nc_1(M), n = 0, 1, 2 \ldots$ is the class of some quantum bundle.

Kodaira has already studied manifolds of this type–viz.

THEOREM (Kodaira) 15.1.2. If M is a compact complex manifold admitting a Kähler form which may be covered by an open bounded domain in C^n then its first Chern class contains a Kähler form.

If M may be covered by a bounded homogeneous domain these spaces have been classified by Harish–Chandra. M is then a product of spaces $\Gamma \backslash G/K = \Gamma \backslash N$, where G is a noncompact simple Lie group with trivial center and K is a maximal connected compact Lie group. Γ is a discrete subgroup of G.

As we saw in Theorem 10.1.7 all line bundles over C-spaces are homogeneous. In the present case $M = \Gamma \backslash N$ where N is a symmetric bounded domain we find a similar type of classification by automorphic factors.

DEFINITION 15.1.3. An *automorphic factor* is a smooth map $J : N \times \Gamma \to C^*$ (or $GL(m, C)$) which satisfies: (i) $J(z, \gamma\delta) = J(z\gamma, \delta)J(z, \gamma)$ and (ii) J is holomorphic in z.

DEFINITION 15.1.4. A C(resp C^m)-valued holomorphic function on N is called a *Γ-automorphic form* for J if $f(\gamma z) = J(z, \gamma) f(z)$.

DEFINITION 15.1.5. Two automorphic factors J, J' are called *equivalent* if there is a nonvanishing holomorphic function f on N such that $J'(z, \gamma) = J(z, \gamma) f(z\gamma) f(z)$ for all (z, γ) in $N \times \Gamma$.

Clearly if J and J' are automorphic factors so is JJ' and J^{-1}. Thus under the equivalence relation the equivalence classes form a group, $\mathscr{A}$.

Given an automorphic factor we can define a line bundle $E(J)$ by the equivalence relation $(z, \mathfrak{z}) \sim (z\gamma, J(z,\gamma)\mathfrak{z})$ on $N \times C$.

THEOREM 15.1.6. $E(J)$ and $E(J')$ are equivalent line bundles iff J and J' are equivalent automorphic factors.

Proof. It is an elementary check.

Now a line bundle E over $\Gamma\backslash N$ is defined by an automorphic factor iff p^*E over N is analytically trivial where $p: N \to M$. N as a homogeneous bounded domain is a Stein manifold; and every complex line bundle over N is analytically trivial. Thus we have

THEOREM 15.1.7. Every complex line bundle, in particular every quantum bundle, over M is defined by an automorphic factor; and the group of equivalence classes of line bundles is isomorphic to $\mathscr{A}$.

EXAMPLE 15.1.8. The classic example is $M = \Gamma\backslash\mathscr{P}$ where $\mathscr{P}$ is the Poincaré upper half plane. The element

$$g = \begin{pmatrix} ab \\ cd \end{pmatrix}$$

in the group $G = SL(2, R)$ acts on z in $\mathscr{P} = SL(2, R)/SO(2)$ by $z \to (az + b)/(cz + d)$. If we set $J(g,z) = cz + d$ then we can check that (i) $J(gg', z) = J(g, g')J(g', z)$ and (ii) $\mathfrak{x}(k) = J(k, i) = \exp(i\vartheta)$ if

$$k = \begin{pmatrix} \cos\vartheta & -\sin\vartheta \\ \sin\vartheta & \operatorname{cox}\vartheta \end{pmatrix} \in SO(2).$$

If we map $G \to \mathscr{P}$ by taking $g \to z(g) = (ai + b)/(ci + d)$ and define $\varphi(g)$ by $\exp(i\varphi(g)) = \mathfrak{x}(g)/\overline{\mathfrak{x}(g)}$ where $\mathfrak{x}(g) = ci + d$, then $z(g), \bar{z}(g)$ and $\varphi(g)$ form a coordinate system for G. The complex Lie algebra $\mathfrak{g}^{\mathbb{C}}$ has a basis

$$H = -(\partial/\partial\varphi),$$

$$E = \exp(-i\varphi)\frac{\partial}{\partial\varphi} + 2y\exp(-i\varphi)\partial/\partial z$$

and

$$\bar{E} = \exp(i\varphi)\frac{\partial}{\partial\varphi} + 2y\exp(i\varphi)\partial/\partial\bar{z},$$

where $y = \operatorname{Im} z(g)$. The Casimir operator is then

$$\mathfrak{c} = \tfrac{1}{4}(E\bar{E} + \bar{E}E) - \tfrac{1}{2}H^2 = 2y^2 \frac{\partial^2}{\partial z \partial \bar{z}} + y\frac{\partial}{\partial \varphi}\left(\frac{\partial}{\partial z} + \frac{\partial}{\partial \bar{z}}\right).$$

A holomorphic function on $\mathscr{P}$ is a Γ-automorphic form of weight $-(2+n)$ if for

$$\gamma = \begin{pmatrix} ab \\ cd \end{pmatrix} \in \Gamma \subset G$$

we have $f(\gamma(z)) = f(z)(cz+d)^{2+n}$. Let $A(\Gamma, n+2)$ denote this space of automorphic forms.

The Lie algebra description of a bounded symmetric domain N goes as follows. If $\mathfrak{g}^{\mathbb{C}} = \mathfrak{p}^{\mathbb{C}} + \mathfrak{k}^{\mathbb{C}}$ is the Cartan decomposition and $\mathfrak{p}^{\mathbb{C}} = \mathfrak{n}^+ + \mathfrak{n}^-$ where $\mathfrak{n}^{\pm}$ are abelian subalgebras of $\mathfrak{g}^{\mathbb{C}}$, then there is a subset Ψ of positive complementary roots such that $\mathfrak{n}^+ = \sum_{\alpha \in \Psi} e_\alpha$ and $\mathfrak{n}^- = \sum_{\alpha \in \Psi} e_{\bar{\alpha}}$; we choose e_α so that $\langle e_\alpha, e_{\bar{\alpha}} \rangle = 1$ for all α in Ψ. Then $e_{\bar{\alpha}} = \bar{e}_\alpha$ where $\bar{\ }$ is conjugation of $\mathfrak{g}^{\mathbb{C}}$ with respect to the real form $\mathfrak{g}$. If $\Delta^+ = \Psi \cup \Theta$ where $\Theta = \{$positive roots α of $\mathfrak{g}^{\mathbb{C}}$ with e_α in $\mathfrak{k}^{\mathbb{C}}\}$ then we let $\{\alpha_1, \ldots, \alpha_l\}$ denote the set of simple roots and we assume that $\{\alpha_1, \ldots, \alpha_s\}$ are in Ψ. s here is equal to the number of simple factors of $\mathfrak{g}^{\mathbb{C}}$.

THEOREM 15.1.9. If $\Psi = \{\alpha_1, \ldots, \alpha_N\}$ and if we set $a = \sum_{i=1}^N \alpha_i$ then $\langle e_{\alpha_i}, e_{\bar{\alpha}_i} \rangle = \alpha_i$ and thus $\sum [e_{\alpha_i}, e_{\bar{\alpha}_i}] = a$ and $\langle a, \alpha_i \rangle = \frac{1}{2}$ for $i = 1, \ldots, N$.

Proof. Since $\langle e_{\alpha_i}, e_{\bar{\alpha}_i} \rangle = 1$ we have $\langle [e_{\alpha_i}, e_{\bar{\alpha}_i}], Y \rangle = \langle e_{\bar{\alpha}_i}[Y, e_{\alpha_i} \rangle = = \alpha_i(Y)$ for all Y in $\mathfrak{h}^{\mathbb{C}}$. The rest is left to the reader.

Let V be a finite dimensional complex vector space and let j be a $GL(V)$-valued automorphic factor on $G \times G/K$, i.e. a smooth mapping $j: G \times G/K \to GL(V)$ such that $j(st, x) = j(s, tx) j(t, x)$ for s, t in G and x in $N = G/K$. Let $A^r(\Gamma, N, j)$ denote the space of all V-valued smooth forms φ of degree r on N such that $(\varphi \cdot L_\gamma)_x = j(x, \gamma)\varphi_x$ where L_γ denotes left translation by γ in Γ. Of course $A^r(\Gamma, N, j)$ is isomorphic to the space of all smooth forms of degree r with values in the sections of E_j.

In the case that N is a symmetric bounded domain in C^N we assume that j is holomorphic in x. Then $A^r(\Gamma, N, j)$ decomposes into a direct sum $\sum A^{p,q}(\Gamma, N, j)$ which allows us to consider the d''-cohomology groups of vector valued forms $H^{p,q}_{d''}(\Gamma, N, j)$.

We will also consider the case where $j = \rho$ is a representation of G on V.

Finally these two cases are tied together by considering the case where $J_\tau(s,x)=\tau(j(s,x))$ where τ is a holomorphic representation of $K^{\mathbb{C}}$. In this case we refer to J_τ as the canonical automorphic factor of type τ.

Let ρ be an irreducible representation of $G^{\mathbb{C}}$ in V with highest weight Λ and eigenspace V_Λ; i.e. $S_1=\{v\in V\,|\,\rho(X)v=0 \text{ for all } X \text{ in } \mathfrak{n}^+\}$. Then take τ to be the representation of $K^{\mathbb{C}}$ induced by that of ρ taking $K^{\mathbb{C}}$ into S_1.

THEOREM (Matsushima–Murakami) 15.1.10. $H^{0,q}(\Gamma,N,J_\tau)\simeq H^{0,q}_{d''}(\Gamma,N,\rho)$ for $q=0,\dots,\mathfrak{n}=\dim_C(N)$.

THEOREM (Matsushima–Murakami) 15.1.11. Let Λ be the highest weight of ρ and let q_ρ be the number of roots $\alpha\in\Psi$ such that $\langle\Lambda,\alpha\rangle>0$. Then $H^{0,q}_{d''}(\Gamma,N,\rho)=\{0\}$ for $q=0,1,\dots,q_\rho-1$.

COROLLARY 15.1.12. If $\langle\Lambda,\gamma_i\rangle>0$ for $i=1,\dots,s$, where γ_i are the simple roots of Ψ then $H^{0,q}_{d''}(\Gamma,N,\rho)=\{0\}$ for $q<n=\dim_C(N)$.

COROLLARLY 15.1.13. Let $\mathfrak{g}^{\mathbb{C}}$ be simple and let τ be a holomorphic representation of $K^{\mathbb{C}}$ in a complex vector space S with highest weight Λ. Let σ be the one dimensional representation of $K^{\mathbb{C}}$ given by $\sigma(t)=\det(\mathrm{ad}(t))$ for t in $K^{\mathbb{C}}$. Then

(i) if $\langle\Lambda,\gamma_1\rangle>0$ then $H^{0,q}_{d''}(\Gamma,N,J_\tau)=\{0\}$ for $q<n$;
(ii) if $r>-2\langle\Lambda,\gamma_1\rangle$ then $H^{0,q}_{d''}(\Gamma,N,J_{G^{-r}\otimes\tau})=\{0\}$ for $q<n$.

The proof of this corollary we leave to the reader; part (i) follows directly from the Theorem and part (ii) follows from part (i) and Theorem 15.1.9.

Recognizing that in the Example 15.1.8$(cz+d)$ is just the Jacobian of γ at z leads to

DEFINITION 15.1.14. A holomorphic function f on M is called an *automorphic form* with respect to Γ of weight r if for all m in M and γ in Γ $f(\gamma m)=$ $=j_\gamma^{-r}(m)f(m)$ where $j_\gamma(m)$ is the Jacobian of γ at m.

THEOREM 15.1.15. The vector space of forms of weight r is isomorphic to $H^0(\Gamma\backslash N,\mathcal{O}(K^r))$.

By Kodaira's vanishing theorem $\dim H^0(\Gamma\backslash N,\ \mathcal{O}(K^r))=\chi(\Gamma\backslash N,r)$ for $r\geqq 2$. By the Hirzebruch proportionality theorem plus the Borel–Weil theorem we are able to calculate the dimension of the space of automorphic forms. Namely we have

THEOREM 15.1.16. $\dim H^0(\Gamma\backslash N, \mathcal{O}(K^r)) = \chi(\Gamma\backslash N)\chi(N^d, \mathcal{O}(K^r))$ where N^d is the Cartan dual to N and $\chi(N^d, \mathcal{O}(K^r))$ is given by the Borel–Weil theorem.

Let $\{\alpha_1, \dots, \alpha_n\}$ be the set of positive roots determining the complex structure of $N^d = G^u/K$. Set $b = \sum_{i=1}^n \alpha_i$ and $n = \dim_{\mathbb{C}} G^u/K$. Then

COROLLARY 15.1.17. $\dim H^0(\Gamma\backslash N, \mathcal{O}(K^r)) = (-1)^n \chi(\Gamma\backslash N) \deg(G^u, \ (r-1)b)$ for $r \geqq 2$ where $\deg(G^u, (r-1)b)$ is the degree of the irreducible representation of G with highest weight $(r-1)b$.

EXAMPLE 15.1.8 cont. Let $\mathscr{P} = N$ be the upper-half plane. The Cartan dual is then $N^d = CP(1) = SU(2)/SO(2)$. By Weyl's degree formula we have $\deg(SU(2), (r-1)b) = 2r - 1$. Thus we have

$$\dim H^0(M, \mathcal{O}(K^r)) = (-1)\chi(\Gamma\backslash N)(2r-1).$$

In the case that N is irreducible there is only one simple root in Ψ; let γ_1 denote this root. Let $\delta = \frac{1}{2}\sum_{\alpha>0} \alpha$. And let $a = \sum_{\alpha\in\Psi} \alpha$ denote the total sum of the complementary positive roots. Then Hirzebruch has shown that

THEOREM (Hirzebruch) 15.1.15. If N is irreducible then

$$\chi(M) = \left(\frac{-2\pi}{n!}\right)^{-n} \frac{\prod_{\alpha\in\Psi} \langle \delta, \alpha \rangle}{(2\langle a, \gamma_1 \rangle)^n} \mathrm{Vol}(M),$$

where Vol(M) is the volume of M with respect to the Bergmann metric on M.

In Chapter 18 we return to the question of calculating the dimension of the space of automorphic forms. The idea that we develop there connects the theory of representations of non-compact semisimple with automorphic forms. Namely we find that the discrete series representation T_n of $G = SL(2, R)$ is realized in the space of analytic functions $f(z)$ on $\mathscr{P}$ which satisfy $\int_{\mathscr{P}} |f(z)|^2 y^{n-1} \,dx\,dy < \infty$. If we consider the space of all analytic functions on $\mathscr{P}$, then the elements in this space left invariant by $T_n(\gamma)$ – i.e. $T_n(\gamma) f(z) = f(az+b)(cz+d)^{-n-1} = f(z)$ are the automorphic forms of Example 15.1.8. The next step is to decompose $L^2(\Gamma\backslash G) = \sum M(n,r)$, $n \in Z$, $r \in R$ where $M(n,r)$ is the space of smooth functions $u(g)$ on G which satisfy (i) $u(\gamma g) = u(g)$, $u(gk) = u(g)\chi^n(k)$ and (iii) $cu = ru$ in the notation of that example.

THEOREM 15.1.16. $A(\Gamma, n+2)$ is isomorphic to $M(-(n+2), n(n+2)/8)$ by the mapping $f(z) \to \chi^{-(n+2)}(g) f(z)$.

Thus in the (n,r) plane only the spectrum on the parabola $8r=n^2+2n$ corresponds to automorphic forms. We leave it to the reader to show that r is given by the spectra of the Schrödinger equation on the upper half plane – i.e. $u\in A(\mathscr{P})$ with $u(\gamma(z))=u(z)$ and $y^2(\partial^2/\partial x^2+\partial^2/\partial y^2)u=ru$.

15.2. Bounded symmetric domains and holomorphic discrete series

A bounded domain D in a bounded open connected subset of C^N. D is called symmetric if each x in D is an isolated fixed point of an involutive holomorphic diffeomorphism $s_x\colon D\to D$. Let G be the connected covering group of the identity component of the group $\mathrm{Hol}(D)$ of holomorphic diffeomorphsims of D onto itself. Then G is transitive on D and $D=G/K$, where $K=\{g\in G\,|\,g,0=0\}$. Let σ be the automorphism $\sigma(g)=s_0gs_0$ in $\mathrm{Hol}(D)$. If $\mathfrak{g}_0$ is the Lie algebra of $\mathrm{Hol}(D)$ then $\dot\sigma=\mathrm{d}\sigma_e$ induces the Cartan-like decomposition $\mathfrak{g}_0=\mathfrak{k}_0\oplus\mathfrak{p}_0$ where $\mathfrak{k}_0$ is the $+1$ eigenspace of σ_e and $\mathfrak{p}_0$ is the -1 eigenspace of $\dot\sigma$.

If $\pi\colon G\to D$ is the usual map, then we may identify $\mathfrak{p}_0$ with the real tangent space $T_0(D)$ by the map $Y\in\mathfrak{p}_0\to\mathrm{d}\pi_eY_e$. If T_z (resp. $\bar T_z$) denotes the holomorphic (resp. antiholomorphic) tangent space of D at z, then we set $\mathfrak{p}_\pm=\{Y\in\mathfrak{p}_0\,|\,\mathrm{d}\pi_eY_e\in\{T_0/T_0\}$.

Theorem (Wolf–Tirao) 15.2.1. Let τ be a continuous representation of K on a finite dimensional vector space and let $E_\tau\to D$ be the associated G-homogeneous vector bundle. Then $E_\tau\to D$ is a G-homogeneous holomorphic vector bundle for which the holomorphic sections s over open U in D, $s\colon U\to G\times_\tau V$, $s(g,0)=[g,f_s(g)]$, are characterized by:

$$Xf_s+\tau(X)f_s=0 \text{ for all } X \text{ in } \mathfrak{k}_0 \text{ (i.e. } f_s \text{ is a section);}$$
$$Xf_s=0 \text{ for all } X \text{ in } \mathfrak{p}_+ \text{ (i.e. } f_s \text{ is holomorphic).}$$

Every bounded domain has a Kähler structure given by the Bergman–kernel function; so D is a Hermitian symmetric space of the noncompact type (i.e. $\mathfrak{g}_0$ is semisimple without compact factors). Conversely, if N is a Hermitian symmetric space of noncompact type then there is a bounded symmetric domain D and a holomorphic diffeomorphism of N onto D. Viz. let $\mathfrak{g}_0$ be the Lie algebra of $\mathrm{Hol}(N)_0$ with Cartan decomposition $\mathfrak{g}_0=\mathfrak{k}_0\oplus\mathfrak{p}_0$. Let $\mathfrak{g},\mathfrak{k}$ and $\mathfrak{p}$ denote the complexifications. Then there is a subset $P_n\subset\Delta$ such that $\mathfrak{p}_\pm=\sum_{\alpha\in P_n}\mathfrak{g}^{\pm\alpha}$. We can choose the ordering on Δ such that

$P_n \subset \Delta^+$. Here $\mathfrak{p}_\pm$ are abelian subspaces of $\mathfrak{p}$. If we let $P_\pm$, $K^{\mathbb{C}}$ denote the subgroups of $G^{\mathbb{C}}$, where $G^{\mathbb{C}}$ is the simply connected Lie group with Lie algebra $\mathfrak{g}$ and G_0, K_0 denote the subgroups of $G^{\mathbb{C}}$ corresponding to $\mathfrak{g}_0, \mathfrak{k}_0$, then one can show that $(q, k, p) \to qkp$ is a holomorphic diffeomorphism of $P_- \times K^{\mathbb{C}} \times P_+$ onto an open submanifold of $G^{\mathbb{C}}$ containing G_0. For g in G let $\zeta(g)$ denote the unique element in P_- such that $g \in \zeta(g) K^{\mathbb{C}} P_-$. Since $P_\pm$ are simply connected abelian Lie groups we let $\log : P_\pm \to \mathfrak{p}_\pm$ be the inverse of the exponential map. Then one has:

(i) $xK_0 \to \log \zeta(g)$ for g in G_0 is a holomorphic diffeomorphism of G_0/K_0 onto a bounded domain D in the complex vector space $\mathfrak{p}_+$;

(ii) If G is the universal covering group of $\mathrm{Hol}(N)_0$, $N = G/K$, and $g \to \bar{g}$: $G \to G_0$ is the covering homomorphism, then ψ: $gK \to \log \zeta(\bar{g})$: $G/K \to D$ is a holomorphic diffeomorphism of G/K onto D.

THEOREM 15.2.2. If τ is a representation of K on V, then the holomorphic vector bundle $E_\tau \to G/K$ is holomorphically equivalent to the trivial bundle $D \times V \to D$, i.e. there is a holomorphic diffeomorphism Ψ such that

$$\begin{array}{ccc} G \times_\tau V & \xrightarrow{\Psi} & D \times V \\ \downarrow & & \downarrow \\ G/K & \xrightarrow{\psi} & D \end{array}$$

where Ψ is given by Φ: $G \to GL(V)$, $\Phi(gk) = \Phi(g)\tau(k)$ and $\Psi(g, k) = (\psi(gK), \Phi(g)v)$.

Let $N = G/K$ be a Hermitian symmetric space of noncompact type: let Λ be the highest weight of an irreducible representation τ_Λ of K on V and let $E_\Lambda \to G/K$ be the homogeneous holomorphic vector bundle over N induced by τ_Λ. E_Λ has a K-invariant inner product (,) iff Λ is a real linear functional – i.e. $\Lambda(h'_\alpha)$ is real for all α in Δ. In this case there is an invariant Hermitian structure on $E_\Lambda \to G/K$. Let $H_\Lambda = H_2^{0,0}(E_\Lambda)$ denote the Hilbert space of all holomorphic sections of $E_\Lambda \to G/K$ for which

$$\int_{G/K} (s(x), s(x))\, dx < \infty.$$

Let Π_Λ denote the representation on H^Λ given by

$$(\Pi_\Lambda(g)s)(x) = gs(g^{-1}x), \qquad g \in G, x \in G/K.$$

Since τ is unitary the weights of τ are imaginary on $\mathfrak{h}$ and hence real on $\mathfrak{h}^+ = i\mathfrak{h}$. Using the ordering on $(\mathfrak{h}^+)^*$ we can speak of the highest weight of τ. The following properties of Λ are known:

(a) Λ is integral; so $\xi_\Lambda(h) = \exp \Lambda(\log h)$ defines a character of $T^{\mathbb{C}} = \exp \mathfrak{h}^{\mathbb{C}}$;

(b) Λ is dominant with respect to $\mathfrak{k}$ – i.e. $\langle \Lambda, \alpha \rangle \geqq 0$ for all positive compact roots α;

(c) the weight space of Λ is one dimensional;

(d) τ is determined up to unitary equivalence by Λ.

THEOREM (Harish–Chandra) 15.2.3. $H_\Lambda \neq \{0\}$ iff $\Lambda(h'_\beta) + \delta(h'_\beta) < 0$ for all noncompact positive roots β, where $\delta = \frac{1}{2}\sum_{\alpha > 0} \alpha$; π_Λ is irreducible and the matrix coefficients of π_Λ are square integrable.

EXAMPLE 15.2.4. Let

$$G = SU(1,1) = \left\{ \begin{pmatrix} \alpha & \beta \\ \bar{\beta} & \bar{\alpha} \end{pmatrix} \Big| |\alpha|^2 - |\beta|^2 = 1 \right\}.$$

G acts as the group of analytic automorphisms on the disk $D = \{|z| < 1\}$ as $g, z \to zg = (\bar{\alpha}z + \beta)/(\bar{\beta}z + \alpha)$. Here $\mathfrak{h}$ = maximal abelian subalgebra of $\mathfrak{k}$

$$\text{and} \quad \mathfrak{h}^{\mathbb{C}} = \left\{ \begin{pmatrix} a & 0 \\ 0 & -a \end{pmatrix} \right\}.$$

Thus

$$\Lambda \begin{pmatrix} a & 0 \\ 0 & -a \end{pmatrix} = -na \qquad \text{and} \qquad \alpha \begin{pmatrix} a & 0 \\ 0 & -a \end{pmatrix} = 2a.$$

$\delta = \frac{1}{2}\alpha$ so

$$\delta \begin{pmatrix} a & 0 \\ 0 & -a \end{pmatrix} = a.$$

Λ integral means n is an integer. Clearly Λ is dominant with respect to $\mathfrak{k}$ since the only positive root α is noncompact. If

$$\Lambda' \begin{pmatrix} 1 & 0 \\ 0 & -1 \end{pmatrix} = c' \quad \text{and} \quad \Lambda'' \begin{pmatrix} 1 & 0 \\ 0 & -1 \end{pmatrix} = c''$$

then $\langle \Lambda', \Lambda'' \rangle = cc'c''$ for some positive constant c. Thus $\langle \Lambda + \delta, \alpha \rangle =$ $= c(-n+1)2$. And Harish–Chandra's condition that $\langle \Lambda + \delta, \alpha \rangle < 0$

is the requirement that $n > 1$. Thus

$$\xi_\Lambda\begin{pmatrix} a & 0 \\ 0 & -a \end{pmatrix} = a^{-n} \text{ with } n \geqq 2.$$

Of course this representation is just the discrete series of Bargmann where he realized G acting on the holomorphic functions f on D for which $\int_D |f(z)|^2 (1 - |z|^2)^{n-2} dx\, dy < \infty$. We leave it to the reader to write out the isomorphism.

THEOREM (Harish–Chandra). The degree of the square integrable holomorphic discrete series is given by

$$d_\Pi = \left| \prod_{\alpha \in \Delta^+} (\Lambda(h'_\alpha) + \delta(h'_\alpha))/\delta(h'_\alpha) \right|,$$

where Δ^+ is the set of all positive roots of $\mathfrak{g}$ with respect to $\mathfrak{h}$.

PROBLEMS

EXERCISE 15.1. Let Γ act freely on $N = G/K$. Let $J(g, z)$ be a $GL(V)$-valued automorphic factor. Let $\tau_J(k) = J(k, z_0)$ where $K = \{g \in G \mid gz_0 = z_0\}$. Then as noted earlier τ_J is a representation of K. Define the action of $\Gamma \times K$ on G by $(\gamma, k)g = \gamma g k^{-1}$. Let $E(\tau_J)$ be the vector bundle given by $G \times V$ modulo the equivalence relation $(g, v) \sim ((\gamma, k)g, \tau_J(k)u)$. Show that $E(J)$ and $E(\tau_J)$ are differentiably equivalent vector bundles over $M = \Gamma \backslash N$.

Chapter 16

Thermodynamics of Homogeneous Spaces

16.1. Density matrices and partition functions

The fundamental construct for quantum statistical mechanics as formulated by von Neuman and Dirac is the density matrix

$$\rho(q,q') = \sum w_k \psi_k(q)\psi_k^*(q')$$

for a quantum mechanical system.

DEFINITION 16.1.1. Given an observable A the *expected value* of this observable is defined by $\langle A \rangle = \mathrm{Tr}(A\rho)/\mathrm{Tr}(\rho)$ where trace means integration over the diagonal.

DEFINITION 16.1.2. $\vartheta(\beta) = \mathrm{Tr}(\rho)$ is called the *partition function* of the system.

The canonical ensemble is defined by taking for the weights $w_k = -\exp(-\beta E_k)$ where $H\psi_k = E_k\psi_k$. β can be interpreted as $1/kT$, where T = temperature, k = Boltzmann's constant.

The partition function determines all the thermodynamic properties of the system. E.g. the average *internal energy* is $U = (-\partial \ln \vartheta(\beta))/\partial\beta$, the *Helmholtz free energy* F is defined by $\vartheta(\beta) = \exp(-F\beta)$. The entropy is defined by $S(T) = kT(\partial \ln \vartheta/\partial T) + k \ln \vartheta(\beta)$.

EXAMPLE 16.1.3. The harmonic oscillator is specified by $H = (1/2m)(p^2 + m^2\omega^2 q^2)$ where $p = -i\partial/\partial q$. The eigen functions are $\psi_k = (m\omega/\pi\hbar)^{1/4} \times (k!)^{1/2} D_k(q\sqrt{2m\omega/\hbar})$ with eigenvalues $E_k = k + \frac{1}{2}, k = 0,1,\ldots$. Using the Mehler identity we find

$$\rho(q,q') =$$

$$= (1/\sqrt{\pi}) \sum_{k=0}^{\infty} \exp(-\beta(k+\tfrac{1}{2}))\frac{\psi_k(q)\psi_k(q')}{2^k k!}\exp\left(-\left(\frac{q^2 - q'^2}{2}\right)\right) =$$

$$= \frac{1}{\sqrt{2\pi \sinh\beta}}\left[\exp\left(-\frac{1}{4}\tanh\left(\frac{\beta}{2}\right)(q+q') - \frac{1}{4}\coth\left(\frac{\beta}{2}\right)(q-q')\right)\right].$$

Thus taking the trace gives the partition function

$$\vartheta(\beta) = \mathrm{Tr}(\rho) = \tfrac{1}{2}\sinh\left(\frac{\beta}{2}\right) = \frac{e^{\beta/2}}{e^{\beta} - 1}$$

which is Planck's formula.

The average internal energy U is easily checked to be $U = \frac{1}{2} + (1/(e^{\beta} - 1))$. Using the expansion

$$\frac{x}{e^x - 1} = 1 - \frac{x}{2} - \sum_{n=1}^{\infty} (-1)^n B_{2n} x^{2n}/(2n)!,$$

where B_{2n} are Bernoulli numbers, the high temperature limit of U is

$$U \sim \frac{1}{\beta} - \frac{1}{\beta} \sum_{n=1}^{\infty} (-1)^n / B_{2n} \beta^{2n}/(2n)!$$

as $\beta \to 0^+$.

The high and low temperature limits of ρ are easily checked to be $\rho(q, q') \sim (1/\sqrt{2\pi\beta}) \exp(q - q')^2/2\beta$ as $\beta \to 0$ and $\rho(q, q') \sim (1/\sqrt{\pi}) \exp(-(q^2 + q'^2))$ as $\beta \to \infty$, using the respective sides of the Mehler identity.

The diagonal term is given by $\rho(q, q) \sim \exp(-\tanh(\beta/2) q^2)$ which has the high temperature limit $\rho(q, q) \sim \exp(-q^2/2)$.

This asymptotic relationship is an example of the formal statement

$$\begin{aligned} \mathrm{Tr}(\rho) &\sim \int \exp(-\beta(T(p) + V(q)) \quad dp\, dq \\ &= c \int \exp(-\beta V(q))\, dq \quad \text{as} \quad \beta \to 0^+, \end{aligned}$$

where $T(p) = (1/2m)p^2$ is the kinetic energy and $V(q) = q^2$ is the potential energy. We formalize this comment as follows:

Definition 16.1.4. The *limiting principle* of quantum statistical mechanics states that in the high temperature limit the quantum mechanical partition function should have the classical partition function as the limiting value.

Example 16.1.5. Consider the rigid rotator with Hamiltonian $H = (1/2I)(p_\vartheta^2 + (p_\varphi^2/\sin^2\vartheta))$. Then the classical partition function is

$$\begin{aligned} \vartheta(\beta)_C &= \frac{1}{h^2} \int\int \exp(-\beta H)\, dp_\vartheta\, dp_\vartheta\, d\vartheta\, d\varphi = \\ &= \frac{1}{h^2} \frac{2\pi I}{\beta} \int_0^{\pi} \int_0^{2\pi} \sin\vartheta\, d\vartheta\, d\varphi = 2I/\hbar^2 \beta. \end{aligned}$$

We will see shortly that the limiting principle holds for this example.

EXAMPLE 16.1.6. In the one dimensional case, since

$$\frac{1}{\sqrt{2\pi\beta}}=\frac{1}{2\pi}\int \exp\left(-\beta\frac{p^2}{2}\right)\mathrm{d}p$$

we see that

$$\frac{1}{\sqrt{2\pi\beta}}\int \exp(-\beta V(q))\,\mathrm{d}q=\frac{1}{2\pi}\int \exp(-\beta(p^2+V(q))\,\mathrm{d}p\,\mathrm{d}q.$$

Thus the limiting principle of quantum statistical mechanics is the requirement that

$$\sum \exp(-E_k\beta)\sim\frac{1}{2\pi}\int_0^\beta \exp(-\lambda\beta)\,\mathrm{d}B(\lambda)\quad\text{as}\quad \beta\to 0^+,$$

where $B(\lambda)$ is the area of the region $(p^2/2)+V(q)\leqq\lambda$. And by the Tauberian theorem the limiting principle states that

$$N(\lambda)=\sum_{E_k\leqq\lambda}1\sim\frac{1}{2\pi}B(\lambda)\quad\text{as}\quad\lambda\to\infty.$$

Since $\exp(-\beta H)\psi_k=\exp(-\beta E_k)\psi_k$ we can write the density matrix as $\rho(\beta)=\sum\exp(-\beta H)\psi_k(q)\psi_k^*(q')$. We see then that formally we have

$$\frac{\partial\rho}{\partial\beta}=-H\rho$$

and

$$\lim_{\beta\to 0}\rho(\beta)=\sum\psi_k(q)\psi_k^*(q')=\delta(q-q').$$

DEFINITION 16.1.7. This differential equation is called *Bloch's equation.* Under interesting situations this is a parabolic equation.

EXAMPLE 16.1.8. Consider the free particle on the interval $0\leqq q\leqq 1$ with $\psi(0)=\psi(1)=0$. Then Bloch's equation is

$$\frac{\partial\rho}{\partial\beta}=\frac{1}{2}\frac{\partial^2}{\partial q^2}\rho$$

which is just the diffusion equation. Note in this case that

$$\rho(q, q') = 2 \sum_{k=1}^{\infty} \exp(-\beta\pi^2 k^2/2) \sin(k\pi q) \sin(k\pi q')$$

$$= \frac{1}{\sqrt{2\pi\beta}} \sum_{l} \exp(-(q' - q + 2l)^2/2\beta) - \exp(-(q' + q + 2l)^2/2\beta)$$

$$= \frac{1}{2}\left\{\vartheta_3\left(\frac{q-q'}{2}, \frac{\beta}{2}\right) - \vartheta_3\left(\frac{q+q'}{2}, \frac{\beta}{2}\right)\right\},$$

where ϑ_3 is one of the Jacobi theta functions. Again the high and low temperature limits are easily found.

We return now to Example 16.5. The kinetic energy of the rigid rotator with an axis of symmetry (but no spin) is $T = (I/2)\{(\mathrm{d}\vartheta/\mathrm{d}t)^2 + \sin^2\vartheta \times (\mathrm{d}\varphi/\mathrm{d}t)^2\}$. The associated Schrödinger equation is $H\psi_\lambda = (\hbar^2/2I)\Delta\psi_\lambda = E_\lambda\psi_\lambda$ where

$$\Delta = \frac{1}{\sin\vartheta}\frac{\partial \sin\vartheta}{\partial\vartheta}\frac{\partial}{\partial\vartheta} + \frac{1}{\sin^2\vartheta}\frac{\partial^2}{\varphi^2}$$

is just the Laplace–Beltrami operator on S^2. The energy levels are $E_\lambda = (\hbar^2/2I)l(l+1), l = 0, 1, \ldots$ with multiplicities $(2l+1)$. The rotational partition function is thus

$$\vartheta_{\mathrm{Rot}}(\beta) = \sum_{l=0}^{\infty} (2l+1) \exp\left(-l\left(\frac{l+1}{2I}\right)\beta\hbar^2\right).$$

We reduce this example to the case just treated as follows. If we restrict the motion of the rotator to rotation in the plane $\vartheta = \pi/2$ and if we let the resulting manifold $M = S^1$ have length L, then the Schrödinger equation becomes

$$\frac{-\hbar^2}{2I}\left(\frac{2\pi}{L}\right)^2 \frac{\partial^2\psi}{\partial\varphi^2} = E_l\psi,$$

where

$$E_l = \frac{\hbar^2}{2I}\left(\frac{2\pi}{L}\right)^2 l^2$$

for l in Z with multiplicity 2 for $E_l \neq 0$. Thus we see that the planar partition function is

$$\vartheta_p(\beta) = \vartheta_3(\beta, 0).$$

The analogue of Mehler's identity is just the Poisson summation formula

$$\sum_{m\in\mathbb{Z}} \exp(-\pi\beta m^2) = \frac{1}{\sqrt{\beta}} \sum_{m\in\mathbb{Z}} \exp(-\pi m^2/\beta).$$

Thus the high temperature limit of the planar rotator is easily seen to be

$$\vartheta_p(\beta) = \frac{2I}{\hbar} L(4\pi\beta)^{-1/2} + \mathcal{O}(e^{-1/\beta}) \quad \text{as} \quad \beta \to 0^+.$$

This example provides a hint at the many relationships that exist between quantum statistical mechanics and analytic number theory. E.g. the classical Riemann zeta function

$$\zeta(2s) = \sum_{n=1}^{\infty} n^{-2s}$$

is given by

$$\zeta(2s) = \frac{\pi}{\Gamma(s)} \int_0 \beta^{s-1} [\vartheta_p(\beta) - \tfrac{1}{2}] \, d\beta.$$

The zeta function is just one example of a Dirichlet L-series:

$$L(s, \mathfrak{X}) = \sum_{n=1}^{\infty} \mathfrak{X}(n) n^{-s},$$

where $|\mathfrak{X}(n)| = 0, 1$ and $\mathfrak{X}(mn) = \mathfrak{X}(m)\mathfrak{X}(n)$ is a Dirichlet character. Clearly $\zeta(s) = L(s, 1)$. Another example of an L-series is

$$\beta(s) = \sum_{n=0}^{\infty} (-1)^n (2n+1)^{-s}.$$

L-series arise in solid state physics as Madelung sums. For various planar lattices this sum is of the form

$$\sum_{(m,n)\neq(0,0)} (m^2 + n^2)^{-s} = 4\zeta(s)\beta(s).$$

As an exercise the reader should show that this lattice sum may be written as a Dirichlet L-series.

One final example arises in Chapter 19 where we consider the vacuum energy momentum tensor $\langle T_{\mu\nu}\rangle$ for a geometry of parallel plates divided into rectangular solids; then the 'Casimir'-renormalized value is

$$\langle T_{00}\rangle_c = -(2\pi)^{-2} \sum_{n_1, n_2, n_3} (a^2 n_1^2 + b^2 n_2^2 + c^2 n_3^2)^{-2}.$$

In the case $a = \infty, b = c = L$, then $\langle T_{00}\rangle_c$ is an Ewald or Madelung lattice sum:

$$\langle T_{00}\rangle_c = -\tfrac{1}{3} L^{-4} \beta(2),$$

where $\beta(2)$ is Catalan's constant.

16.2. Epstein Zeta Functions

The planar rotator is a very interesting example since it is also the partition function for the free particle in a box D of length L, where I is replaced by the particle mass μ.

Here the Schrödinger equation is

$$\frac{-\hbar^2}{2\mu} \frac{\partial^2 \psi}{\partial x^2} = \lambda \psi,$$

say with boundary conditions

$$\psi(x) = \psi(x + nL), \qquad n \in Z.$$

The wave functions are then

$$\psi(x) = A \exp\left[i \frac{2\pi \hbar n}{L} \frac{x}{\hbar} \right]$$

with eigenvalues

$$\lambda_n = n^2 2\pi^2 \hbar^2 / \mu L^2, \qquad n \in Z.$$

If we take for the boundary conditions

$$\psi(x) = 0 \text{ on boundary } B$$

then $\psi(x) = A \sin(n\pi x/L)$ with associated energy eigenvalues

$$\lambda_n = n^2 \pi^2 \hbar^2 / 2\mu L^2,$$

where $n = 1, 2, 3 \ldots.$

The series $\sum_{n=1}^{\infty}\psi_n(x)\psi_n(y)/\lambda_n^s$ for $s\in\mathbb{C}$ was first studied by Epstein in 1903 and is called the Epstein zeta function. Minakshisundaram in 1949 M28 returned to the study of this zeta function using the Green's function for the heat equation or Bloch's equation. Viz., letting $\rho(x,y;\beta)$ denote the kernel for Bloch's

$$\frac{-\hbar^2}{2\mu}\frac{\partial^2}{\partial x^2}\rho(x,y;\beta)=\frac{-\partial}{\partial\beta}\rho(x,y;\beta)$$

with $\lim_{\beta\to 0^+}\rho(x,y;\beta)=\delta(x;y)$, then it is easily shown that

$$\rho(x,y;\beta)=\sqrt{\frac{\mu}{2\mu\hbar^2\beta}}\exp\left[-\frac{\mu}{2\hbar^2\beta}(x-y)^2\right]-g(x,y;\beta).$$

After seeing that

$$|g(x,y;\beta)|\leqq\frac{C(y)}{\beta^{1/2}}\exp(-l_y^2/4\beta)$$

(v. M28, K3, etc.) for all x in D where l_y is the minimum distance between y and points on boundary B it follows that:

THEOREM (Minakshisundaram) 16.2.1. The series $Z(x,y;s)=\sum_{n=1}^{\infty}\psi_n(x)\psi_n(y)/\lambda_n^s$ converges uniformly in x and y for all $s\in\mathbb{C}$ such that Re(s) is large. $Z(x,y;s)$ extends to an entire function of s if $x\neq y$ are in D, and has 'trivial zeros' at $s=0,-1,-2,\ldots$.

The series $Z(x,x;s)=\sum_{n=1}^{\infty}\psi_n^2(x)/\lambda_n^s$ represents a meromorphic function of s, with a simple pole at $s=\frac{1}{2}$ and residue $1/\Gamma(\frac{1}{2})2\sqrt{\pi}$, and has 'trivial zeros' at $s=0,-1,-2,\ldots$.

The proof follows from the above facts about the kernel of the density matrix and the representation

$$\Gamma(s)Z(x,y;s)=\int_0^{\infty}\beta^{s-1}\rho(x,y;\beta)\,\mathrm{d}\beta.$$

As a corollary of these results we have

COROLLARY (Carleman's Asymptotic Formulae) 16.2.2

(1) $$\sum_{\lambda_n\leqq\lambda}\psi_n^2(x)\sim\frac{1}{2\sqrt{\pi}\,\Gamma(\frac{3}{2})}\sim\lambda^{1/2}\quad\text{as}\quad\lambda\to\infty.$$

(2) and noting that the number $N(\lambda)$ of eigenvalues $\lambda_n \leqq \lambda$ is $N(\lambda) = \sum_{\lambda_n \leqq \lambda} \int \psi_n^2(x)\,\mathrm{d}x$, it follows that

$$N(\lambda) \sim \frac{L}{2\sqrt{\pi}\,\Gamma(\frac{3}{2})} \lambda^{1/2} \quad \text{as} \quad \lambda \to \infty .$$

16.3. Asymtotes of the Density Matrix

The high temperature limit of the density matrix $\rho(\beta, m, m')$ for the rigid rotator should be of the form

$$\rho(\beta, x, y) \sim \frac{\exp(-g^2/2)}{(2\pi\beta)} (\sqrt{k}\,g/\sin\sqrt{k}\,g),$$

where $k = 1/r^2$ and $g(x, y)$ is the Riemann distance function on $S^2(r)$. Here we assume that x, y are nonconjugate (i.e. $g(x, y) < \pi r$). To demonstrate this fact requires a digression into more differential geometry. Let s denote the length of the geodesic and let N be the hypersurface orthogonal to the unique shortest geodesic $\gamma = \overline{mm_1}$. Here $0 \leqq s \leqq g(m, m_1)$. The Jacobi field $Y(s)$ is the field of speeds $Y(s) = \mathrm{d}\gamma_\alpha(s)/\mathrm{d}\alpha$ along some geodesic variation γ_α, $0 \leqq \alpha \leqq 1$, $\gamma_0 = \gamma$. Along γ, $Y(s)$ satisfies

$$\frac{\nabla}{\mathrm{d}s}\frac{\nabla}{\mathrm{d}s} Y(s) + R_{\gamma(s)}(Y(s), \dot{\gamma}(s))\dot{\gamma}(s) = 0,$$

where $R(X, Y)Z$ is the curvature form of M. Consider the Jacobi fields Y orthogonal to γ – i.e. $(Y, \dot{\gamma}) = 0$. Then the curvature form becomes

$$R_{\gamma(s)}(Y(s), \dot{\gamma}(s))\dot{\gamma}(s) = K_i^j Y^i,$$

where

$$K_i^j = R_{lil}^j = -\frac{1}{2}\frac{\partial^2 g_{ll}}{\partial x^i \partial x^j}.$$

Let $K = \{K_i^j, i, j \leqq n-1\}$. The Jacobi field equation becomes

$$\ddot{Y} + KY = 0. \qquad (*)$$

Theorem 16.3.1. Let $\mathcal{Y}$ be an arbitrary $(n-1) \times (n-1)$ matrix formed from $n-1$ independent Jacobi fields satisfying $(*)$. Then the matrix

$$\mathfrak{z}(t) = \mathcal{Y}(t) \int_0^t (\mathcal{Y}^* \mathcal{Y})^{-1}(s)\,\mathrm{d}s$$

also satisfies $(*)$ with $\mathfrak{z}(0) = 0$, $\dot{\mathfrak{z}}(0) = I$.

The proof is just an elementary check.

COROLLARY 16.3.2. Let $\Psi(m, m') = \det \mathfrak{z}(g(m, m')$. Then $\Psi(x, y) = \Psi(y, x)$.

Let $S_{m,m'}(z) = g^2(m, z) + g^2(z, m')$ be the action of the piecewise geodesic $m \to z \to m'$. If m, m' are not conjugate along same shortest geodesic then the Hession of $S(z)$ is non-singular at $z_0 = \gamma(g/2)$.

THEOREM 16.3.3.

$$\det \text{Hess}\left(\frac{S_{mm'}(z)}{2}\right) = 4g^{1-n}\frac{\Psi(m, m')}{\Psi(m, z_0)\,\Psi(z_0, m')}.$$

Using Varadhan's estimate V2 and these results on Jacobi fields it follows that:

THEOREM (Molchanov) 16.3.4. For any compact set $D \subset M$ there is a constant $\varepsilon > 0$ such that whenever $g(m, m') < \varepsilon$ uniformly in m and m' we have

$$\rho(\beta, m, m') \sim \frac{\exp^{-g^2(m,m')/2\beta)}}{(2\pi\beta)^{n/2}} H(m, m'),$$

$n = \dim M, H(m, m') = g^{(n-1)/2}(m, m')\,\Psi^{-1/2}(m, m')$.

If M is a simply connected symmetric space, then $K_i^j = \delta_{ij}\lambda_j$ where λ_j are principal curvatures of any geodesic hypersurface N orthogonal to $\overline{mm'}$. (*) is then a system of $n-1$ equations with constant coefficients. It follows that

$$\det \mathfrak{z}(g(m, m')) = g^{k_0} \prod_{i=1}^{k^+} \frac{\sin(\sqrt{\lambda_i^+}\, g)}{\sqrt{\lambda_i^+}} \prod_{j=1}^{k^-} \frac{\sin h(\sqrt{-\lambda_j}\, g)}{\sqrt{-\lambda_j}}$$

where k_0, k^+, k^- are the number of zero, positive, and negative principal curvatures; here $k_0 + k^+ + k^- = n - 1$.

COROLLARY 16.3.5. If M is a symmetric simply connected manifold, then

$$\rho(\beta, m, m') \sim \frac{e^{-g^2(m,m')/2\beta}}{(2\pi\beta)^{n/2}} \prod_{i=1}^{k^+} \left(\frac{\sqrt{\lambda_i^+}\, g}{\sin(\sqrt{\lambda_i^+}\, g)}\right)^{1/2} \times$$

$$\times \prod_{j=1}^{k^-} \left(\frac{\sqrt{-\lambda_j}\, g}{\sinh(\sqrt{-\lambda_j} g)}\right)^{1/2}.$$

In fact if M is of nonpositive curvature, then the asymptotic formulae hold for all m, m'. In particular for symmetric spaces of negative curvature

$$\rho(\beta, m, m') \sim \frac{e^{-g^2(m,m')/2\beta}}{(2\pi\beta)^{n/2}} \prod_{i=1}^{n} \left(\frac{\sqrt{-\lambda_i g}}{\sinh(\sqrt{-\lambda_i g})} \right)^{1/2}$$

for all m, m'.

We note that the expansion for the rigid rotator is as desired. And, furthermore, it follows from Molchanov's theorem that the high temperature limit in the presence of a potential energy V is as conjectured:

COROLLARY 16.3.6. If the Bloch equation is of the form $\partial\rho/\partial\beta = H\rho$ where $H = \frac{1}{2}\Delta + V$ then

$$\rho(\beta, x, y) \sim \frac{\exp(-g^2/2\beta + A(x, y))}{(2\pi\beta)^{n/2}} H(x, y),$$

where $A(x, y) = \int_0^g (V, \dot{\gamma}(s))\,\mathrm{d}s$ is the work of field V along the geodesic $\gamma_{x,y}$.

The Molchanov expansion is just the first term of a generalized expansion that goes back to Hadamard and has been used by several physicists – notably de Witt. We turn now to a review of Hadamard's method of parametrices in the context of vector bundles. This development will be needed in Chapter 19.

Let (M, g) be a Riemannian manifold of dimension n. Let E be a complex vector bundle over M. Let $\mathscr{V}$ denote the volume bundle over M – i.e. the vector bundle of rank one associated to the frame bundle. Let E^* denote the dual vector bundle i.e. the fiber bundle with fiber the dual space to that of E. The bundle $E' = E^* \otimes \mathscr{V}$ is the bundle isomorphic to $\mathrm{Hom}(E, \mathscr{V})$ i.e. if s and s' are smooth cross sections of E and E', the section $m \in M \to \langle s(m), s'(m) \rangle_m$ belongs to $\mathscr{V}_m$. Let $S(E)$ denote the space of smooth cross-sections of E. The covariant derivative of Chapter 3 is generalized to be a map

$$L: S(E) \to S(T^*M \otimes E)$$

which satisfies $L(fs) = \mathrm{d}f \otimes s + fLs$.

DEFINITION 16.3.7. The generalized Laplacian is the map $\Delta: S(E) \to S(E)$ which satisfies $\Delta(fs) = (\Delta f)s - 2\mathrm{Tr}(\mathrm{d}f \otimes Ls) + f\Delta s$, where Δf is the usual Laplacian on functions, $\Delta = -(\det g)^{-1/2}\,\partial_i (\det g)^{1/2} g^{ij}\,\partial_j$ and the trace is with respect to g – i.e.

$$\mathrm{Tr}(\mathrm{d}f \otimes Ls) = g^{ij}\,\partial_i f L_j s.$$

$(\nabla^2 s)(X,Y) = (\nabla_Y(\nabla s))(X) = \nabla_Y(\nabla_X s) - \nabla_{D_Y X} s$

$(\nabla^2(fs))(X,Y) = (D^2 f)(X,Y).s + (df \otimes s)(X,Y) + (df \otimes s)(Y,X) + f.(\nabla^2 s)(X,Y)$

The generalized Laplacian is then a formal differential operator of second order.

DEFINITION 16.3.8. The symbol σ_Δ of Δ is the function $\sigma_\Delta(m,\alpha)u = -\frac{1}{2}\Delta(f^2 s)(m)$ where $f \in A(M)$ has $f(m)=0$ and $\mathrm{d}f = \alpha \in T_m^* M$ and $u \in E_m$ with $s(m)=u$.

We leave it to the reader to show that

$$\sigma_\Delta(m,\alpha)\,u = (g^{ij}(m)\,\alpha_i\,\alpha_j)\,u.$$

Thus if g is nondegenerate, Δ is an elliptic operator.

DEFINITION 16.3.9. Let $E \boxtimes E'$ denote the bundle whose base is $M \times M$ and whose fiber is $E_x \otimes E'_y$.

DEFINITION 16.3.10. A parametrix P for a generalized Laplacian Δ is a distribution in $\mathscr{D}'(E \boxtimes E' \,|\, U)$ where U is a neighbourhood of $M \times M$ for which there is a $k \geqq 0$ and an element Q in $C^k(E \boxtimes E' \,|\, U)$ such that $\Delta_m P(m,m') = \delta(m,m') + Q(m,m')$.

The Hadamard construction of a parametrix is as follows. In $M \times M$ consider a neighbourhood U, normal for the Riemannian structure (M,g). This means that for (m,m') in U there is a unique geodesic call it $\overline{mm}$, joining m and m'. Covariant derivation L in E defines a parallel transport along $\overline{mm'}$. This provides an isomorphism $E_{m'} \simeq E_m$ for (m,m') in U and so defines a local section ϑ of $E \boxtimes E^*$.

Taking components g_{ij} of g in the system of normal coordinates we set $G(m,m') = \det g_{ij}(m))^{-1/4}$.

DEFINITION 16.3.11. Define the sequence of kernels $u_k(m,m')$ by

$$u_k(m,m') = g(m,m')\vartheta(m,m')v_g,$$

where v_g is the Riemannian measure on (M,g) and

$$u_i(m,m')\left(i - \frac{g(m,m')}{G(m,m')}\frac{\partial}{\partial r} G(m,m') \right) + g(m,m')\frac{L(m)}{\mathrm{d}r}u_i(m,m') +$$

$$+ \Delta(m)u_{i-1}(m,m') = 0$$

when written in the polar geodesic coordinates for the orthonormal frame of $T_{m'}M$.

THEOREM 16.3.12.

$$P(\beta, m, m') = (2\sqrt{\pi})^{-n}\beta^{-n/2}\exp\left(-\frac{g^2(m,m')}{4\beta}\right)\times$$
$$\times \sum_{i=0}^{N} \beta^i u_i(m,m'))$$

is a parametrix for Bloch's equation $(\partial/\partial\beta) + \Delta$ for $N > n/2$.

In particular if $\tilde{P}$ is a smooth extension of P to $]0, \infty) \times \times M \times M$ we have $s = \lim_{\beta\to 0} \langle \tilde{P}(\beta, ., m'), s(m')\rangle$.

This theorem is just an extension of the Minakshisundaram–Pleijel theorem to bundles. A second example of this theorem is to take for the generalized Laplacian, the Laplacian Δ_p on the bundle of p-forms $E = A^p(M)$. This case is of interest to physicists for the Proca equations to describe spin 1 massive particles is $\Delta\varphi = -m^2\varphi$ and $\delta\varphi = 0$ where φ is a 1-form and $\Delta = d\delta + \delta d$. For future reference in Chapter 19 we note that this field equation arises from the Lagrangian $L = -\frac{1}{2}(\varphi_\mu \Delta\varphi^\mu + m^2\varphi_\mu\varphi^\mu)$. However the condition $\delta\varphi = 0$ does not; it is just a restriction to consider only divergence free forms. Finally if $m = 0$ in this case we must add the restriction that φ is not exact.

THEOREM 16.3.13. For the Laplacian Δ_p on p-forms if u_i^p are the components in the parametrix then:

(i) $\mathrm{Tr}(u_i^p) = (-1)^{p(n-p)}\mathrm{Tr}(u_i^{n-p})$ for $p = 0, 1, \ldots n$ and for every i;

(ii) $$\left(\frac{1}{2\sqrt{\pi}}\right)^n \sum_{p=0}^{n} (-1)^p \int_M \mathrm{Tr}(u_i^p) = \begin{cases} 0 & i \neq n/2 \\ \mathfrak{X}(M) & i = n/2 \\ & n \text{ even;} \end{cases}$$

(iii) if $\dim M = 4m$, $\int_M \mathrm{Tr}(*_m u_{2m}^{2m}(m,m'))_{m=m'} = 16^m\pi^{2m}\,\mathrm{Sgn}(M)$,

where $*$ is the Hodge star operator on forms and $\mathrm{Sgn}(M)$ is the Hirzebruch signature (v. H22).

The fundamental solution $\rho(\beta, m, m')$ of Bloch's equation was studied for compact manifolds M by Kotake and others. $\rho(\beta, m, m')$ belongs to $S(R^+)\otimes S(E \boxtimes E')$. It is constructed by using the parametrix $\rho(\beta, m, m') = P(\beta, m, m') + W(\beta, m, m')$. Here W is given by

$$W(\beta, m, m') = \int_0^\beta du \int_M \tilde{P}(\beta - u, m, n)R(u, n, m')$$

and R is selected so that ρ satisfies Bloch's equation. One checks quickly that this means that

$$\operatorname{Tr}(\rho(\beta, m, m)) = \operatorname{Tr} P(\beta, m, m) + \operatorname{Tr} W(\beta, m, m)$$

and we have

THEOREM 16.3.14. The high temperature limit of the trace of the kernal of the density matrix is given by

$$\operatorname{Tr} \rho(\beta, m, m) \sim (2\sqrt{\pi})^{-n} \beta^{-n/2} \sum_{i=0}^{\infty} \beta^i \operatorname{Tr} u_i(m).$$

Since $\sum_{i=0}^{\infty} \exp(-\lambda_i \beta) = \int_M \rho(\beta, m, m) v(m)$, we have by integrating the expansion in theorem

THEOREM 16.3.15. $\vartheta(\beta) = \exp(-\lambda_i \beta)$ has the asymptotic high temperature expansion

$$\vartheta(\beta) \sim (4\pi\beta)^{-n/2}(a_0 + a_1\beta + \cdots),$$

where $a_i = \int_M u_i(m, m) v_g(m)$ are spectral invariants – i.e. $u_i(m, m)$ can be computed invariantly in any local coordinates as smooth functions of m, the symbol σ_Δ and derivatives of σ_Δ.

For a proof of this result see Gilkey G13

In particular in the case that E is the trivial bundle and Δ is the Laplacian on functions on M we have

(i) $\quad a_0 = \operatorname{Vol}(M, g)$,

(ii) $\quad a_1 = \frac{1}{6}\int_M K_g v_g$,

where K_g is the scalar curvature of (M, g). By the Gauss–Bonnet theorem $\int_M K_g v_g = 2\pi \mathfrak{X}(M)$ if $\dim M = 2$. Thus $a_1 = (\pi/3)\mathfrak{X}(M)$ if $\dim M = 2$.

(iii) $\quad a_3 = \frac{1}{360}\int_M (2|R|^2 - 2|\mathrm{Ric}|^2 + 5K^2) v_g$,

where Ric is the Ricci curvature and R is the Riemann curvature.

EXAMPLE 16.3.16. In particular we have for the rigid rotator on $M = S^2$ since $K_g = 2$, $|\mathrm{Ric}|^2 = 2$, $|R|^2 = 4$, $\operatorname{Vol}(S^2, g) = 4$,

$$\vartheta_{\mathrm{Rot}}(\beta) \sim \frac{1}{\beta}\left(1 + \frac{\beta}{3} + \frac{\beta^2}{15} + \cdots\right);$$

and for the constrained rigid rotator $M = S^1$ since $|K|^2 = |R|^2 = |\mathrm{Ric}|^2 = 0$, $\vartheta(\beta) \sim (1/\sqrt{\beta}) + \mathcal{O}(e^{-1/\beta})$. These two results will be generalized in the next chapter.

Chapter 18 deals with the generalization of the Poisson summation formula which on $M = R^l/\Gamma$ has the form

$$\sum \exp(-4\pi^2(n,\alpha)\beta) = \frac{\mathrm{Vol}(M)}{(2\pi\beta)^{l/2}} \exp(-(n,\alpha)/2\beta),$$

where Γ is the group of parallel translations generated by $x \to x + a_i e_i$ where $\{e_i\}$ is an orthonormal basis on R^l. Here $a = (a_1, \ldots, a_l)$ and $n = (n_1, \cdots, n_l)$. The high temperature limit is

$$\vartheta_M(\beta) = \frac{\mathrm{Vol}(M)}{(2\pi\beta)^{l/2}} + \mathcal{O}(\exp(-\delta/\beta)).$$

In the case $M = R^2/\Gamma$ we are able to reconstruct Γ from a knowledge of the geodesics and the high temperature limit of the partition function. That is, in the language of physicists, we are able to determine or reconstruct spectroscopically.

The reconstruction works as follows. The inner product $(n, a) = \sum a_i n_i$ is just the lengths of periodic geodesics on M from x to γx. Let $\Lambda = \{|\gamma|\}$ denote the collection of distance $|\gamma| = g(x, \gamma x)$. Let $\gamma_1 = \min_{\gamma\in\Lambda} |\gamma|$. Then we can take γ_1 as one of the sides of the fundamental lattice parallelogram. Delete from Λ the values $k\gamma_1$ where $k = 1, 2, \ldots$. Let $|\gamma_2|$ be the minimum of the remaining numbers; and take γ_2 to be the second side of the parallelogram. Since we know the volume of M by the high temperature limit, we have reconstructed M and also Γ.

Uniqueness of reconstruction fails in higher dimensions – in particular for $M = R^{16}/\Gamma$ in an example due to Milnor; v. Berger *et al.* B8.

16.4. Zeta functions on compact Lie groups

The following objects are related: the zeta function $\zeta(s) = \sum \lambda_n^{-s}$, its distribution kernel $Z(p, q; s) = \sum \lambda_n^{-s} \psi_n(p)\psi_n(q)$; the theta function $\vartheta(\beta) = \sum \exp(-\lambda_n\beta)$, its distribution kernel $\Theta(p, q; \beta) = \sum \exp(-\lambda_n\beta)\psi_n(p)\psi_n(q)$; the resolvent kernel $R(p, q; \tau) = \sum \psi_n(p)\psi_n(q)/(\lambda_n + \tau)$ and the trace $R(\tau) = \sum 1/(\lambda_n + \tau)$ of the resolvent; all go back to Epstein, Carleman, Bochner, Weyl, Pleijel, Minakshisundaram *et al.* These objects of study are of course all interrelated: e.g.

$$Z(p, q; s) = \int (-\tau)^{-s} R(p, q; \tau)\, d\tau,$$

where the contour encloses all the poles, $\tau = -\lambda_n$, of the resolvent.

$$Z(p, q; s) = \frac{1}{\Gamma(s)} \int_0^\infty \Theta(p, q; \beta)\beta^{s-1}\, d\beta.$$

Everything can be reduced to the study of the kernel of Bloch's equation because

$$\Theta(x, x; \beta) = \mathrm{Tr}\,(\rho(\beta; x, x)).$$

We recall the theorem of Minakshisundaram and Pleijel:

THEOREM 16.4.1. Let M be a compact connected smooth Riemannian manifold with Laplace-Beltrami operator Δ_M and eigenvalue problem

$$\frac{-\hbar^2}{2\mu}\Delta_M\psi - \lambda\psi = 0$$

with eigenvalues and eigenfunctions $\{\lambda_n, \psi_n\}$. Then $Z(p, q; s)$ converges uniformly in p and q for all complex s such that $\mathrm{Re}(s)$ is large. The series can be continued arbitrarily far to the left of the abscissa of convergence giving rise to an entire function $Z(p, q; s)$ for $p \neq q$, with zeros at nonpositive integers; and for $p = q$ a meromorphic function $Z(p, p; s)$ with simple poles at

$$s = \tfrac{1}{2}d - n,\ n = 0, 1, 2, \cdots \quad \text{if } d \text{ is odd}$$

and

$$s = \tfrac{1}{2}d, \tfrac{1}{2}d - 1, \cdots, 2, 1 \quad \text{if } d \text{ is even.}$$

The residues at these poles are Riemannian invariants; and $Z(p, p; s)$ has zeroes at nonpositive integers for d odd.

COROLLARY (Carleman's Formulae) 16.4.2.

$$\sum_{\lambda_n \leq \lambda} \psi_n^2(p) \sim \frac{(2\mu\lambda)^{d/2}}{(4\pi\hbar^2)^{d/2}\,\Gamma\left(\dfrac{d}{2} + 1\right)} \quad \text{as} \quad \lambda \to \infty$$

and

$$\sum_{\lambda_n \leq \lambda} \psi_n(p)\psi_n(q) = o(\lambda) \quad \text{for } p \neq q.$$

THEOREM 16.4.3. The Dirichlet series $\zeta(s) = \sum \lambda_n^{-s}$ can be continued analytically to the left of its abscissa of convergence and the function so obtained can be written in the form

$$\zeta(s) = \frac{1}{(4\pi)^{d/2}} \sum_{k=0}^{m} \frac{\int_M u_k(p,p)\,dv(p)}{\Gamma\left(\frac{d}{2} - k\right)\left(s - \frac{d}{2} + k\right)} + R_n(s),$$

where $m = n$ if d is odd and $m = (d/2) - 1$ if d is even and where $R_n(s)$ is regular in the half-plane $Re(s) > (d/2)n - 2$. Here n is the order of the expansion

$$\Theta_n(p, q; \beta) = \frac{1}{(4\pi)^{d/2}} e^{-r^2/4\beta} \times \beta^{d/2} \times (u_0 + u_1 + \cdots + u_n\beta^n).$$

where we have set $\hbar^2/2\mu = 1$.

Thus $\Gamma(s)\zeta(s)$ has simple poles at $-k/2$ for $k \geqq -d$ and

$$\operatorname{Re}s(\Gamma(s)\zeta(s))|_{-k/2} = \int_M u_{(k+d)/2}(p,p)\,dv(p) = a_{(k+d)/2}.$$

Corollary 16.4.4. If $N(\lambda)$ is the number of eigenvalues $\lambda_n < \lambda$, then

$$N(\lambda) \sim \frac{\operatorname{Vol}(M)(2\mu\lambda)^{d/2}}{(4\pi\hbar^2)^{d/2}\Gamma\left(\frac{d}{2}+1\right)}.$$

16.5. Ising Models

The Ising model has been studied in statistical mechanics since it is one of the simplest systems which can be treated analytically and which undergoes a phase transition (in the two-dimensional case). The model is a lattice Γ of spins – originally the scalar spins μ_x, $x \in \Gamma$ assumed two orientations: $+1$ (spin up) and -1 (spin down); and each pair of nearest neighbor sites are joined by a bond. The energy of the configuration

$$\{\mu\} = \{\mu_x | x \in \Gamma\} \quad \text{is} \quad E(\{\mu\}) = -J \sum_{\text{bonds}} \mu_x\mu_y - \mathscr{H} \sum_{x\in\Gamma} \mu_x.$$

Here J is a coupling constant and $\mathscr{H}$ is the external magnetic field. The statistical mechanical problem is to evaluate the partition function

$$Z = \sum_{\{\mu\}} \exp(-\beta E(\{\mu\})).$$

EXAMPLE 16.5.1. In one dimension the case with $\mathscr{H}=0$ has

$$Z_N=\exp\left(v\sum_{i=1}^{N-1}\mu_i\mu_{i+1}\right),$$

where $v=\beta J$. It is easy to see that $Z=2(2\cosh v)^{N-1}$. The free energy F per spin in the 'thermodynamic limit' is

$$-F=\frac{1}{\beta}\lim_{N\to\infty}\frac{\log}{N}\left(Z_N\right)=\frac{1}{\beta}\log(2\cosh v)$$

which is analytic in v. Thus there is no phase transition in this one dimensional case.

EXAMPLE 16.5.2. Consider the case with $\mathscr{H}=0$ in two dimensions. This partition function can be expressed, as discovered by van der Waerden, as a sum over a finite group. If we let G denote the group of paths on Γ with coefficients in Z_2 i.e.

$$G=\underbrace{Z_2\times\cdots\times Z_2}_{B},$$

where $B=$ number of bonds on Γ, and if we let H denote the subgroup of G of closed paths (i.e. cycles), then

$$Z=\sum_{\text{spins}}\prod_{\text{bonds}}e^{v\mu_x\mu_y},$$

where $v=J\beta$. Setting c^* to be the root of $\sinh 2c\sinh 2c^*=1$; then

$$\prod_{\text{bonds}}(e^{c^*}+\mu_x\mu_y e^{-c^*})=\sum_{g\in G}e^{(B-n)c^*-nc^*}\prod_{\text{bonds of } g}\mu_x\mu_y,$$

where $n=n(g)$ is the number of bonds of $g\in G$. If A is the number of sites in the lattice we have

$$Z(\{\mu\})=2^A(\tfrac{1}{2}\sinh 2c)^{1/2B}\sum_{g\in H}f(g),$$

where $f(g)=e^{(B-n)c^*-nc^*}$.

Consider now a compact Lie group G with a fixed fundamental weight λ. For each $L=1,2,\ldots$ let $\mathscr{H}_L$ and Π_L be the space and representation with maximal weight $L\lambda$. For fixed L we let $\mathscr{H}_\alpha$ be a copy of $\mathscr{H}_L$ for $\alpha\in\Lambda$, where Λ

is a finite subset of Z^ν. We set $\mathscr{H}_\Lambda = \otimes_{\alpha\in\Lambda}\mathscr{H}_\alpha$. On $\mathscr{H}_\Lambda$ we define operators $S_\alpha(X)(\alpha\in\Lambda, X\in\mathfrak{g})$ to be the tensor product of $\Pi_L(X)$ in the αth factor and 1 in the other factors. Fix a basis $X_1,\ldots,X_m$ of $\mathfrak{g}$ and a function H of $|\Lambda|m$-vectors $S_{\alpha,i}$, $\alpha\in\Lambda$, $i=1,\ldots,m$ which is multiaffine, i.e. a sum of monomials which are of degree zero or one in the variables at each 'site'. Let $d_L = \dim\mathscr{H}_L$. Then the lattice partition function is defined as

$$Z_Q(\nu) = d_L^{-|\Lambda|}\,\mathrm{Tr}\,(\exp(-H(\nu S_\alpha(X_L)/L)).$$

If $\mathfrak{h}$ is the Cartan subalgebra of $\mathfrak{g}$ we extend λ to $\mathfrak{g}$ by setting $\lambda=0$ on $\mathfrak{h}$. We let γ be the coadjoint orbit in $\mathfrak{g}^*$ containing λ and we let $\mathrm{d}\tilde{\mu}(.)$ be the probability measure on γ given by the Haar measure $\mathrm{d}g$ on G. For each $\alpha\in\Lambda$ and copy γ_α of γ we set $\Gamma^{|\Lambda|} = \prod_\alpha \gamma_\alpha$. The classical partition function is defined as

$$Z_{\mathrm{cl}}(\nu) = \int_{\Gamma^{|\Lambda|}} \exp(-H(\nu \mathrm{l}_\alpha(X_i))\prod_\alpha \mathrm{d}\tilde{\mu}(\mathrm{l}_\alpha).$$

THEOREM (Simon–Lieb) 16.5.3.

$$Z_{\mathrm{cl}}(\nu) \leqq Z_Q(\nu) \leqq Z_{\mathrm{cl}}(\nu(1+aL^{-1})),$$

where $a = 4(\lambda,\delta)/(\delta,\delta)$ where $\delta = \frac{1}{2}\sum_{i=1}^r \lambda_i$.

In particular in the case $G = SO(3)$ we have

$$Z_{\mathrm{cl}}(\nu) = \int \prod_{\alpha\in\Lambda} \frac{\mathrm{d}\Omega}{4\pi}(S_\alpha)\exp(-H(\nu S_\alpha)),$$

where $\mathrm{d}\Omega$ is the usual measure on $S^2 \subset R^3$ while

$$Z_Q(\nu) = (2l+1)^{-|\Lambda|}\mathrm{Tr}(\exp(-H(\nu L_\alpha/l))$$

for $l = \frac{1}{2}, 1, \frac{3}{2}, \ldots$ and $\{L_\alpha\}$ is a family of independent spin l quantum spins.

The theorem of Lieb then demonstrates the convergence of Z_Q to Z_{cl} as $l\to\infty$ in a sufficiently strong way that one can interchange the $l\to\infty$ and the $|\Lambda|\to\infty$ limit in the free energy per unit volume. For details, v. Lieb L11a.

The basis of the proof of the Simon–Lieb result is to define a set of coherent states on the Lie group G based on the maximal weight vectors. These coherent states are parametrized by the limit manifold which is the associated orbit in $\mathfrak{g}^*$. Viz. let $P(e)$ be the projection onto the maximal weight vector for irreducible representation π of G. For g in G set $P(g) = \pi(g)P(e)\pi(g)^{-1}$. Let $\mathrm{d}\mu$ denote the Haar measure on G normalized so the

$\int_G \mathrm{d}\mu = d = \dim(\pi)$. Then by Schur's lemma $\int P(g)\,\mathrm{d}\mu(g) = 1$. We say a map from a measure space (G, μ) to $P(g)$, an orthogonal projection on Hilbert space H, is a *family of coherent projections* if $\dim P(g) = 1$ for all g in G and $\int P(g)\,\mathrm{d}\mu(g) = 1$. Coherent projections always arise from coherent vectors – i.e. if $P(g)$ is a family of coherent projections, then there is a measurable family $\Psi(g)$ of unit vectors so that $P(g) = (\Psi(g), .)\ \Psi(g)$. Furthermore the coherent vectors $\Psi(g)$ are complete. We return to this topic in Chapter 20.

Problems

Exercise 16.1.

(i) Show that $\dfrac{rL(x)}{\mathrm{d}r}\vartheta(m, m') = 0$,

(ii) $$\frac{1}{G}\frac{\partial G}{\partial r} = -\tfrac{1}{2}\det(g)^{-1/2}\frac{\partial}{\partial r}(\det(g)^{-1/2});$$

and

(iii) $$\Delta_m g^2(m, m') = -2n + \frac{4g}{G}\frac{\partial G}{\partial r}.$$

Exercise (Perelomov) 16.2. Consider the Schrödinger equation on a compact manifold M of dimension d:

$$H\psi_\lambda = \lambda\psi_\lambda,$$

where $H = -\Delta + u$. Let

$$\vartheta(\beta) = \operatorname{Tr}(e^{-\beta H}) = \sum_\lambda e^{-\lambda\beta}.$$

and let G be the Green's function for the heat equation

$$\frac{\partial G}{\partial\beta}(x, y, \beta) + HG(x, y, \beta) = \delta(\beta)\delta(x - y).$$

If $u = 0$, we set $G = G_0$.

(a) Show that $F(x, y, \beta) = G_0^{-1}G$ asymptotically satisfies

$$\frac{\partial F}{\partial\beta} + \beta^{-1}(x - y)\nabla_x F(x, y, \beta) - \Delta_x F(x, y, \beta) + u(x)F(x, y, \beta) = 0.$$

(b) Show that in the asymptotic expansion

$$\vartheta(\beta) \sim (4\pi\beta)^{-d/2}(1 + a_1\beta + a_2\beta^2 + \cdots)$$

the coefficients a_k are given by

$$a_k = \int F_k(x, x)\,dx,$$

where $F_k(x, x)$ are polynomials in u and the derivatives of u: viz.

$$a_k = \frac{(-1)^k}{k!} \int P_k(u, u_i, u_{ij}, \ldots)\,dx;$$

where

$$P_1 = u, P_2 = u^2, P_3 = u^3 + \tfrac{1}{2}(\nabla u)^2, \quad \text{etc.}$$

(c) Show the coefficient a_k is equal to $(-1)^k/k!$ times the kth conserved integral of Korteweg–de Vries equation:

$$U_t = 6UU_x - U_{xxx},$$

viewing the potential as a one-parameter family which satisfies the $K\,dV$ equation.

EXERCISE (Dikii) 16.3. For the Sturm–Liouville problem

$$-\psi'' + u(x)\psi = \lambda\psi, \quad \psi(0) = \psi(\pi) = 0$$

with $u(x)$ infinitely differentiable.

(a) Show that

$$z(s) = \sum_{\lambda} \lambda^{-s}$$

has poles at $s = \frac{1}{2}, -\frac{1}{2}, -\frac{3}{2}, \ldots$;

(b) Show that

$$\sum_n \frac{1}{\lambda_n + z} \sim \frac{\pi}{2\sqrt{z}} + \frac{1}{2z} + \sum_{k=1}^{\infty} \frac{m_k}{z^{k+1/2}}$$

for large z where the m_k are certain combinations of terms of the form

$$\int_0^{\pi} u^{(k_1)}(x) \ldots u^{(k_n)}(x)\,dx$$

with

$$k_1 + \cdots + k_n + 2n = 2k;$$

(c) Evaluate the residues of the poles of

$$z(\beta) = \sum_{n=1}^{\infty} \lambda_n^{-s}.$$

(d) Show that

$$\sum_{n=1}^{\infty} e^{\lambda_n \beta} \sim \frac{1}{\sqrt{4\pi\beta}} \sum_{m=0} \frac{(-\beta)^m}{(2m-3)\dots 3.1} H_{m-1} - \tfrac{1}{2} \sum_{p=0}^{\infty} \frac{(-t)^p}{p!} L_p, \text{ as } \beta \to 0$$

where H_m is the 'Hamiltonian'

$$H_m = \int_0^{\pi} I_m[u(x), u'(x), \dots]\,\mathrm{d}x$$

and

$$L_p = \mu_0^p + \sum_{i=1}^{\infty} (\mu_{\alpha i-1}^p + \mu_{\alpha i}^p - 2\lambda_i).$$

For supplementary reading, cf. McKean–van Moerbeke [M12] (and references therein), Hermann [H9]; etc.

Exercise 16.4. The asymptotic solution of Bloch's equation was derived as early as 1940 by Husimi; cf also work of Dirac in 1934 and Copson in 1948. Viz. for the equation

$$\partial\rho/\partial\beta = \{\tfrac{1}{2}\partial^2/\partial q^2 - V(q)\}\rho$$

assume that the solution is of the form

$$\rho = \frac{1}{\sqrt{2\pi\beta}} \exp(-(q-q')^2/2\beta - \beta U(qq'|\beta)).$$

Expand U as $U = U_0 + U_1\beta + \cdots$ substitute and equate coefficients; show that $U_0 = (1/q - q') \int_{q'}^{q} V(q)\,\mathrm{d}q$; so for $q \to q'$, $U_0 = V(q')$; $U_1 = 1/[2(q-q')^2]\ \{V(q) + V(q') - 2U_0\}$; so for $q \to q'$, $U_1 = \frac{1}{12}V''(q')$; similarly find U_2; thus show that

$$\rho(q, q', \beta) \sim \frac{1}{\sqrt{2\pi\beta}} \exp\left[\frac{-(q-q')^2}{2\beta} \frac{-\beta}{q-q'} \int_q^q V(q)\,\mathrm{d}q - \cdots\right]$$

and

$$\rho(q, q, \beta) \sim \frac{1}{\sqrt{2\pi\beta}} \exp\left[-V(q)\beta - \frac{V''(q)}{12}\beta^2 \cdots\right].$$

Chapter 17

Quantum Statistical Mechanics

17.1. Quantum statistical mechanics on compact symmetric spaces

Let G be a compact real connected Lie group of dimension N and Lie algebra $\mathfrak{g}$. We assume there is a biinvariant Riemannian structure on G. Let T be a maximal torus of G with Lie algebra $\mathfrak{h}$. Let $n = \dim T$. As we have discussed before there exists an orthonormal basis for $\mathfrak{g}$ given by $H_1, \ldots, H_n$ (the basis for $\mathfrak{h}$) and X_r, Y_r for $1 \leqq r \leqq (N-n)/2$ which satisfy the relationships

(i) $[H, X_r] = -2\prod \alpha_r(H) Y_r$ for H in $\mathfrak{h}$;

(ii) $[H, Y_r] = 2\prod \alpha_r(H) X$;

(iii) $[Y_r, X_r] = h_r'$,

where $\langle H, h_r' \rangle = 2\prod \alpha_r(H)$. With respect to this basis we have

THEOREM 17.1.1. The Laplacian on G is of the form

$$-\Delta = \sum_{j=1}^{n} H_j^2 + \sum_r (X_r^2 + Y_r^2).$$

DEFINITION 17.1.2. Let U be an irreducible representation of group G on space V and let K be a closed subgroup of G. Assume that U is class one with respect to K and K is massive–i.e. there is only one invariant element v in V with respect to U. The function $v(g) = (U(g)v, v)$ is the *zonal spherical function* corresponding to v.

In the present case we have

THEOREM 17.1.3. The zonal spherical functions on the compact symmetric space $M = G/H$ are given by

$$\Phi(h.\lambda) = (-i)^{(N-n)/2} \prod_{\alpha \geqq 0} \frac{(\alpha, \delta) \sum_{w \in W} \det w \exp(i(\lambda, wh)}{\prod_{\alpha > 0} (\alpha, \lambda) j(h)},$$

where

$$j(h) = \prod_{\alpha > 0} \sin \pi(\alpha, h).$$

THEOREM 17.1.4. $\Phi(h, \lambda)$ are eigenfunctions of Δ_M with eigenvalue $-(\lambda, \lambda) - (\delta, \delta)$ where $\delta = \sum_{\alpha \geqq 0} \alpha$.

THEOREM 17.1.5. Consider the Bloch equation $\frac{1}{2}\Delta_M \rho = \partial\rho/\partial\beta$ on a compact symmetric space $M = G/K$ and let $\Delta_M \psi_\lambda = E_\lambda \psi_\lambda$. Then $E_\lambda = -(\lambda,\lambda) - (\delta,\delta)$ and the multiplicity of E_λ is Mult $(E_\lambda) = n(\lambda) d(\lambda)$ where $n(\lambda) = \int_K \mathfrak{X}^{(\lambda)}(k) \mathrm{d}k$ (with $\int_K \mathrm{d}k = 1$) and $d(\lambda) = \prod_{j=1}^n [(a/2) + \lambda), \alpha_j]$ where $a = \sum_{i=1}^r \alpha_i$, sum over the simple roots.

Proof. Let $E_G(\beta, g)$ be the fundamental solution of the Bloch equation $\partial\rho/\partial\beta = \Delta_G \rho$ on G. Since E_G is invariant by inner automorphisms of G, E_G has an expansion in terms of characters (see Exercise 1.4) of the form

$$E_G(\beta, g) = \sum_{\lambda \in \hat{G}} a_\lambda(\beta) \mathfrak{X}_\lambda(g).$$

Substituting this into the Bloch equation we find that

$$a'_\lambda(\beta) = -E_\lambda a_\lambda(\beta).$$

Thus $a_\lambda(\beta) = a_\lambda(0) \exp(-E_\lambda \beta)$ where

$$a_\lambda(0) = \lim_{\beta \to 0} \int_G E(\beta, g) \mathfrak{X}_\lambda(g^{-1}) \mathrm{d}g = \mathfrak{X}_\lambda(e) = \mathrm{d}(\lambda).$$

Checking that

$$E_{G/K}(\beta, gK) = \int_K E_G(\beta, gk) \,\mathrm{d}k$$

we have

$$E_{G/K}(\beta, gK) = \sum_{\lambda \in \hat{G}_1} \mathrm{d}(\lambda) \exp(-E_\lambda \beta) \varphi_\lambda(gK), \qquad (*)$$

where

$$\varphi_\lambda(gK) = \int_K \mathfrak{X}_\lambda(gk) \,\mathrm{d}k.$$

Here we have set $E_{G/K}(\beta, y^{-1}xK, K) = E_{G/K}(\beta, gK)$. Thus using $g = e$ in $(*)$ we have

$$E_{G/K}(\beta, gK, gK) = \sum_{\lambda \in \hat{G}_1} d(\lambda) \dim Z(V_\lambda) \exp(-E_\lambda \beta),$$

where V_λ is the representation space of U_λ and $Z(V_\lambda)$ is the subspace of V_λ of elements v such that $U_\lambda(k)v = v$.

By Corollary 18.1.5 to the Frobenius Reciprocity Theorem we have $n(\lambda) = \int_K \mathfrak{X}_\lambda(k)\,dk$. And by the Weyl character formula we have

$$d(\lambda) = \prod_{\alpha\in\Delta^+} \frac{(\lambda+\delta,\alpha)}{(\delta,\alpha)}.$$

Thus the result follows.

Corollary 17.1.6. If G/K is a symmetric space, then $\dim Z(V_\lambda) = 1$ and

$$E_{G/K}(\beta, gK, gK) = \sum_{\lambda\in\hat{G}_1} d(\lambda)\exp(-E_\lambda\beta).$$

A more refined result can be stated by using the following theorem of Cartan.

Let G be a compact connected Lie group with involutive automorphism σ. Let K be an open subgroup of $G^\sigma = \{g\in G\,|\,\sigma(g) = g\}$. The Lie algebra $\mathfrak{g}$ of G decomposes under σ as $\mathfrak{g} = \mathfrak{k} + \mathfrak{s}$. We choose a maximal abelian subspace $\mathfrak{a}$ of $\mathfrak{s}$ and $\Delta_\mathfrak{a}^+$ a positive $\mathfrak{a}^\mathbb{C}$-root system of $\mathfrak{g}^\mathbb{C}$. Set $\mathfrak{m} = \{X\in\mathfrak{k}\,|\,[X,\mathfrak{a}] = 0\}$ and let $\mathfrak{t}$ be a Cartan subalgebra of $\mathfrak{m}$. Then $\mathfrak{h} = \mathfrak{t} + \mathfrak{a}$ is a Cartan subalgebra of $\mathfrak{g}$. Any choice of a positive $\mathfrak{t}^\mathbb{C}$-root system on $\mathfrak{m}^\mathbb{C}$ specifies a choice of a positive root system Δ^+ of $(\mathfrak{g}^\mathbb{C}, \mathfrak{h}^\mathbb{C})$ such that $\Delta_\mathfrak{a}^+ = \{\alpha|\mathfrak{a}\,|\,\alpha\in\Delta^+ \alpha|\mathfrak{a} \neq 0\}$. Define

$$\Lambda^+ = \left\{\lambda\in i\mathfrak{a}^* \,\middle|\, \frac{(\lambda,\alpha)}{(\alpha,\alpha)} \in\mathbb{Z}, \text{ is} \geqq 0 \text{ for all } \alpha\in\Delta_\mathfrak{a}^+\right\}.$$

Theorem (Cartan) 17.1.7. If G/K is a symmetric space of rank one with G simply connected and K connected, then there is a bijection $\hat{G}_1 \to \Lambda^+$ where $[T]\in\hat{G}_1 \to$ highest weight relative to $(\mathfrak{h}, \Delta^+)$.

Corollary 17.1.8. If M is a simply connected symmetric space, then the partition function is given by

$$\vartheta_{G/K}(\beta) = \sum_{\lambda\in\Lambda^+} \mathrm{d}(\lambda)\exp(-\beta E_\lambda),$$

where $E_\lambda = \|\lambda+\delta\|^2 - \|\delta\|^2 = \|\lambda+\delta_\mathfrak{a}\|^2 - \|\delta_\mathfrak{a}\|^2$ where

$$\|\alpha\|^2 = (\alpha,\alpha) \quad \text{and} \quad \alpha_\mathfrak{a} = \alpha|\mathfrak{a}.$$

We also note here that Eskin has developed a closed form expression for the density matrix on a compact Lie group. Since $\rho(\beta, x, y) = \rho(\beta, y^{-1}x, e) = \rho(\beta, y^{-1}x)$ we see that $g \to \rho(\beta, g)$ is invariant under inner automorphisms of maximal torus T. Thus $\rho(\beta, .)$ is determined by its restriction to T. Thus we state

THEOREM (Exkin) 17.1.9. The density matrix on a compact semisimple Lie group G is given by

$$\rho(\beta, h) = \frac{\mathrm{Vol}(\mathfrak{t}/\Gamma)\Delta^{1/2}\exp(4\pi^2(\delta,\delta)\beta)}{2^{(N+n)/2}\pi^{N/2}(i)^{(N-n)}\prod_r(\delta, h'_r)\beta^{n/2}}\sum_{\gamma\in\Gamma} \times$$

$$\times \frac{L(h)\exp(-(h,h)/4\beta)(h+\gamma)}{j(H+\gamma)}$$

for $h \in T'$ (regular elements in T) where Γ is the integral lattice in $\mathfrak{t}$, Δ is the discriminant of the scalar product on $\mathfrak{t}$ with respect to R^n.

$$j(H) = \prod_r \exp(\pi i \alpha_r(H)) - \exp(-\pi i \alpha_r(H))$$

and $L(h) = \prod_r L(h, h'_r)$ where $L(h, h'_r)$ is the derivation at h along h'_r.

COROLLARY 17.1.10. The diagonals of the density matrices on symmetric spaces G/K are given by

$$E_{G/K}(\beta, xK, yK) = \int_G E_G(\beta, gk)\,\mathrm{d}k = \frac{1}{|W|}\int_T E_G(\beta, u)|j(u)|^2\,\mathrm{d}u.$$

Rather than expanding these integrals for the high temperature asymptotic limits we turn to an approach which connects to the work in quantum field theory in Chapter 19.

The first interesting examples of the Minakshisundaram-like expansion of the partition function were given by Fowler and his student Mulholland for the rigid rotator in 1928. Some twenty years later Minakshisundaram treated the more general case of the arbitrary n sphere. However, only recently did Wolf and Cahn realize that the original work by Mulholland provides a solution to the asymptotic expansion for the general compact symmetric space of rank one. Viz. we are able to generate formulas for all the coefficients in the high temperature expansion

$$\vartheta(\beta) = (4\pi\beta)^{-n/2}(a_0 + a_1\beta + a_2\beta^2 + \cdots) \quad \text{as} \quad \beta \to 0.$$

Before examining the asymptotics we use Theorem 17.1.7 and the theory of Lie algebras to derive the partition functions for the symmetric spaces $M = G/K$, rank one. These were first derived by Cahn and Wolf, whose work we follow closely.

Consider first the odd dimension spheres $S^{2n-1} = SO(2n)/SO(2n-1) n \geqq 1$. We have already treated $n = 1$. In case $n = 2$ $G = SO(4)$ has Dynkin diagram $D_2 : \mathring{\alpha}_1 \mathring{\alpha}_2$. Thus $\Delta^+ = \{\alpha_1, \alpha_2\}$, $\Delta_{\mathfrak{a}}^+ = \{\alpha\}$ where $\delta = \delta_{\mathfrak{a}} = \alpha = \frac{1}{2}(\alpha_1 + \alpha_2)$. Thus $\Lambda^+ = \{m\alpha : m \geqq 0 \text{ integer}\}$ and $E_{m\alpha} = \| m\alpha + \delta_{\mathfrak{a}} \|^2 - \| \delta_{\mathfrak{a}} \|^2 = (m^2 + 2m) \| \delta_{\mathfrak{a}} \|$ and we calculate

$$\mathrm{d}(m\alpha) = \frac{\langle m\alpha + \delta, \alpha_1 \rangle}{\langle \delta, \alpha_1 \rangle} \cdot \frac{\langle m\alpha + \delta, \alpha_2 \rangle}{\langle \delta, \alpha_2 \rangle} = (m+1)^2.$$

Using the negative of the Killing form to specify the Riemannian metrics of S^3 and $P^3(R)$, the tables at the end of Bourbaki $B19$ show $\langle \alpha_i, \alpha_i \rangle = \frac{1}{2}$, so $\| \delta_{\mathfrak{a}} \|^2 = \frac{1}{4} \langle \alpha_1 + \alpha_2, \alpha_1 + \alpha_2 \rangle = \frac{1}{4}$, and

$$\vartheta_{S^3}(\beta) = \sum_{m=0}^{\infty} (m+1)^2 e^{-\beta(m^2 + 2m)/4}.$$

It is classical that $T_{m\alpha}(-I) = 1$ just when m is even, so also

$$\vartheta_{P^3(R)}(\beta) = \sum_{r=0}^{\infty} (2r+1)^2 e^{-\beta(r^2 + r)}.$$

This case was treated also by C. deWitt in D6.

Now we assume $n \geqq 3$ in (2.1) so that $G = SO(2n)$ is a simple group of type D_n, and denote its Dynkin diagram

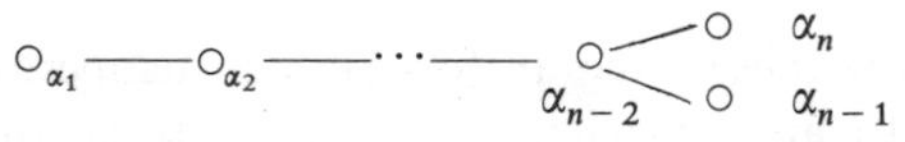

with

$$\Delta_{\mathfrak{a}}^+ = \{\alpha\}, \quad \alpha_1|_{\mathfrak{a}} = \alpha \quad \text{and} \quad \alpha_i|_{\mathfrak{a}} = 0 \quad \text{for} \quad i > 1.$$

Relative to an appropriate positive multiple of the Cartan–Killing form, $i\mathfrak{z}^*$ has orthonormal basis $\{\varepsilon_1, \ldots, \varepsilon_n\}$ such that

$$\alpha_i = \varepsilon_i - \varepsilon_{i+1} \quad \text{for} \quad 1 \leqq i < n \text{ and } \alpha_n = \varepsilon_{n-1} + \varepsilon_n.$$

Thus Δ^+ consists of the roots $\varepsilon_i \pm \varepsilon_j$ for $1 \leqq i < j \leqq n$, and so $\alpha = \varepsilon_1$ and $\varepsilon_1 \pm \varepsilon_j (1 < j \leqq n)$ are the roots that restrict to α.

Now

$$\Lambda^+ = \{m\varepsilon_1 \mid m \geqq 0 \text{ integer}\}, \ \delta_{\mathfrak{a}} = (n-1)\varepsilon_1 \text{ and } \delta = \sum_{j=1}^{n-1} (n-j)\varepsilon_j.$$

If $1 \leqq i < j \leqq n$ then $\langle \delta, \varepsilon_i \pm \varepsilon_j \rangle = \{(n-i) \pm (n-j)\} \| \varepsilon_1 \|^2$, so

$$\frac{\langle m\varepsilon_1 + \delta, \varepsilon_i \pm \varepsilon_j \rangle}{\langle \delta, \varepsilon_i \pm \varepsilon_j \rangle} = 1 \quad \text{if} \quad i > 1, = \frac{m + (n-1) \pm (n-j)}{(n-1) \pm (n-j)} \quad \text{if} \quad i = 1.$$

That gives us

$$\delta(m\varepsilon_1) = \prod_{j=2}^{n} \frac{m + 2n - j - 1}{2n - j - 1} \cdot \frac{m - 1 + j}{j - 1} = \frac{m + n - 1}{n - 1} \prod_{k=1}^{2n-3} \frac{m + k}{k}.$$

Recall from the tables at the end of Bourbaki B19 that $\| \varepsilon_1 \|^2 = (1/4)(m-1)$. Now

$$E_{m\varepsilon_1} = \| m\varepsilon_1 + \delta_{\mathfrak{a}} \|^2 - \| \delta_{\mathfrak{a}} \|^2 = \{m^2 + 2m(n-1)\}/4(n-1).$$

Now Corollary 17.1.8 gives us

THEOREM 17.1.11. Let S^{2n-1} denote the sphere of odd dimension $2n - 1$, $n \geqq 3$, with Riemannian metric of constant positive curvature induced by the negative of the Cartan–Killing form of $SO(2n)$. It has partition function

$$\vartheta_{S^{n-1}}(\beta) = \sum_{m=0}^{\infty} \left\{ \frac{m + n - 1}{n - 1} \cdot \prod_{k=1}^{2n-3} \frac{m + k}{k} \right\} e^{-\beta\{m^2 + 2m(n-1)\}/4(n-1)}. \quad (*)$$

The real projective space $P^{2n-1}(R) = S^{2n-1}/\{\pm I\}$ has partition function given by summing the summands of (*) whose representations $[T_{m\varepsilon_1}]$ occur in $L_2(P^{2n-1}(R))$, that is the one with vector fixed under the subgroup $SO(2n-1) \cup (-I_{2_n}) \cdot SO(2n-1)$. These are the $[T_{m\varepsilon_1}]$ whose kernel contains $-I_{2n}$, which are easily seen to be the ones for which m is even.

COROLLARY 17.1.12. Let $P^{2n-1}(R)$ denote the real projective space of odd dimension $2n - 1$, $n \geqq 3$, with Riemannian metric of constant positive curvature induced by the negative of the Cartan–Killing form of $SO(2n)$. It has partition function

$$\vartheta_{P^{2n-1}(R)}(\beta) = \sum_{r=0}^{\infty} \left\{ \frac{2r + n - 1}{n - 1} \cdot \prod_{k=1}^{2n-3} \frac{2r + k}{k} \right\} e^{-\beta\{r^2 + r(n-1)\}/(n-1)}.$$

We now treat the case of partition functions of spheres and real projective spaces of even dimension $2n$.

$$S^{2n} = SO(2n+1)/SO(2n), \quad n \geqq 1,$$

and

$$P^{2n}(R) = S^{2n}/\{\pm I\} = SO(2n+1)/SO(2n) \times O(1).$$

$G = SO(2n+1)$ has Dynkin diagram

$$B_n : \circ_{\alpha_1} \text{---} \circ_{\alpha_2} \text{---} \cdots \text{---} \circ_{\alpha_{n-1}} \text{---} \circ_{\alpha_n}$$

with

$$\Delta_{\mathfrak{a}}^{+} = \{\alpha\}, \quad \alpha_1|_{\mathfrak{a}} = \alpha \quad \text{and} \quad \alpha_i|_{\mathfrak{a}} = 0 \quad \text{for} \quad i > 1.$$

Relative to an appropriate multiple of the Cartan–Killing form, $i\mathfrak{z}^*$ has orthonormal basis $\{\varepsilon_1, \ldots, \varepsilon_n\}$ such that

$$\alpha_i = \varepsilon_i - \varepsilon_{i+1} \quad \text{for} \quad 1 \leqq i < n \quad \text{and} \quad \alpha_n\text{-}\varepsilon_n.$$

Thus Δ^+ consists of the roots $\varepsilon_i \pm \varepsilon_j$ for $1 \leqq i < j < n$ and the roots ε_k for $1 \leqq k \leqq n$, and so $\alpha = \varepsilon_1$, and ε_1 and $\varepsilon_1 \pm \varepsilon_j$ $(2 \leqq j \leqq n)$ are the roots restricting to α.

Now

$$\Lambda^+ = \{m\varepsilon_1 \mid m \geqq 0 \text{ integer}\}, \quad \delta_{\mathfrak{a}} = \frac{2n-1}{2}\varepsilon_1 \quad \text{and}$$

$$\delta = \frac{1}{2}\sum_{j=1}^{n} (2n-2j+1)\varepsilon_j.$$

If $1 \leqq i < j \leqq n$ then $\langle \delta, \varepsilon_i \pm \varepsilon_j \rangle = \frac{1}{2}\{(2n-2i+1) \pm (2n-2j+1)\} \|\varepsilon_1\|^2$, so

$$\frac{\langle m\varepsilon_1 + \delta, \varepsilon_i \pm \varepsilon_j \rangle}{\langle \delta, \varepsilon_i \pm \varepsilon_j \rangle} = 1 \quad \text{if} \quad i > 1,$$

$$= \frac{2m + (2n-1) \pm (2n-2j+1)}{(2n-1) \pm (2n-2j+1)} \quad \text{if} \quad i = 1.$$

If $1 \leqq k \leqq n$ then $\langle \delta, \varepsilon_k \rangle = \frac{1}{2}(2n-2k+1)\|\varepsilon_1\|^2$, so

$$\frac{\langle m\varepsilon_1 + \delta, \varepsilon_k \rangle}{\langle \delta, \varepsilon_k \rangle} = 1 \quad \text{if} \quad k > 1, = \frac{2m+2n-1}{2n-1} \quad \text{if} \quad k = 1.$$

That gives us

$$\mathrm{d}(m\varepsilon_1) = \frac{2m+2n-1}{2n-1} \cdot \prod_{j=2}^{n} \frac{m+2n-j}{2n-j} \cdot \frac{m+j-1}{j-1} =$$

$$= \frac{2m+2n-1}{2n-1} \prod_{k=1}^{2n-2} \frac{m+k}{k}.$$

Recall from the tables of Bourbaki B19 that $\|\varepsilon_1\|^2 = 1/4n - 2$. Then

$$E_{m\varepsilon_1} = \|m\varepsilon_1 + \rho_{\mathfrak{a}}\|^2 - \|\rho_{\mathfrak{a}}\|^2 = \{m^2 + m(2n-1)\}/(4n-2).$$

Thus we have

THEOREM 17.1.13. Let S^{2n} denote the sphere of even dimension $2n$ with Riemannian metric of constant positive curvature induced by the negative of the Cartan–Killing form of $SO(2n+1)$. It has partition function

$$\vartheta_{S^{2n}}(\beta) = \sum_{m=0}^{\infty} \left\{ \frac{2m+2n-1}{2n-1} \prod_{k=1}^{2n-2} \frac{m+k}{k} \right\} e^{-\beta\{m^2+m(2n-1)\}/(4n-2)}.$$

COROLLARY 17.1.14. Let $P^{2n}(R)$ denote the real projective space of even dimension $2n$ with Riemannian metric of constant positive curvature induced by the negative of the Cartan–Killing form $SO(2n+1)$. It has partition function

$$\vartheta_{P^{2n}(R)}(\beta) = \sum_{r=0}^{\infty} \left\{ \frac{4r+2n-1}{2n-1} \prod_{k=1}^{2n-2} \frac{2r+k}{k} \right\} e^{-\beta\{2r^2+r(2n-1)\}/(2n-1)}$$

We now consider the case of partition functions on the complex projective spaces

$$P^n(C) = U(n+1)/U(n) \times U(1) = SU(n+1)/S(U(n) \times U(1))$$

of complex dimension n, real dimension $2n$. Since $P^1(C)$ is the sphere S^2, which we already considered, we will work under the hypothesis $n > 1$. The case $n = 1$ already treated will show the conclusion is valid in general.

$$G = SU(n+1)/\{e^{2\pi ik/(n+1)}I\}$$

has Dynkin diagram

$$A_n : \circ_{\alpha_1} \text{———} \circ_{\alpha_2} \text{———} \cdots \text{———} \circ_{\alpha_n}$$

with

$$\Delta_{\mathfrak{a}}^{+} = \{\alpha, 2\alpha\}, \alpha_1|_{\mathfrak{a}} = \alpha = \alpha_n|_{\mathfrak{a}}, \alpha_i|_{\mathfrak{a}} = 0 \quad \text{for} \quad 1 < i < n.$$

Relative to appropriate multiple of the Cartan–Killing form, $i\mathfrak{z}^*$ is isometric to the hyperplane

$$\left\{ \sum_{1}^{n+1} x^i \varepsilon_i \,\middle|\, \sum x^i = 0 \right\}$$

in an R^{n+1} with orthonormal basis $\{\varepsilon_1, \ldots, \varepsilon_{n+1}\}$, where $\alpha_i = \varepsilon_i - \varepsilon_{i+1}$. Thus Δ^+ consists of the roots $\alpha_i + \alpha_{i+1} + \cdots + \alpha_j = \varepsilon_i - \varepsilon_{j+1}$ for $1 \leqq i \leqq j \leqq n$, and so

$\alpha = \frac{1}{2}(\varepsilon_1 - \varepsilon_{n+1})$, and $\varepsilon_1 - \varepsilon_j$ and $\varepsilon_j - \varepsilon_{n+1} (2 \leqq j \leqq n)$ restrict to α;

$2\alpha = \varepsilon_1 - \varepsilon_{n+1}$, and $\varepsilon_1 - \varepsilon_{n+1}$ is the only root restricting to 2α.

Now

$$\Lambda^+ = \{2m\alpha = m(\varepsilon_1 - \varepsilon_{n+1}) \mid m \geqq 0 \text{ integer}\} \text{ and}$$

$$\delta_{\mathfrak{a}} = n\alpha = \frac{n}{2}(\varepsilon_1 - \varepsilon_{n+1})$$

and

$$\delta = \frac{1}{2}\sum_{i \leqq j}(\varepsilon_i - \varepsilon_{j+1}) = \frac{1}{2}\sum_{j=1}^{n+1}(n - 2j + 2)\varepsilon_j.$$

Thus

$$\frac{\langle m(\varepsilon_1 - \varepsilon_{n+1}) + \delta, \varepsilon_i - \varepsilon_{j+1}\rangle}{\langle \delta, \varepsilon_i - \varepsilon_{j+1}\rangle}$$

has value 1 if $1 < i \leqq j < n$, $m + j/j$ if $1 = i \leqq j < n$, $(m + n + 1 - i)/(n + 1 - i)$ if $1 < i \leqq j = n$, and $(2m + n)/n$ if $1 = i < j = n$. So

$$\mathrm{d}(m(\varepsilon_1 - \varepsilon_{n+1})) = \frac{2m+n}{n}\prod_{k=1}^{n-1}\left(\frac{m+k}{k}\right)^2.$$

Recall from the tables at the end of Bourbaki B19 that $\|\varepsilon_i\|^2 = (1/2)(n+1)$, so $\|\alpha\|^2 = (1/4)(n+1)$. Now

$$E_m(\varepsilon_1 - \varepsilon_{n+1}) = \|2m\alpha + \delta_{\mathfrak{a}}\|^2 - \|\delta_{\mathfrak{a}}\|^2 = (m^2 + mn)/(n+1).$$

THEOREM 17.1.15. *Let $P^n(C)$ denote the complex projective n-space with Riemannian metric induced by the negative of the Cartan–Killing form of $SU(n+1)$. It has partition function*

$$\vartheta_{P^n(C)}(\beta) = \sum_{m=0}^{\infty}\left\{\frac{2m+n}{n}\prod_{k=1}^{n-1}\left(\frac{m+k}{k}\right)^2\right\}e^{-\beta\{m^2+mn\}/(n+1)}.$$

We now consider the partition functions on the quaternionic projective spaces

$$P^{n-1}(Q) = \mathrm{Sp}(n)/\mathrm{Sp}(n-1) \times \mathrm{Sp}(1), \qquad n \geqq 2,$$

of real dimension $4(n-1)$. Here note that $P^1(Q) = S^4$.

$G = \mathrm{Sp}(n)/\{\pm I\}$ has Dynkin diagram

C_n: 

with

$$\Delta_{\mathfrak{a}}^+ = \{\alpha, 2\alpha\}, \qquad \alpha_2|_{\mathfrak{a}} = \alpha, \qquad \alpha_i|_{\mathfrak{a}} = 0 \qquad \text{for} \qquad i \neq 2.$$

Relative to an appropriate multiple of the Cartan–Killing form, $i\mathfrak{z}^*$ has orthonormal basis $\{\varepsilon_1, \ldots, \varepsilon_n\}$ such that

$$\alpha_i = \varepsilon_i - \varepsilon_{i+1} \quad \text{for} \quad 1 \leqq i \leqq n-1 \quad \text{and} \quad \alpha_n = 2\varepsilon_n.$$

Thus Δ^+ consists of the roots $\varepsilon_i \pm \varepsilon_j$ for $1 \leqq i < j \leqq n$ and $2\varepsilon_k$ for $1 \leqq k \leqq n$, and so

$$\alpha = \tfrac{1}{2}(\varepsilon_1 + \varepsilon_2), \text{ and } \varepsilon_1 \pm \varepsilon_j \text{ and } \varepsilon_2 \pm \varepsilon_j (3 \leqq j \leqq n) \text{ restrict to } \alpha,$$
$$2\alpha = \varepsilon_1 + \varepsilon_2, \text{ and } 2\varepsilon_1, \varepsilon_1 + \varepsilon_2 \text{ and } 2\varepsilon_2 \text{ restrict to } 2\alpha.$$

Now

$$\Lambda^+ = \{2m\alpha = m(\varepsilon_1 + \varepsilon_2) : m \geq 0 \text{ integer}\} \text{ and}$$
$$\delta_{\mathfrak{a}} = (2n-1)\alpha = \frac{2n-1}{2}(\varepsilon_1 + \varepsilon_2),$$

and

$$\delta = \tfrac{1}{2}\left\{ \sum_{i<j} (\varepsilon_i - \varepsilon_j) + \sum_{i<j} (\varepsilon_i + \varepsilon_j) + 2 \sum_k \varepsilon_k \right\} = \sum_{l=1}^{n} (n - l + 1)\varepsilon_l.$$

Calculate $\langle \delta, \varepsilon_i \pm \varepsilon_j \rangle = \{(n-i+1) \pm (n-j+1)\} \| \varepsilon_1 \|^2$ and $\langle \delta, 2\varepsilon_k \rangle = 2(n-k+1) \| \varepsilon_1 \|^2$. Now, for $1 \leqq i \leqq j \leqq n$,

$$\frac{\langle m(\varepsilon_1 + \varepsilon_2) + \delta, \varepsilon_1 + \varepsilon_j \rangle}{\langle \delta, \varepsilon_i + \varepsilon_j \rangle} = 1 \quad \text{for} \quad 3 \leqq i, = \frac{m + 2n - j}{2n - j} \quad \text{for}$$

$$2 \leqq i < j,$$
$$= \frac{m + 2n - j + 1}{2n - j + 1} \quad \text{for} \quad 1 = i \quad \text{and}$$
$$3 \leqq j, = \frac{2m + 2n - i - j + 2}{2n - i - j + 2} \quad \text{for} \quad j \leqq 2;$$

and, for $1 \leqq i < j \leqq n$,

$$\frac{\langle m(\varepsilon_1 + \varepsilon_2) + \delta, \varepsilon_i - \varepsilon_j \rangle}{\langle \delta, \varepsilon_i - \varepsilon_j \rangle} = 1 \quad \text{if} \quad i > 2 \quad \text{or} \quad (i,j) = (1,2), =$$
$$= \frac{m + j - i}{j - i} \quad \text{if} \quad 1 \leqq i \leqq 2 < j.$$

Thus

$$\mathrm{d}(m(\varepsilon_1 + \varepsilon_2)) = \frac{2m + 2n - 1}{2n - 1} \cdot \prod_{r=2}^{2n-2} \frac{m + r}{r} \cdot \prod_{s=1}^{2n-3} \frac{m + s}{s}.$$

Recall from the tables at the end of Bourbaki B19 that $\|\varepsilon_i\|^2 = 1/4(n+1)$. Now $\|\alpha\|^2 = 1/8(n+1)$ and

$$E_m(\varepsilon_1+\varepsilon_2) = \|m(\varepsilon_1+\varepsilon_2)+\delta_{\mathfrak{a}}\|^2 - \|\delta_{\mathfrak{a}}\|^2 = (m^2+m(2n-1))/2(n+1).$$

THEOREM 17.1.16. Let $P^{n-1}(Q)$ denote the quaternionic projective $n-1$ space, with Riemannian metric induced by the negative of the Cartan–Killing form of Sp(n). It has partition function

$$\vartheta_{P^{n-1}(Q)}(\beta) = \sum_{m=0}^{\infty} \left\{ \frac{2m+2n-1}{2n-1} \cdot \prod_{r=2}^{2n-2} \frac{m+r}{r} \cdot \prod_{s=1}^{2n-3} \frac{m+s}{s} \right\} e^{-\beta(m^2+2mn-m)/2(n+1)}.$$

Notice that the case $n=2$ is $P^1(Q) = S^4$, which agrees with the earlier case, and has the partition function

$$\vartheta_{S^4}(\beta) = \sum_{m=0}^{\infty} \left\{ \frac{2m+3}{3} \cdot \frac{m+2}{2} \cdot \frac{m+1}{1} \right\} e^{-\beta(m^2+3m)/6}.$$

Finally, we work out the partition function for the Cayley projective plane

$P^2(\mathrm{Cay}) = F_4/\mathrm{Spin}(9)$, real dimension 16.
$G = F_4$ has Dynkin diagram 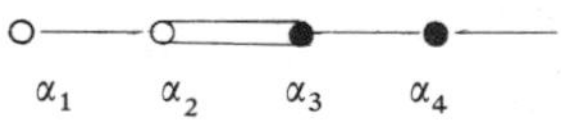 with $\Delta_{\mathfrak{a}}^+ = \{\alpha, 2\alpha\}$

where $\alpha_4|_{\mathfrak{a}} = \alpha$ and the other three $\alpha_i|_{\mathfrak{a}} = 0$. Relative to an appropriate multiple of the Cartan–Killing form, $i\mathfrak{z}^*$ has orthonormal basis $\{\varepsilon_1, \varepsilon_2, \varepsilon_3, \varepsilon_4\}$ with

$$\alpha_1 = \varepsilon_2 - \varepsilon_3, \alpha_2 = \varepsilon_3 - \varepsilon_4, \alpha_3 = \varepsilon_4 \quad \text{and}$$
$$\alpha_4 = \tfrac{1}{2}(\varepsilon_1 - \varepsilon_2 - \varepsilon_3 - \varepsilon_4).$$

Thus Δ^+ consists of the roots

$$\varepsilon_i (1 \leqq i \leqq 4), \varepsilon_i \pm \varepsilon_j (1 \leqq i < j \leqq 4), \tfrac{1}{2}(\varepsilon_1 \pm \varepsilon_2 \pm \varepsilon_3 \pm \varepsilon_4).$$

Now

$\alpha = \frac{1}{2}\varepsilon_1$ and the $\frac{1}{2}(\varepsilon_1 \pm \varepsilon_2 \pm \varepsilon_3 \pm \varepsilon_4)$ are the roots restricting to α,
$2\alpha = \varepsilon_1$ and the roots restricting to it are $\varepsilon_1, \varepsilon_1 \pm \varepsilon_2, \varepsilon_1 \pm \varepsilon_3, \varepsilon_1 \pm \varepsilon_4$.

Thus

$$\Lambda^+ = \{m\varepsilon_1 \mid m \geqq 0 \text{ integer}\},\ \delta_\alpha = 11\alpha = \tfrac{11}{2}\varepsilon_1,\ \delta = \tfrac{1}{2}(11\varepsilon_1 + 5\varepsilon_2 + 3\varepsilon_3 + \varepsilon_4).$$

Now calculate

$$\mathrm{d}(m\varepsilon_1) = \frac{2m+11}{11} \cdot \prod_{q=1}^{3} \frac{m+q}{q} \cdot \prod_{r=4}^{7} \left(\frac{m+r}{r}\right)^2 \cdot \prod_{s=8}^{10} \frac{m+s}{s}.$$

From the tables at the end of Bourbaki B19, $\|\varepsilon_1\|^2 = 1/18$, so $\|\alpha\|^2 = 1/72$, and thus

$$E_m\varepsilon_1 = \|(2m+11)\alpha\|^2 - \|11\alpha\|^2 = (m^2 + 11m)/18.$$

THEOREM 17.1.17. Let $P^2(\mathrm{Cay})$ denote the Cayley projective plane with Riemannian metric induced by the negative of the Cartan–Killing form of F_4. It has partition function

$$\vartheta_{P^2(\mathrm{Cay})}(\beta) = \sum_{m=0}^{\infty} \left\{ \frac{2m+11}{11} \cdot \prod_{q=1}^{3} \frac{m+q}{q} \cdot \right.$$

$$\left. \cdot \prod_{r=4}^{7} \left(\frac{m+r}{r}\right)^2 \cdot \prod_{s=8}^{10} \frac{m+s}{s} \right\} e^{-\beta(m^2+11m)/18}.$$

We turn now to the high temperature asymptotics of the partition functions on compact symmetric spaces of rank one. As noted above Cahn and Wolf have shown that Mulholland's results extend to these homogeneous spaces. We review this work now.

THEOREM (Mulholland). 17.1.18. Let $f(t) = \sum_{s\in Z} \exp(-s^2 t)$. Then $f(t) = \pi^{1/2} t^{-1/2} + 0(e^{-1/t})$ as $t \to 0$ and

$$(-1)^k f^{(k)}(t) = \sum_{s\in Z} \exp(-s^2 t) = \frac{(1)}{2}\frac{(3)}{2}\frac{(2k-1)}{2} \times$$

$$\times\, t^{\frac{-(2k+1)}{2}} + 0(e^{-1/t}) \quad \text{as} \quad t \to 0.$$

Proof. If we apply the Poisson summation formula to $\exp(-s^2 t)$ we have

$$\sum_{s\in Z} \exp(-s^2 t) = \pi^{1/2} t^{-1/2} \sum_{s\in Z} \exp - \pi^2 s^2/t.$$

Thus the first part follows by noting that $\sum_{s\neq 0}\exp(-\pi^2 s^2/t)$ is $0(e^{-1/t})$. We now take derivatives with respect to t to derive the rest.

DEFINITION 17.1.19. We set $B_0=1$, $B_k=(\frac{1}{2})(\frac{3}{2})\cdots(2k-1)/2$ for $k\geqq 1$.

We note that $\pi^{1/2}\ B_k=\Gamma((2k+1)/2)$.

Mulholland in M39 showed that

THEOREM 17.1.20. Let $g(t)=\sum_{j=0}^{\infty}(2j+1)e^{-(j+1/2)^2 t}$. Then

$$g(t)=\frac{1}{t}+c_0+c_1 t+\frac{c_2}{2!}t^2+\cdots+\frac{c_n}{n!}t^n+0(t^{n+1})$$

and

$$g^{(k)}(t)=\frac{(-1)^k k!}{t^{k+1}}+c_k+c_{k+1}t+\cdots+\frac{c_{k+n}}{n!}t^n+0(t^{n+1})t\downarrow 0$$

with

$$c_n=\frac{(-1)^n}{(n+1)}B_{2n+2}(1-2^{-2n-1}),$$

B_n is the nth Bernoulli number.

Before proceeding we wish to note that $\sum_{n=0}^{\infty}(c_n/n!)t^n$ is not convergent. Therefore these are asymptotic series.

THEOREM 17.1.21. Let $g_1(t)=\sum_{j=0}^{\infty}(4j+1)e^{-(2j+1/2)^2 t}$ and $g_2(t)=\sum_{j=0}^{\infty}(4j+3)e^{-(2j+3/2)^2 t}$. Then $g_1(t)+g_2(t)=g(t)$ and

$$g_i^{(k)}(t)=\frac{1}{2}\left(\frac{1}{t}+c_0+c_1 t+\cdots+\frac{c_n}{n!}t^n+0(t^{n+1})\right),\quad i=1,2$$

$$g_i^{(k)}(t)=\frac{1}{2}\left(\frac{(-1)^k k!}{t^{k+1}}+c_k+c_{k+1}t+\cdots+\frac{c_{k+n}}{n!}t^n+0(t^{n+1})\right),\quad i=1,2$$

as t 0.

THEOREM 17.1.22. Let $h(t)=\sum_{j=0}^{\infty}2je^{-j^2 t}$. Then

$$h(t)=\frac{1}{t}+d_0+d_1 t+\frac{d_2}{2!}t^2+\cdots+\frac{d_n}{n!}t^n+0(t^{n+1})$$

and

$$h^{(k)}(t)=\frac{(-1)^k k!}{t^{k+1}}+d_k+d_{k+1}+\cdots+\frac{d_{n+k}}{n!}t^n+0(t^{n+1})$$

as $t\downarrow 0$ with

$$d_n = \frac{(-1)^n}{n+1} B_{2n+2}.$$

Using Mulholland's results we are now able to calculate the coefficients a_n for the high temperature asymptotics of the partition functions just derived. Again we are following Cahn and Wolf's development.

From Theorem 17.1.18 we have

$M = S^1$.

$$\vartheta_M(\beta) = f\left(\frac{4\pi^2}{l^2}\beta\right).$$

Consequently

$$\vartheta_M(\beta) = \pi^{1/2}\left(\frac{4\pi^2}{l^2}\beta\right)^{-1/2} + 0(e^{-1/\beta})$$
$$= l(4\pi\beta)^{-1/2} + 0(e^{-1/\beta}).$$

Thus we conclude $a_0 = l$ and $a_m = 0$, $m \geqq 1$.

$M = S^3$.

$$\begin{aligned}
\vartheta_M(\beta) &= \sum_{m=0}^{\infty} (m+1)^2 e^{-\beta(m^2+2m)/4} \quad \text{if} \quad p = m+1 \\
&= \sum_{p=0}^{\infty} p^2 e^{-\beta(p^2-1)/4} \\
&= \tfrac{1}{2} e^{\beta/4} \sum_{p=-\infty}^{\infty} p^2 e^{-p^2\beta/4} \\
&= \tfrac{1}{2} e^{\beta/4}(-1) f'\left(\frac{\beta}{4}\right) \\
&= 2\pi^{1/2} t^{-3/2} e^{\beta/4} + ES = \frac{16\pi^2 e^{\beta/4}}{(4\pi t)^{3/2}} + ES,
\end{aligned}$$

therefore $a_m = 16\pi^2/4^m m!$. ES is an error which is exponentially small as $\beta \to 0$.

$M = P^3(R)$.

$$\begin{aligned}
\vartheta_M(\beta) &= \sum_{r=0}^{\infty} (2r+1)^2 e^{-(r^2+r)\beta} \\
&= \sum_{r=0}^{\infty} (2r+1)^2 e^{-[(2r+1)^2-1]\beta/4} \\
&= \tfrac{1}{2} e^{\beta/4} \left[\sum_{s\in\mathbb{Z}} s^2 e^{-s^2\beta/4} - \sum_{s\in 2\mathbb{Z}} s^2 e^{-s^2\beta/4} \right] \\
&= \tfrac{1}{2} e^{\beta/4} \left[(-1) f'\left(\frac{\beta}{4}\right) + 4 f'(\beta) \right] \\
&= \frac{\pi^{1/2}}{2} e^{\beta/4} [4\beta^{-3/2} - 2\beta^{-3/2}] + ES \\
&= \pi^{1/2} t^{-3/2} e^{\beta/4} + ES = \frac{8\pi^2 e^{\beta/4}}{(4\pi\beta)^{3/2}} + ES
\end{aligned}$$

so $a_m = 8\pi^2/4^m m!$.

$M = S^{2n-1} \cdot n \geqq 3$.

$$\vartheta_M(\beta) = \sum_{m=0}^{\infty} \left\{ \frac{m+n-1}{n-1} \prod_{k=0}^{2n-3} \frac{m+k}{k} \right\} e^{-\beta\{m^2+2m(n-1)\}/4(n-1)}.$$

We will let $s = m + n - 1$. Then

$$\begin{aligned}
\vartheta_M(\beta) &= \frac{1}{(n-1)(2n-3)!} \sum_{s=n-1}^{\infty} \left\{ \prod_{j=0}^{n-2} (s^2 - j^2) \right\} e^{-\{s^2-(n-1)^2\}\beta/4(n-1)} \\
&= \frac{e^{(n-1)/4}}{2(n-1)(2n-3)!} \sum_{s\in\mathbb{Z}} \left\{ \prod_{j=0}^{n-2} (s^2 - j^2) \right\} e^{-s^2\beta/4(n-1)}
\end{aligned}$$

We now define $\alpha_{k,n}$ by

$$\prod_{j=0}^{n-2} (s^2 - j^2) = \sum_{k=0}^{n-1} \alpha_{k,n} s^{2k}.$$

Then

$$\vartheta_M(\beta) = \frac{e^{(n-1)\beta/4}}{(2n-2)!} \sum_{s\in\mathbb{Z}} \sum_{k=0}^{n-1} \alpha_{k,n} s^{2k} e^{-s^2\beta/4(n-1)}$$

$$= \frac{e^{(n-1)\beta/4}}{(2n-2)!} \sum_{k=0}^{n-1} \alpha_{k,n}{}^{(-1)^k f^{(k)}}\left(\frac{\beta}{4(n-1)}\right)$$

$$= \frac{\pi^{1/2} e^{(n-1)\beta/4}}{(2n-2)!} \sum_{k=0}^{n-1} \alpha_{k,n} b_k \left(\frac{\beta}{4(n-1)}\right)^{-(2k+1)/2} + ES.$$

Now by convolving the series we conclude that

$$a_m = \frac{2^{2n-1}\pi^n}{(2n-2)!} \sum_{k=0}^{m} \frac{(n-1)^{n-1/2}}{k!} \alpha_{n-1-k,n} b_{n-1-k} 4^{n-1/2-2k}$$

if $m < n$

and

$$a_m = \frac{2^{2n-1}\pi^n}{(2n-2)!} \sum_{k=m-n}^{m} \frac{(n-1)^{m-1/2}}{k!} \alpha_{m-k-1,n} b_{n-k-1} 4^{m-1/2-2k}$$

if $m \geqq n$.

$M = P^{2n-1}(R)$. To compute the asymptotic expansion for the projective spaces we take $s = r + ((n-1)/2)$. Then

$$\frac{2r+n-1}{n-1} \prod_{j=1}^{2n-3} \frac{2r+k}{k} = \frac{2}{(2n-2)!} \prod_{j=0}^{n-2} 4(s^2 - (j/2)^2) =$$

$$= \frac{4^{n-1/2}}{(2n-2)!} \sum_{j=0}^{n-1} \alpha'_{j,n} s^{2j}.$$

$$\vartheta_M(\beta) = \frac{4^{n-1}}{(2n-2)!} e^{(n-1)\beta/4} \sum_{s \in 1/2\mathbb{Z}} \sum_{j=0}^{n-1} \alpha'_{j,n} s^{2j} e^{-s^2\beta/n-1}$$

$$= \frac{4^{n-1}}{(2n-2)!} e^{(n-1)\beta/4} \sum_{j=0}^{n-1} \alpha'_{j,n} 4^{-j} f^{(j)}\left(\frac{\beta}{4(n-1)}\right)$$

$$= \frac{4^{n-1}\pi^{1/2}}{(2n-2)!} e^{(n-1)\beta/4} \sum_{j=0}^{n-1} \alpha'_{j,n} b_j 4^{-j} \left(\frac{\beta}{4(n-1)}\right)^{-(2j+1)/2} +$$

$$+ ES.$$

Therefore

$$a_m = \frac{4^{n-1}\pi^n}{(2n-2)!} \sum_{j=0}^{m} \frac{(n-1)^{n-1/2}}{j!} \alpha'_{n-1-j,n} b_{n-1-j} 4^{-j-1/2}$$

if $m < n$

and

$$a_m = \frac{4^{n-1}\pi^n}{(2n-2)!} \sum_{j=m-n}^{m} \frac{(n-1)^{m-1/2}}{j!} \alpha'_{m-j-1,n} b_{m-j-1} 4^{-j-1/2}$$

if $m \geqq n$.

$M = S^{2n}$. As in the preceding cases we will change variables to utilize the Mulholland theorem. Recall that

$$\vartheta_M(\beta) = \sum_{m=0}^{\infty} \left\{ \frac{2m+2n-1}{2n-1} \prod_{k=1}^{2n-2} \frac{m+k}{k} \right\} e^{-\beta\{m^2+m(2n-1)\}/4n-2}.$$

We will let $s = m + (n - 1/2)$.

Then

$$\frac{2m+2n-1}{2n-1} \prod_{k=1}^{2n-2} \frac{m+k}{k} = \frac{2s}{(2n-1)!} \prod_{j=1/2}^{n-3/2} (s^2 - j^2) =$$

$$= \frac{2s}{(2n-1)!} \sum_{j=0}^{n-1} \beta_{j,n} s^{2j},$$

where the product runs through the half-integers which are not integers. Also $\beta(m^2 + m(2n-1)\}/4n - 2 = \beta\{s^2 - (n-1/2)^2\}/4n - 2$. Therefore

$$\vartheta_M(\beta) = \frac{e^{(n-1/2)\beta/4}}{(2n-1)!} \sum_{s \geqq 1/2} \sum_{j=0}^{n-1} \beta_{j,n} 2s^{2j+1} e^{-s^2\beta/4(n-1/2)}$$

$$= \frac{e^{(n-1/2)\beta/4}}{(2n-1)!} \sum_{j=0}^{n-1} \beta_{j,n}(-1)^j g^{(j)}\left(\frac{\beta}{4(n-1/2)}\right).$$

Thus

$$a_m = \frac{(4\pi)^2}{(2n-1)!} \sum_{k=0}^{m} \frac{(n-1-m+k)!}{k!} 4^{n-m-1}(n-1/2)^{m-n+2k+1} \beta_{n-1-m+k,n}$$

if $m < n$

and

$$a_m = \frac{(4\pi)^n}{(2n-1)!} \left(\sum_{k=0}^{n-1} \frac{k!}{(m-n+k+1)!} (n-1/2)^{m-n+2k+2} \beta_{k,n} 4^{n-m} + \right.$$

$$\left. + \sum_{k=0}^{m-n} \sum_{j=0}^{n-1} \frac{(-1)^j c_{j+k} \beta_{j,n} (n-1/2)^{m-n-2k}}{k!(m-n-k)!} 4^{n-m} \right)$$

if $m \geqq n$.

$M = P^{2n}(R)$. Recall that

$$\vartheta_M(\beta) = \sum_{r=0}^{\infty} \left\{ \frac{4r+2n-1}{2n-1} \prod_{k=1}^{2n-2} \frac{2r+k}{k} \right\} e^{-\beta\{2r^2+r(2n-1)\}/2n-1}.$$

We will let

$$s = r + \left(\frac{2n-1}{4}\right).$$

Then

$$\frac{4r+2n-1}{2n-1} \prod_{k=1}^{2n-2} \frac{2r+k}{k} = \frac{4s}{(2n-1)!} \prod_{j=1/2}^{n-3/2} \left(4s^2 - \frac{j^2}{4}\right) =$$

$$= \frac{4s}{(2n-1)!} \sum_{j=0}^{n-1} \beta'_{j,n} (2s)^{2j}$$

and

$$\beta\{2r^2 + r(2n-1)\}/(2n-1) = \beta\left\{2s^2 - 2\left(\frac{2n-1}{4}\right)^2\right\} \Big/ (2n-1).$$

Then

$$\vartheta_M(\beta) = \frac{e^{\beta((2n-1)/8)}}{(2n-1)!} \sum_{s \geqq 1/2} \sum_{j=0}^{n-1} \beta'_{j,n} 4s(2s^{2j} e^{-2s^2\beta/2n-1}$$

$$= \frac{e^{\beta(n-1/2)/4}}{(2n-1)!} \sum_{j=0}^{n-1} \beta'_{j,n} (-1)^j g_i^{(j)} \left(\frac{\beta}{2(2n-1)}\right),$$

where $i = 1$ if n is odd and $i = 2$ if n is even.

Thus

$$a_m = \frac{(4\pi)^n}{2(2n-1)!} \sum_{k=0}^{m} \frac{(n-1-m+k)!}{k!} 4^{n-m-1} (n-1/2)^{n-m+2k+1} \beta'_{n-1-m+k,n}$$

if $m < n$ and

$$a_m = \frac{(4\pi)^m}{2(2n-1)!} \left(\sum_{k=0}^{n-1} \frac{k!}{(m-n+k+1)!} (n-1/2)^{m-n+2k+2} \beta'_{k,n} 4^{n-m} + \right.$$

$$\left. + \sum_{k=0}^{m-n} \sum_{j=0}^{n-1} \frac{(-1)^j c_{j+k} \beta'_{j,n} (n-1/2)^{m-n-2k}}{k!(m-n-k)!} 4^{n-m} \right) \text{ if } \quad m \geqq n.$$

We will have to treat $P^n(\mathbb{C})$ with 2 separate arguments according to whether n is odd or even. The reason for this division is that when n is even $\delta_{\mathfrak{a}} \in \Lambda^+$ while when n is odd $\rho_{\mathfrak{a}} \notin \Lambda^+$. The treatment of the 2 cases will then differ only in that when n is even we will use Theorem 17.1.20 while for n odd we will use Theorem 17.1.22.

$M = P^n(\mathbb{C})$, n odd.

$$\vartheta_M(\beta) = \sum_{m=0}^{\infty} \left\{ \frac{2m+n}{n} \prod_{k=1}^{n-1} \left(\frac{m+k}{k} \right)^2 \right\} e^{-\{m^2+mn\}/(n+1)}.$$

Let $s = m + (n/2)$. Then

$$\frac{2m+n}{n} \prod_{k=1}^{n-1} \left(\frac{m+k}{k} \right)^2 = \frac{2s}{n!(n-1)!} \prod_{j=1/2}^{n/2-1} (s^2 - j^2)^2 = \frac{2s}{n!(n-1)!} \sum_{k=0}^{n-2} \gamma_{k,n} s^{2k}$$

and

$$\{m^2 + mn\}/n + 1 = \left\{ s^2 - \left(\frac{n}{2} \right)^2 \right\} \Big/ n + 1.$$

Therefore

$$\vartheta_M(\beta) = \frac{e^{n^2/4(n+1)}}{n!(n-1)!} \prod_{k=0}^{n-2} \gamma_{k,n} \sum_{s \geqq 1/2} 2s^{2k+1} \, e^{-s^2\beta/(n+1)}$$

$$= \frac{e^{n^2/}4(n+1)}{n!(n-1)!} \sum_{k=0}^{n-2} (-1)^k \gamma_{k,n} g^{(k)}\{\beta/(n+1)\}.$$

Thus

$$a_m = \frac{(4\pi)^{n-1}}{n!(n-1)!} \sum_{k=0}^{m} (n+1)^{n-1-m} \left(\frac{n}{2} \right)^{2k} \frac{(n-m+k-2)!}{k!} \gamma_{n-m+k-2,n}$$

if $m < n-1$ and

$$a_m = \frac{(4\pi)^{n-1}}{n!(n-1)!} \left(\sum_{k=0}^{n-2} \frac{k!}{(m-n+2+k)!} \left(\frac{n}{2} \right)^{2(m-n+2+k)} \gamma_{k,n} (n+1)^{n-m-1} + \right.$$

$$\left. + \sum_{k=0}^{m-n+1} \left(\frac{n^2}{4(n+1)} \right)^k \frac{1}{k!} \sum_{j=0}^{n-2} \frac{(-1)^j \gamma_{j,n} C_{m-n+1-k+j}}{(m-n+1-k)!} (n+1)^{m-n+1-k} \right)$$

if $m > n-1$

$M = P^n(\mathbb{C})$, n even. Since n is even we will let $n = 2n_0$. If $s = m + n_0$ then we will write

$$\prod_{k=1}^{n-1}(m+k)^2=\prod_{k=0}^{n_0-1}(s^2-j^2)^2=\sum_{k=0}^{n-2}\gamma_{k,n}s^{2k}.$$

Then

$$\begin{aligned}\vartheta_M(\beta)&=\frac{e^{\beta n_0^2/(n+1)}}{n!(n-1)!}\sum_{k=0}^{n-2}\gamma_{k,n}\sum_{s\geqq 0}2s^{2k+1}e^{-s^2\beta/(n+1)}\\&=\frac{e^{\beta n_0^2/(n+1)}}{n!n-1!}\sum_{k=0}^{n-2}(-1)^k\gamma_{k,n}h^{(k)}(\beta/(n+1)).\end{aligned}$$

Thus

$$a_m=\frac{(4\pi)^{n-1}}{n!(n-1)!}\sum_{k=0}^{m}(n+1)^{n-1-m}n_0^{2k}\frac{(n-m+k-2)!}{k!}\gamma_{n-m+k-2,n}$$

if $m<n-1$

and

$$a_m=\frac{(4\pi)^{n-1}}{n!(n-1)!}\left(\sum_{k=0}^{n-2}\frac{k!}{(m-n+2+k)!}n_0^{2(m-n+2+k)}\gamma_{k,n}(n+1)^{n-m-1}+\right.$$

$$\left.+\sum_{k=0}^{m-n-1}\left(\frac{n_0^2}{n+1}\right)^k\frac{1}{k!}\sum_{j=0}^{n-2}\frac{(-1)^j\gamma_{j,n}d_{m-n+1-k+j}}{(m-n+1-k)!}(n+1)^{m-n+1-k}\right)$$

if $m>n-1$.

If $M=P^{n-1}(Q), n\geqq 2$ then recall that

$$\vartheta_M(\beta)=\sum_{m=0}^{\infty}\frac{2m+2n-1}{2n-1}\prod_{r=2}^{2n-2}\frac{m+r}{r}\prod_{P=1}^{2n-2}\frac{m+P}{P}e^{-\beta(m^2+2mn-m)/2(n+1)}.$$

We will let $s=m+(n-1/2)$. Then

$$\begin{aligned}\frac{2m+2n-1}{2n-1}\prod_{r=2}^{2n-2}\frac{m+r}{r}\prod_{P=1}^{2n-3}\frac{m+P}{P}&=\frac{2s}{(2n-1)!(2n-3)!}\times\\&\times\prod_{j=1/2}^{n-3/2}(s^2-j^2)\prod_{j=1/2}^{n-5/2}(s^2-j^2)\\&=\frac{2s}{(2n-1)!(2n-3)!}\sum_{k=0}^{2n-3}\delta_{k,n}s^{2k}\end{aligned}$$

$$\beta(m^2+2mn-m)/2(n+1)=\beta(s^2-(n-1/2)^2)/2(n+1).$$

Therefore

$$\vartheta_M(\beta) = \frac{e^{\beta(n-1/2)^2/2(n+1)}}{(2n-1)!(2n-3)!} \sum_{k=0}^{2n-3} \sum_{s \geqq 1/2} \delta_{k,n} 2s^{2k+1} e^{-s^2\beta}$$

$$= \frac{e^{\beta(n-1/2)^2/2(n+1)}}{(2n-1)!(2n-3)!} \sum_{k=0}^{2n-3} \delta_{k,n} (-1)^k g^{(k)}(\beta).$$

Thus

$$a_m = \frac{(4\pi)^{2n-2}}{(2n-1)!(2n-3)!} \sum_{k=0}^{m} \left(\frac{(n-1/2)^2}{2(n+1)}\right)^k \frac{(2n-3-m+k)!}{k!} \delta_{2n-3-m+k,n}$$

if $m < 2n-2$

and

$$a_m = \frac{(4\pi)^{2n-2}}{(2n-1)!(2n-3)!} \left(\sum_{k=0}^{2n-3} \left(\frac{(n-1/2)^2}{2(n+1)}\right)^{2(m+2n-3-k)} \times \right.$$

$$\times \frac{k!}{(m+2n-3-k)!} \delta_{k,n} + \sum_{k=0}^{m-2n+2} \frac{(n-1/2)^{2k}}{2^k (n+1)^k k!} \times$$

$$\left. \times \sum_{j=0}^{2n-3} \frac{(-1)^j \delta_{j,n} c_{j+m-k}}{(m-k)!} \right) \quad \text{if} \quad m \geqq 2n-2.$$

We will deal with the Cayley projective plane by letting $s = m + \frac{11}{2}$. Then

$$d(m\varepsilon_1) = \frac{3!}{11!7!} 2s(s^2 - (\tfrac{1}{2})^2)^2 (s^2 - (\tfrac{3}{2})^2)^2 (s^2 - (\tfrac{5}{2})^2)(s^2 - (\tfrac{7}{2})^2)(s^2 - (\tfrac{9}{2})^2) =$$

$$= \frac{3!}{11!7!} 2s \sum_{j=0}^{7} \eta_j s^{2j}$$

with

$$\eta_7 = 1, \quad \eta_6 = -\frac{170}{4}, \quad \eta_5 = \frac{10\,437}{16}, \quad \eta = -\frac{262\,075}{64}, \quad \eta_3 = \frac{2\,858\,418}{256},$$

$$\eta_2 = -\frac{13\,020\,525}{1024}, \quad \eta_1 = \frac{18\,455\,239}{4096}, \quad \eta_0 = -\frac{8\,037\,225}{16\,384}.$$

$t(m^2 + 11m)/18 = t(s^2 - (11/2)^2)/18$ so

$$\vartheta_{P^2(\mathrm{Cay})}(\beta) = \frac{3!}{7!11!} e^{(121/72)\beta} \sum_{j=0}^{7} \sum_{s \geqq 1/2} \eta_j 2s^{2j+1} e^{-s^2\beta}$$

$$= \frac{3!}{7!11!} e^{(121/72)\beta} \sum_{j=0}^{7} \eta_j (-1)^j g^{(j)}(\beta).$$

Therefore

$$a_m = \frac{3!}{7!11!}(4\pi)^8 \sum_{k=0}^{m} \left(\frac{121}{72}\right)^k \eta_{7-m+k} \frac{(7-m+k)!}{k!} \quad \text{if} \quad m \leqq 7$$

$$a_m = \frac{3!}{7!11!}(4\pi)^8 \left(\sum_{k=0}^{7} \left(\frac{121}{72}\right)^{(m+7-k)} \eta_k \frac{k!}{(m+7-k)!} + \right.$$

$$\left. + \sum_{k=0}^{m-8} \frac{(121/72)^k}{k!} \sum_{j=0}^{7} \frac{(-1)^j n_j c_{j+m-k}}{(m-k)!} \right), \quad \text{if} \quad m > 7.$$

17.2. Zeta functions on compact Lie groups

Using the explicit formula for the kernel $\rho_G(\beta; g)$ on a compact Lie group G, Ěskin [E4] studied the zeta function

$$Z(g; s) = \frac{1}{\Gamma(s)} \int_0^\infty [\rho_G(\beta; g) - 1]\beta^{s-1} \, d\beta.$$

As in 16.2.1 he found that $Z(g; s)$ for $g \neq e$ is an entire function of s; and for $g = e$, $Z(e; s)$ is a meromorphic function of s with poles at

$$\frac{N}{2}, \frac{N}{2} - 1, \ldots 1 \text{ for } N \text{ even}$$

and at

$$\frac{N}{2}, \frac{N}{2} - 1, \ldots \text{for } N \text{ odd}$$

where $N = \dim G$.

Theorem (Eskin) 17.2.1. At these poles the residues are

$$\operatorname{Res} Z(e; s)|_{(N/2)-k} = \frac{\pi^{2k-N+(n/2)} V_\Gamma 2^{2k} \langle \delta, \delta \rangle^k}{2^{N+r} k! \prod_{\alpha \geqq 0} \langle \alpha, \delta \rangle \Gamma\left(\frac{N}{2} - k\right)}.$$

Problems

Exercise 17.1. Let G be a compact connected, simply connected semi-simple Lie group with Lie algebra $\mathfrak{g}$. Let L be the lattice of integral weights; $|W|$ is the order of the Weyl group;

$$\delta = \frac{1}{2} \sum_{\alpha > 0} \alpha; \text{ and } F(\Lambda) = \prod_{\alpha > 0} (\Lambda, \alpha) / \prod_{\alpha > 0} (\delta, \alpha),$$

where (,) denotes the Killing form. Show that if rank $G = 1$, the partition function

$$Z(\beta) = \frac{1}{|W|} \sum_{\Lambda \in L} F^2(\Lambda) \exp[-(|\Lambda|^2 - /\delta|^2)\beta]$$

has the form

$$Z(\beta) = \frac{\mathrm{Vol}(G)}{(4\pi\beta)\dfrac{\dim G}{2}} \times \exp(|\delta|^2 \beta) + ES,$$

where ES is an exponentially small error as $\beta \to 0$. Viz. show in this case that $F(\Lambda) = \Lambda$, $L = Z$, $|\Lambda|^2 = 2\Lambda^2$ and

$$Z(\beta) = \tfrac{1}{2} e^{|\delta|^2 \beta} \sum_{\Lambda \in Z} \Lambda^2 \exp(-|\Lambda|^2 \beta).$$

Since

$$\sum \Lambda^2 e^{-2\Lambda^2\beta} = \frac{-d}{d\beta}\left(\frac{1}{2} \sum_{\Lambda \in Z} e^{-2\Lambda^2\beta}\right),$$

use Poisson summation formula to complete the proof.

Extend this proof to rank n case.

Chapter 18

Selberg Trace Theory

18.1. The Selberg Trace formula

In example 3.34 we studied the classical mechanical system given by geodesic flow on the hyperbolic plane $\mathfrak{P} = SL(2, R)/SO(2)$. We want to examine the partition function for quantum statistical physics of these spaces. $\mathfrak{P}$ is noncompact, thus we will simplify the situation by studying $M = \Gamma\backslash\mathfrak{P}$ where Γ is a discrete subgroup of $SL(2, R)$ chosen so that M is compact. This classical example was the original case studied by Maas, Selberg and others. More recently these and related spaces have been studied by Dowker and others in quantum field theory as we shall show in the next chapter.

We will examine the general case $M = \Gamma\backslash G/K$ where G/K is a symmetric space of rank one. In this case G is a connected semisimple Lie group with finite center and K is a maximal compact subgroup. Γ is just the fundamental group of M. The general set up is to also consider finite dimensional unitary representations T of Γ. Let $\mathfrak{X}$ denote the character of T.

If $\mathfrak{g}$ and $\mathfrak{k}$ denote the Lie algebras of G and K we let $\mathfrak{g} = \mathfrak{k} \oplus \mathfrak{p}$ be the Cartan decomposition. Let $\mathfrak{a}_{\mathfrak{p}}$ be a maximal abelian subspace of $\mathfrak{p}$. Thus we have an Iwasawa decomposition $\mathfrak{g} = \mathfrak{k} \oplus \mathfrak{a}_{\mathfrak{p}} \oplus \mathfrak{n}$ or $G = KA_{\mathfrak{p}}N$. We assume that $\dim \mathfrak{a}_{\mathfrak{p}} = 1$. Λ denotes the real dual of $\mathfrak{a}_{\mathfrak{p}}$. If τ is the involution determined by $\mathfrak{k}$ and B is the Killing form we set $|X|^2 = -B(X, \tau X)$ for X in $\mathfrak{g}$ and $\sigma(g) = |X|$ for $g = k \exp X, k \in K, X$ in $\mathfrak{p}$.

Let $\mathfrak{a}$ denote a Cartan subalgebra of $\mathfrak{g}$ and let Δ denote the set of roots of $\mathfrak{g}^{\mathbb{C}}, \mathfrak{a}^{\mathbb{C}}$. Let $P_+ = \{\alpha \in \Delta^+ \mid \alpha \not\equiv 0 \text{ on } \mathfrak{a}_{\mathfrak{p}}\}$, where Δ^+ is the set of positive roots. Let $\delta = \frac{1}{2}\sum_{\sigma \in P_+} \alpha$. Let $\sum$ denote the set of restrictions to $\mathfrak{a}_{\mathfrak{p}}$ of elements in P_+. Since $\dim \mathfrak{a}_{\mathfrak{p}} = 1$ there is an element b in $\sum$ such that $2b$ is the only possible element in $\sum$. We define H_0 in $\mathfrak{a}_{\mathfrak{p}}$ by $b(H_0) = 1$ and we set $\delta_0 = \delta(H_0)$. For any h in $A_{\mathfrak{p}}$ we set $u(h) = b(\log h)$.

Let $p =$ the number of roots in P_+ whose restriction to is equal to b and let q be the number of remaining elements. Then $(H_0, H_0) = 2p + 8q$, $\delta(H_0) = \frac{1}{2}(p + 2q)$, $\langle \delta, \delta \rangle = \frac{1}{4}(p + 2q)^2 \times (2p + 8q)^{-1}$.

Consider the space $A_0(K\backslash G/K)$ of *spherical* functions in $A(G)$. Consider the associated space $L_1(K\backslash G/K)$. For any g in G let $H(g)\in\mathfrak{a}_\mathfrak{p}$ be the unique element of $\mathfrak{a}_\mathfrak{p}$ such that $g\in K\exp H(q)N$. Then for any λ in $\Lambda^{\mathbb{C}}$ the function $\phi_\lambda(g)=\int_K\exp(i\lambda-\delta)(H(gk))\,\mathrm{d}k$ is the *elementary* spherical function corresponding to λ. For f in $L_1(K\backslash G/K)$ and λ in $\Lambda^{\mathbb{C}}$ the *Harish–Chandra* or *spherical Fourier transform* is defined by

$$\hat{f}(\lambda)=\int_G f(g)\phi_\lambda(g)\,\mathrm{d}g,$$

where dg is the Haar measure in G. The *Abel transform* of f is defined by

$$F_f(a)=\exp\delta(\log a)\int_N f(an)\,\mathrm{d}n.$$

The Abel and Harish–Chandra transforms are related by

$$\hat{f}(\lambda)=\int_{A_\mathfrak{p}}F_f(a)\exp i\lambda(\log a)\,\mathrm{d}a.$$

The inverse transform to $\hat{f}$ is $f(g)=|W|^{-1}\int_\Lambda f(\lambda)\phi_\lambda(g^{-1})c(\lambda)^{-1}\times c(-\lambda)^{-1}\,\mathrm{d}\lambda$ where $c(\lambda)$ is the Harish–Chandra c-function. c is explicitly known.

The unitary representation T induces a representation U of G. U will be a discrete direct sum of irreducible representations occurring with finite multiplicities (v. G10 for a proof of this fact). We are interested in the spherical representations which occur in U. Let $\{U_j\}$ denote this set with their multiplicities $n_j(\mathfrak{X})$. U_j is completely determined by its elementary spherical function ϕ_{ν_j}, $\nu_j\in\Lambda^{\mathbb{C}}$. Since U_j is unitary ϕ_{ν_j} is positive definite and $|\nu_j|^2+|\delta|^2\geqq 0$. Thus ν_j is either purely real – i.e. $\nu_j\in\Lambda$ or purely imaginary $\nu_j\in i\Lambda$.

The tool to examine the partition functions is the Selberg trace formula. We review this theory next. What we find is that, as in the last chapter, the quantum statistical mechanics of the systems on these spaces is completely determined by the representation theory – in the present case by the spherical series representations.

The Selberg trace formula is merely an extension of the Poisson summation formula to noncommutative groups. Viz. let G be a locally compact group, let Γ be a closed subgroup of finite index. Let L be a finite dimensional unitary representation of Γ. Thus the induced representation U^L of G is also finite dimensional. Since the character $\mathfrak{X}_{U^L}=\mathrm{Tr}\,U^L$ is a class function it can be shown to depend only on $\mathfrak{X}_L=\mathrm{Tr}\,L$. Let $\mathfrak{X}_L^0$ be the function on G which is zero outside of Γ and equals $\mathfrak{X}_L$ on Γ. In this case we have

THEOREM 18.1.1.

$$\mathfrak{X}_{U^L} = \frac{1}{|G/\Gamma|} \sum_{y \in G/\Gamma} \mathfrak{X}_L^0(y \times y^{-1}).$$

If G/Γ is infinite we must look at the trace of $U^L(f)$ where

$$U^L(f) = \int_G f\, U(g)\, \mathrm{d}g.$$

If U^L is finite dimensional then $\mathrm{Tr}(U^L(f)) = \int \mathfrak{X}_{U^L}(g) f(g) \mathrm{d}g$.

Let v be the finite invariant measure on G/Γ normalized so that $\int_{G/\Gamma} \mathrm{d}v = 1$.

THEOREM 18.1.2.

$$\mathrm{Tr}(U^L(f)) = \int \sum_{y \in G/\Gamma} \int_\Gamma f(y^{-1}\gamma y) \mathfrak{X}_L(\gamma)\, \mathrm{d}\gamma\, \mathrm{d}v(y).$$

If U^L is completely reducible, say $U^L = \oplus\, n_j M^j$ where $M^j \in \hat{G}$ then clearly we have

$$\mathrm{Tr}\, U^L(f) = \sum n_j\, \mathrm{Tr}\, M^j(f) = \int_{\hat{G}} M(f)\, \mathrm{d}\mu(M).$$

Thus we have

THEOREM (STF Version I) 18.1.3.

$$\int_{G/\Gamma} \int_\Gamma f(y^{-1}\gamma y) \mathfrak{X}_L(\gamma)\, \mathrm{d}\gamma\, \mathrm{d}v(y) = \int_{\hat{G}} \mathrm{Tr}(M(f))\, \mathrm{d}\mu(M).$$

COROLLARY (Poisson Summation Formula) 18.1.4. Let G be a commutative locally compact group with G/Γ having a finite invariant measure. Take $L = I$, the one dimensional identity representation. Then the Selberg trace formula states that

$$\int_\Gamma f(\gamma)\, \mathrm{d}\gamma = \sum_{\gamma \in \Gamma^\perp} \hat{f}(\mathfrak{X}).$$

Proof. First we note that G/Γ has a finite invariant measure iff G/Γ is compact. In this case $U^I = \oplus_{\mathfrak{X} \in \Gamma^\perp \subset \hat{G}} \mathfrak{X}$. Second, since G is commutative the left-hand side of the STF is just

$$\operatorname{Tr}(U^I(f)) = \int_\Gamma f(\gamma)\, d\gamma.$$

Finally the components $M^j(g)$ of U^I are just $\mathfrak{X}(g)I$ and

$$\operatorname{Tr} M^j(f) = \int_G \mathfrak{X}(g) f(g)\, dg = \hat{f}(\mathfrak{X}).$$

Thus we have $\operatorname{Tr}(U^I(f)) = \sum m_j \operatorname{Tr}(M^j(f))$ is as stated.

Corollary 18.1.5. Let G be a compact group; then the multiplicity m_j that M^j occurs in U^L is

$$m_j = \frac{1}{|\Gamma|} \sum_{\gamma \in \Gamma} \overline{\mathfrak{X}_M{}^j(g)} \operatorname{Tr} L(\gamma).$$

Proof. Just take $f(g) = \mathfrak{X}_M{}^j(g)$ and use the Schur orthogonality relations.

A slight extension of the Selberg trace formula is given by dropping the requirement that G/Γ has a finite invariant measure. Let $N(\Gamma)$ = the normalizer of Γ in G, and let $N(\Gamma)_L$ be the subgroup of $N(\Gamma)$ consisting of all automorphisms of leaving $\mathfrak{X}_L$ stable.

Theorem (STF, Version II) 18.1.6. Let G, Γ be as stated above with $G/N(\Gamma)_L$ having a finite invariant measure. Then

$$\int_{G/N(\Gamma)_L} \int_\Gamma f(y^{-1}\gamma y)\mathfrak{X}_L(\gamma)\, d\gamma\, dv = \int_{\hat{G}} \operatorname{Tr}(M(f))\, d\mu(M).$$

Corollary 18.1.7. Let $\Gamma = \{e\}$. Thus $N(\Gamma) = G$ and the STF Version II states that

$$f(e) = \int_{\hat{G}} M(f)\, d\mu(M)$$

– i.e. the Plancherel formula.

The Frobenius reciprocity formula stated in Theorem or Corollary 18.1.5 has a beautiful generalization called the duality theorem, which we develop next.

Let T be an irreducible unitary representation of a locally compact G on

Hilbert space H. The space of vectors v in H such that the function $f(g) = T(g)v$ is in $A(G,H)$ is called the *Gårding space* of H and is denoted H_∞. Clearly H_∞ is invariant under T and $G \times H_\infty \to H_\infty$: $(g,v) \to T(g)v$ is continuous. Let H^*_∞ denote the dual space to H_∞ – i.e. the space of continuous linear functional on H_∞. We define the representation T^* on H^* by $\langle v, T^*(g)f\rangle = \langle T(g^{-1})v, f\rangle$ for v in H_∞ and f in H^*_∞.

Let Γ be a closed subgroup of G such that $\Gamma\backslash G$ is compact with a G-invariant on $\Gamma\backslash G$. Let V be a finite dimensional unitary representation of Γ on Hilbert space M. The induced representation is denoted by $U^V = \mathrm{Ind}_\Gamma^G(V)$; we let L denote the space on which U^V acts. Olshanski O7 has generalized Frobenius reciprocity as follows:

THEOREM (Duality Theorem) 18.1.8. The spaces of continuous linear intertwining maps $\mathrm{Hom}_G(H,L)$ and the $\mathrm{Hom}_\Gamma(M^*,H^*_\infty)$ are isomorphic.

Proof. If $A \in \mathrm{Hom}_G(H,L)$ then A maps H_∞ continuously into L_∞. Since $A(h)$ is a smooth M-valued function on G we can define $J \in \mathrm{Hom}_\Gamma(M^*,H^*)$ by

$$\langle h, J(\xi)\rangle = \langle A(h)(e), \xi\rangle.$$

The reader can check that $J(\xi) \in H^*_\infty$. Clearly $\xi \to J(\xi)$ is linear and an easy calculation shows that $JV^*(\gamma) = T^*(\gamma)J$.

The inverse mapping is given by

$$\langle A(h)(g), \xi\rangle = \langle T(g)h, J(\xi)\rangle.$$

This defines $A: H_\infty \to L$. It only remains to check that A admits a closure. This is left to the reader.

COROLLARY (Frobenius–Cartier) 18.1.8. Let V be a $2n$-dimensional real vector space with a nondegenerate alternating bilinear form B on $V \times V$. Assume V admits a complex structure J where $J^2 v = -v$, $B(Jv, Jv') = B(v,v')$ and $B(v, Jv) \geqq 0$. Let L be a lattice in V such that B takes integral values on $L \times L$. Let $F: V \to R$ satisfy $F(l_1 + l_2) \equiv F(l_1) + F(l_2) + B(l_1, l_2)$ mod 2 where $l_i \in L$. Then the Fock representation $(U_J, \mathscr{F}_J)$, $U_J(i_t e^v) = \mathrm{e}(\lambda t)U(v)$, $(U(v)f)(v') = \exp(-\pi\lambda H(v,v)/2 + H(v,v')f(v+v')$ for f holomorphic on $(V_J, (|))$ where $(f|f') = \int_V \exp(-\pi\lambda H(v,v))\overline{f(v)}f'(v)dv$, is irreducible and the solutions of the $U_J(e^l)\cdot t = \mathrm{e}(\frac{1}{2}F(l))\cdot t$ for every l in L form an e-dimensional subspace Θ of $(F_J)^*_\infty$ – i.e. dimension of space of theta functions or solutions of $t(v) = t(v+l)\exp - \pi[\frac{1}{2}H(l,l) + H(l,v) + iF(l)]$ is the square root of the discriminent of B with respect to $L(e = \sqrt{[L^*:L]})$.

COROLLARY (Gelfand–Graev–Piatetski–Shapiro–Ehrenpreis–Mautner) 18.1.9. Let G be a connected semisimple Lie group with finite center and let Γ be a cocompact discrete subgroup of G. Then $\mathrm{Hom}_G(H, L^2(\Gamma\backslash G)) = (H^*_\infty)^\Gamma$.

Specializing this to the case $G = SL(2, R)$ we have

COROLLARY 18.1.9′. Let $G = SL(2, R)$ and consider the discrete series representation $T_n^+(g)$. Then $N_n^+ = \dim \mathrm{Hom}_G(T_n^+, U^\Gamma) = \dim \{f(z)$ analytic for $\mathrm{Im}\, z > 0$ and $f(\gamma z)(\gamma_{12}z + \gamma_{22})^{-n-1} = f(z)\}$ i.e. the dimension of the space of automorphic forms.

Now the calculation of this multiplicity is provided by the Selberg trace formula. Viz.

THEOREM 18.1.10.

$$N_n^+ = [1 + (-1)^{n-1}\varepsilon] \deg(L) \operatorname{Vol} \frac{(\Gamma\backslash G)n}{\pi^2} - \sum_\gamma \sum_{s=1}^{k-1} \frac{1}{4ki} \frac{\mathrm{Tr}(L(\gamma^s))}{\sin(\pi s/k)} \exp(i\pi sn/k).$$

Here $\varepsilon = \pm 1$ if $L(-e) = \pm I$; and if Γ contains no elliptic elements we have

$$N_n^+ = N_n^- = [1 + (-1)^{n-1}\varepsilon] \deg(L) \frac{\operatorname{Vol}(\Gamma\backslash G)n}{\pi^2}.$$

Thus the Selberg trace formula essentially contains the Riemann–Roch formula.

The next version of the Selberg trace formula is for the case of compact space forms of symmetric spaces of rank one, $M = \Gamma\backslash G/K$, as in Section 18.1. For f in $L^1(G)$ we have the bounded operator $U(f) = \int_G f(g)U(g)\,dg$.

DEFINITION 18.1.11. f is said to be *admissable* if (i) the series $\sum_\gamma f(y^{-1}\gamma x) \times T(\gamma)$ converges absolutely, uniformly on compacts of $G \times G$, to a continuous End (V)-valued function $F(x, y, T)$ and (ii) $U(f)$ is of trace class.

THEOREM (STF Version III) 18.1.12. If f is admissable and if $U = \sum_{\lambda\in\hat{G}} n_\Gamma(\lambda, T)\lambda$, then $\sum_{\lambda\in\hat{G}} n_\Gamma(\lambda, T) \mathrm{Tr}\, U_\lambda(f) = \int_{\Gamma\backslash\hat{G}} \mathrm{Tr}\, F(x, x, T)\,dx$.

If we let C_Γ denote the set of representatives in Γ of the conjugacy classes of elements of Γ, if G_γ is the centralizer of γ in G, $\Gamma_\gamma = \Gamma \cap G_\gamma$, and dx_γ is the Haar measure on G_γ, and if we set $I_\gamma(f) = f_{G\gamma\backslash G} f(g^{-1}\gamma g)\,dx^*_\gamma$ where dx^*_γ is the G-invariant measure on $G_\gamma \backslash G$ normalized so that $dx = dx_\gamma dx_\gamma^*$, then a simple calculation shows

COROLLARY (STF Version IV) 18.1.13.

$$\sum_{\lambda \in \hat{G}} n_\Gamma(\lambda, T) \operatorname{Tr} U_\lambda(f) = \sum_{\gamma \in C_\Gamma} \operatorname{Tr}(T(\gamma)) \operatorname{Vol}(\Gamma_\gamma \backslash G_\gamma) I_\gamma(f).$$

If $f \in \mathfrak{C}_1(K \backslash G/K)$, the Harish–Chandra Schwartz space, then Gangolli and Warner showed that f is admissable. Since f is spherical $U_\lambda(f) = 0$ unless λ is of class one with respect to K. And, in this case, U_λ has associated with it a positive definite elementary spherical function ϕ_λ. In this case $U_\lambda(f) = \hat{f}(\lambda)$. Thus we have

COROLLARY (STF Version V) 18.1.14. If $f \in \mathfrak{C}_1(K \backslash G/K)$ then

$$\sum_{\lambda \in \hat{G}_1} n_\Gamma(\lambda, T) \hat{f}(\lambda) = \sum_{\gamma \in C_\Gamma} \operatorname{Tr}(T(\gamma)) \operatorname{Vol}(\Gamma_\gamma \backslash G_\gamma) I_\gamma(f).$$

DEFINITION 18.1.15. Element g in G is said to be *elliptic* if it is conjugate to some element of K. (It is then semisimple.) We say g in G is *hyperbolic* if it is semisimple, but not elliptic. In all other cases g is called *parabolic*.

The following facts regarding Γ are known. If G/Γ is compact, Γ contains no parabolic elements. Element γ in Γ is elliptic iff it is of finite order. We assume below that Γ contains no nontrivial elliptic elements, i.e. no nontrivial fixed points in G/K. Thus each $\gamma \in \Gamma$, $\gamma \neq e$, is hyperbolic. Every γ in Γ is then conjugate in G to an element of the Cartan subgroup $A =$ $=$ centralizer of $\mathfrak{a}$ in G. Here $A = A_\mathfrak{t} A_\mathfrak{p}$. We choose an element $h(\gamma)$ of A to which γ is conjugate and set $h(\gamma) = h_\mathfrak{t}(\gamma) h_\mathfrak{p}(\gamma)$. We define $u_\gamma = b(\log h_\mathfrak{p}(\gamma))$. Thus $u_\gamma = u(\log h_\mathfrak{p}))$.

In the present case G/K is the simple connected covering manifold of $M = \Gamma \backslash G/K$, and we identify Γ with the fundamental group, $\pi_1(M)$. The free homotopy classes of closed paths on M are in a natural $1-1$ correspondence with the set of conjugacy classes of Γ – i.e. C_Γ. For any γ in C_Γ the corresponding free homotopy class contains a periodic geodesic which we denote by g_γ. g_γ has minimum length among all paths in that class. Let $l(\gamma)$ be the length of g_γ. Any closed path in this homotopy class can be lifted to G/K to a path of equal length, joining x in G/K to γx. If $g(.,.)$ is the Riemannian distance on G/K then it follows that $l(\gamma) = \inf_{x \in G/K} g(x, \gamma x)$. Writing $x = gK$ we have $g(x, \gamma x) = g(gK, \gamma gK) = g(K, g^{-1}\gamma gK) = \sigma(g^{-1}\gamma g)$. Thus $l(\gamma) = \inf_{g \in G} \sigma(g^{-1}\gamma g)$.

THEOREM 18.1.16. In the present set up $l(\gamma)=\sigma(h(\gamma))=|\log h_{\mathfrak{p}}(\gamma)|$.

As we will discuss below an element γ of Γ, $\gamma \neq e$, is called *primitive* if it cannot be expressed as γ_0^n for some γ_0 in Γ, $n>1$. Every $\gamma \neq e$ in Γ can be expressed in the form $\gamma=\gamma_0^j(\gamma)$ for a unique primitive element γ_0.

For any root α define the character ξ_α on A by $\xi_\alpha(h)=\exp \alpha(\log h)$. Let $\varepsilon_R^A(h)=\operatorname{Sgn} \prod_{\alpha\in\Phi_R^+}(1-\xi_\alpha(h)^{-1})$, where Φ_R^+ are the roots of $(\mathfrak{g}^{\mathbb{C}}, \mathfrak{a}^{\mathbb{C}})$ that are real on $\mathfrak{a}$. Set

$$C(h(\gamma))=\varepsilon_R^A(h(\gamma))\xi(h_{\mathfrak{p}}(\gamma))\prod_{\alpha\in P_+}(1-\xi_\alpha(h(\gamma))^{-1})^{-1}$$

COROLLARY (STF Version VI) 18.1.17. If f is admissable, then

$$\sum_{\lambda\in G_1} n_\Gamma(\lambda, T)\hat{f}(\lambda)=$$

$$=\int_{\Gamma\backslash G}\sum \mathfrak{X}(\gamma) f(g^{-1}\gamma g)\,dg=(e)\operatorname{Vol}(G) f(e)\times$$

$$\times\sum_{\gamma\in C_\Gamma\backslash\{e\}}\mathfrak{X}(\gamma)\,|u_\gamma|\, j(\gamma)^{-1}\, C(h(\gamma))\, F_f(h_{\mathfrak{p}}(\gamma)).$$

A particular function f will be studied. Consider Bloch's equation $\partial\rho/\partial\beta=\Delta\rho$ where Δ is the Laplace–Beltrami operator on G/K. The physics of the situation is this. The Hilbert space of the quantum mechanical system has a natural decomposition

$$L^2(\Gamma\backslash G/K)=\sum H(\lambda).$$

The Hamiltonian Δ acts on $H(\lambda)$ with energy eigenvalue $E_\lambda(\Delta)=-((\lambda,\lambda)+(\delta,\delta))$. Thus $\exp(-\beta E_\lambda(\Delta))\dim H(\lambda)$ is the trace of $\exp(-\beta\Delta)$ on $H(\lambda)$; i.e.

$$\vartheta(\beta)=\sum \exp(-\beta E_\lambda(\Delta))\dim H(\lambda)$$

is the trace of $\exp(-\beta\Delta)$ on $L^2(\Gamma\backslash G/K)$. The multiplicity $\dim H(\lambda)$ is just the multiplicity $n_\Gamma(\lambda, T)$ of the spherical representation in the induced representation U.

We must show that the trace exists. Let $E(\beta)$ be the fundamental solution of Bloch's equation. Then, as a function on G, $E(\beta)$ is spherical and non-negative real valued. One has $E(\beta_1+\beta_2)=E(\beta_1)^* E(\beta_2)$. And for each $\beta>0$ $E(\beta)$ is in $L_1(K\backslash G/K)$. One easily checks that $E_\beta(\lambda)=\exp(-((\lambda,\lambda)+(\delta,\delta))\beta)$. Since $E(\beta)$ is integrable $\hat{E}_\beta$ is defined by all λ such that φ_λ is bounded.

THEOREM 18.1.18.
(i) $E(\beta)$ belongs to $\mathfrak{C}_1(K\backslash G/K)$ so $E(\beta)$ is admissable
(ii) $F_{E(\beta)}(a_p) = (4\pi\beta)^{-1/2} \exp - [\beta(\delta, \delta) + |\log a_p|^2/4\beta]$.
Setting $f = E(\beta)$ in the STF, Version III, we have

THEOREM (STF Version VII) 18.1.19.

$$\sum_{\lambda \in \hat{G}_1} n_\Gamma(\lambda, T) \exp - ((\lambda, \lambda) + (\delta, \delta))\beta$$
$$= \sum_{\gamma \in C_\Gamma} \operatorname{Tr} T(\gamma) \operatorname{Vol}(\Gamma_\gamma \backslash G_\gamma) I_\gamma(E(\beta)).$$

Gangolli and his student Eaton have evaluated the high temperature limits of the partition functions on $M = \Gamma\backslash G/K$. Viz. they were able to show that the sum $\sum_{\gamma \neq e}$ goes to zero as $\beta \to 0^+$.

COROLLARY 18.1.20. $\lim_{\beta\to 0} \beta^{n/2} \vartheta(\beta) = C_G \operatorname{Vol}(\Gamma\backslash G) \deg T$ where $C_G = \lim \beta^{n/2} E_\beta(e)$ is a constant that depends only on $G, n = \dim G/K$.
Setting

$$N(r, T) = \sum_{\substack{\lambda \in \hat{G}_1 \\ |E_\lambda| \leq r}} n_\Gamma(\lambda, T)$$

and realizing $\vartheta(\beta) = \int_0^\infty e^{-\beta r}\, dN(r)$ we have that since $E(\beta)$ is admissable that $N(r, T)$ is finite for each r and $\vartheta(\beta)$ exists. Furthermore, by Karamata's Tauberian theorem we have

THEOREM 18.1.21. $r^{-n/2} N(r, T) \sim C_G \Gamma((n/2) + 1) \operatorname{Vol}(\Gamma\backslash G) \deg(T)$ as $r \to \infty$; or $N(r) \sim s \operatorname{Vol}(\Gamma\backslash G/K) M(r)$ where s = order of the center of the intersection of Γ and the center of G and

$$M(r) = \int_{(\lambda,\lambda)+(\delta,\delta) \leq r} |c(\lambda)|^{-2} d\lambda.$$

For simplicity we assume $T = 1$. Using STF Version VI we can rewrite the partition functions as

$$\sum_{\lambda \in \hat{G}_1} N_\Gamma(\lambda, 1) e^{-[(\lambda,\lambda)+(\delta,\delta)]\beta} =$$
$$= E_\beta(e) \operatorname{Vol}(\Gamma\backslash G) + (4\pi\beta)^{-1/2} \sum_{\gamma \in C_\Gamma \backslash \{e\}} \times$$
$$\times\, C(h(\gamma)) e^{-[(\delta,\delta)\beta + l(\gamma)^2/4\beta]},$$

where $l(\gamma) = |\log h_p(\gamma)|$. Clearly this agrees with Molchanov's results. In fact using his Jacobi field approach Molchanov showed that

THEOREM (Molchanov) 18.1.22. Let M be a connected compact n-dimensional manifold with negative curvature and let $M = \Gamma\backslash N$. Then by Varadhan's result we have for any γ in $\Gamma = \pi_1(M)$

$$\lim_{\beta\to 0} -2\beta \ln \vartheta(\beta) = \min_{m\in M} g^2(m, \gamma m),$$

where

$$\vartheta(\beta) = \sum e^{-\lambda_\kappa \beta} = \sum \vartheta_\gamma(\beta)$$

and

$$\vartheta_\gamma(\beta) = \int_M \rho_N(\beta, m, \gamma m)\, \mathrm{d}m$$

where

$$\frac{\partial}{\partial\beta}\rho_N = \frac{\Delta_N}{2}\rho_N$$

and

$$\rho_M = \sum_{\gamma\in\Gamma} \rho_N(\beta, x, \gamma y);$$

and $\vartheta_\gamma(\beta)$ can be defined asymptotically by periodic Jacobi fields. In particular if M is a compact symmetric space with negative curvature the Jacobi fields provide

$$\vartheta_\gamma(\beta) \sim \frac{l(\gamma)\exp\{-l(\gamma)^2/2\beta\}}{\sqrt{2\pi\beta}\sqrt{\prod_{i=1}^{n-1} 2\sinh\left(\sqrt{-k_i}\frac{l(\gamma)}{2}\right)}},$$

where the Jacobi equations are just $\ddot{y} + K_i y = 0$, $K_i\, i = 1, \ldots, n-1$ are the eigenvalues of the curvature form along the geodesic.

One final version of the STF is motivated by the following. If $G = SL(2, R)$, then the compact Cartan subgroup $T \subset G$ has its set of regular elements $T' \sim n \in Z\backslash\{0\}$. We have the Harish–Chandra map $\omega: T' \to \hat{G}_d$. As we mentioned in Chapter 14 when Γ has no elliptic elements, the multiplicities $n_{\omega(n)}(\Gamma)$ of discrete series in $L^2(\Gamma\backslash G)$ is equal the dimension of the space of automorphic forms of weight $|n|+1$. Viz.

$$n_{\omega(n)} = \begin{cases} |n|\ (g-1) & |n| \neq 1 \\ g & |n| = 1 \end{cases}.$$

Here $|n|(g-1) = d_{\omega(n)} \operatorname{Vol}(\Gamma\backslash G)$ where $d_{\omega(n)}$ is the formal degree, for $n \neq 1$ – i.e. iff $\omega(n)$ is in fact an integrable representation. However, more generally we have

THEOREM (Dimension of Space of Automorphic Forms) 18.1.23. Let $T \subset K \subset G$ be as in Theorem 14.1.8 and assume (i) and (ii) from that theorem. Then if Γ is a discrete subgroup such that $\Gamma\backslash G$ is compact and Γ has no elliptic elements then

$$N_{\omega(\lambda+\delta)} = d_{\omega(\lambda+\delta)} \operatorname{Vol}(\Gamma\backslash G).$$

Using B a left invariant metric $\langle | \rangle$ is defined on G by $\langle X | X \rangle = B(X, X)$ for X in $\mathfrak{p}$ and $\langle X | X \rangle = -B(X, X)$ for X in $\mathfrak{k}$. We get a Riemannian metric on $\Gamma\backslash G$ by requiring the projection $\pi_\Gamma : G \to \Gamma\backslash G$ to be a Riemannian submersion. Since $A\,\mathrm{d}k$ preserves the Carten decomposition and the Killing form for k in K, we see that K acts by isometries on the right on G on $\Gamma\backslash G$.

The Casimir element $\mathfrak{C}$ is defined by $\mathfrak{C} = \sum_{i=1}^{r+s} X_i X^i$, $r = \dim \mathfrak{p}$, $s = \dim \mathfrak{k}$, $\{X_i\}$, $i = 1, \ldots, r+s$ a basis for $\mathfrak{g}$ and $\{X^i\}$, $i = 1, \ldots, r+s$ the dual basis. $\mathfrak{C}$ defines a second order differential operator $\tilde{\mathfrak{C}}$ on $A(G)$ by $\tilde{\mathfrak{C}} = \sum \tilde{X}_i \tilde{X}^i$ where $\tilde{X} f(g) = (\mathrm{d}/\mathrm{d}t) f(g \exp tX)|_{t=0}$ for X in $\mathfrak{g}$, and f in $A(G)$. $\tilde{\mathfrak{C}}$ is biinvariant and setting $\tilde{\mathfrak{C}}_\Gamma f \cdot \pi_\Gamma = \mathfrak{C}(f \cdot \pi)$ we have a second order differential operator on $A(\Gamma\backslash G)$. One can check that $\sigma(\mathfrak{C}_\Gamma)(\xi) = -|\xi|^2$ for ξ in T_K^{*M}; so $\mathfrak{C}_\Gamma$ is an elliptic operator. Let $\rho \in \hat{K}$ and let V be the ρ-module. Set $v_\rho(\lambda) =$ = multiplicity of ρ in the eigenspace C^λ of $-\mathfrak{C}_\Gamma$ with eigenvalue λ i.e. $v_\rho(\lambda) = \dim \operatorname{Hom}_K(V, C^\lambda)$. Set $m = \dim G/K$. Set $l = \dim \operatorname{Hom}_{K_m}(V, C)$, for m in M_0. Then by Theorem 1.4.4 we have

$$N(t) = \sum_{\substack{\lambda \in \operatorname{spec}(-\mathfrak{C}_\Gamma) \\ \lambda \leq t}} v_\rho(\lambda) \sim t^{m/2} \operatorname{Vol}(B(m))\, l \operatorname{Vol}(M_0/K)/(2\pi)^m.$$

Let $N = \{g \in G \mid gxK = xK$ for all x in $G\}$. Thus $\operatorname{Vol}(M_0/K) =$ $= \operatorname{Vol} M |\Gamma \cap N| / \operatorname{Vol} K$.

For $\omega \in \hat{G}$ let $n_\Gamma(\omega) =$ multiplicity of ω in $L^2(\Gamma\backslash G)$ and let $[\omega | K : \rho] =$ = multiplicity of ρ in $\omega | K$. Let λ_ω be the value of the Casimir operator on all representation of class ω. This leads to a result due to Gelfand, Gangolli, and Wallach:

THEOREM 18.1.24.

$$\sum_{\substack{\lambda \leqq t \\ \omega \in \hat{G}}} n_\Gamma(\omega)[\omega | K:\rho] \sim t^{m/2} \operatorname{Vol}(B(m)) \dim V^{\Gamma \cap N} \operatorname{Vol} \times$$
$$\times \Gamma \backslash G | \Gamma \cap N | / (2\pi)^m \operatorname{Vol} K.$$

Here Vol $(B(m))$ is the volume of the m-dimensional unit ball.

18.2. THE PARTITION FUNCTION AND THE LENGTH SPECTRA

As we mentioned at the end of Chapter 16, the Selberg trace formula, being a generalization of the Poisson summation formula may have properties relating to the length spectra of $\Gamma\backslash G/K$. For the cases at hand that is actually the case. In the language of physics, we can show that the quantum statistical mechanical partition function is determined by the geodesics.

This requires a brief review of the geometry of $\Gamma\backslash G/K$. Since G/K is the simply connected covering space of M we identify Γ and $\Pi_1(M)$. Using the notation preceding Theorem 18.1.16 we have

THEOREM 18.2.1.

(i) $l(\gamma) = \sigma(h(\gamma)) = |\log h_p(\gamma)|$,

(ii) $l(\gamma) = l(\gamma^{-1})$,

(iii) $l(\gamma^j) = jl(\gamma)$,

(iv) if $\gamma \neq e$, then $\Gamma_\gamma = Z$.

By the last theorem every $\gamma \neq e$ in Γ can be written as $\gamma = \gamma_0^j$ where γ_0 is primitive. j is unique and will be written $j(\gamma)$.

Using these facts and the STF, Version IV, we have

THEOREM 18.2.2.

$$\vartheta(\beta) = E_\beta(e)\operatorname{Vol}(\Gamma\backslash G) + \sum_{\gamma \in C_\Gamma \backslash \{e\}} (4\pi\beta)^{-1/2} \times$$
$$\times C(h(\gamma)) l(\gamma) j(\gamma)^{-1} \exp(-(\delta, \delta)\beta + l(\gamma)^2 \backslash 4\beta).$$

We let m_i denote the *multiplicity* of l_i in $\{l(\gamma), \gamma \in C_\Gamma \backslash \{e\}\}$. It is clear that the lengths $\{l_i\}$, are determined by the partition function $\vartheta(\beta)$. For details of the algorithm to find l_i see Gangolli G4.

Using the Selberg trace formula Duistermaat, Kolk, and Varadarjan were able to examine the spectra of compact locally symmetric manifolds S of negative curvature. Viz. let Spec denote the spectrum of G-invariant differential operators on S which are realized on $M = \Gamma\backslash S$, M compact. The multiplicity $n_\Gamma(\lambda, \mathfrak{X})$ must be replaced by $m(\lambda) = |w\cdot\lambda|^{-1} n_\Gamma(\lambda, \mathfrak{X})$, with w in the Weyl group W. The Selberg trace formula in this case shows that

THEOREM (STF Version IV) 18.2.3.

$$\sum_{\lambda\in\mathrm{Spec}} m(\lambda)\exp(\lambda\cdot\log) = \sum_{\gamma\in C_\Gamma} d(\gamma)v(\gamma)T_\gamma$$

where $v(\gamma) = [G_\gamma : G_\gamma^0]^{-1}\,\mathrm{Vol}(\Gamma_\gamma\backslash G_\gamma)$ = the volume of the submanifolds of M consisting of all closed geodesics in M with free homotopy type of γ. T_γ is a W-invariant tempered distribution. $d(\gamma)$ is a complicated function, v. D31.

COROLLARY 18.2.4. If the split rank of G is one (i.e. $\dim A = 1$) then

$$\sum m(\lambda)\exp(\lambda) = K + \tfrac{1}{2}\sum_{c\in C_\Gamma\backslash\{e\}} l_0(c)|\det(I - P_c)|^{-1/2}(\delta_{\gamma(c)} + \delta_{-\gamma(c)}),$$

where $l_0(c)$ is the primitive length corresponding to c; and P_c is the Poincaré map along the element of c. (For details on the Poincaré map see D31, K13.)

The eighth version of the Selberg trace formula is to write it in the form of the Atiyah–Bott Lefschetz formula. Recall that the set up for this is a compact manifold M, a sequence of vector bundles E_j over M and a complex of differential operators $d_j : S(E_j) \to S(E_{j+1})$. If f is a smooth function $f : M \to M$ with a simple fixed point p, then $\det(1 - \mathrm{d}f_p) \neq 0$. Let $\varphi_j : f^*E \to E$ be a bundle homomorphism and set $T_j(s) = \varphi_j\cdot s\cdot f$. If T_j commutes with d_j then T_j induces a map T_j^* on homology. The theorem states that

$$\sum(-1)^j\,\mathrm{Tr}\,T_j^* = \sum_{p\in F}(-1)^j\frac{\mathrm{Tr}\,\varphi_{j,p}}{|\det(1 - \mathrm{d}f_p)|}$$

and

$$\sum(-1)^j\,\mathrm{Tr}\,T_j = \sum(-1)^j\,\mathrm{Tr}\,T_j^*,$$

where F is the set of fixed points.

Let Γ be a discrete cocompact group of G – i.e. $\Gamma\backslash G$ is compact, where G is a connected semisimple Lie group. Let T be a unitary representation of Γ on V_0. Let $\mathfrak{g} = \mathfrak{k} + \mathfrak{p}$ be the Cartan decomposition and $\mathfrak{h} \subset \mathfrak{g}$ a Cartan subalgebra. Let H be the centralizer of $\mathfrak{h}$ in G and let H' be the regular elements in H.

For h in H' define the action T_h on $\Gamma\backslash G$ by $T_h(\Gamma g) = \Gamma gh$. Define the map $\tilde{T}_h : T_h^* E \to E$ as follows. Take $(\Gamma gh^{-1}, v) \in T_h^* E$, $v \in E$. Corresponding to the principal bundle $\Gamma \to G \to \Gamma\backslash G$ we have a fiber bundle $E \to \Gamma\backslash G$ with fiber

V_0. The principal map P is defined by

$$\begin{array}{ccc} G \times V_0 & \xrightarrow{P} & E \\ \downarrow & & \downarrow p \\ G & \to & \Gamma\backslash G, \end{array}$$

where $p(v) = \Gamma g$. Then define $\tilde{T}_h(\Gamma g h^{-1}, v) = P(gh^{-1}, w), w \in V_0$. If U is the T-induced representation of G acting on V, then one can check that for f in V

$$U(h) f = \tilde{T}_h f T_h.$$

Let $\{\gamma_{j_h}\}$ be a complete set of representatives of disjoint conjugacy classes in Γ such that for each γ_{j_h} there is a g_{j_h} in G such that $g_{j_h}^{-1} \gamma_{j_h} g_{j_h} = h \in H'$. For h in H' we let G_h denote the centralizer of h in G.

An elementary check shows that the disjoint union $\bigcup \Gamma g_{j_h} G_h$ is precisely the set of fixed points in $\Gamma\backslash G$ of the T_h-action. Note also that $\mathrm{d}T_h = \mathrm{Ad}(h^{-1})$. We let N_{j_n} denote the normal bundle to the fixed point set $\Gamma g_{j_h} G_h$.

We let $\Lambda = \{h \in H' \mid T_h \text{ has a fixed point}\}$, A_h = fixed point set of h in Λ. Thus A_h decomposes into a disjoint union $\bigcup A_h^{j_h}$ where $A_h^{j_h} = \Gamma g_{j_h} G_h$ for some g_{j_h} in G. Let $\tilde{T}_h^{j_h}$ denote the action of T_h on fibers of E over $A_h^{j_h}$. Then from STF Formula Version I we have the Atiyah–Bott–Lefschetz version of STF:

THEOREM 18.2.5. For φ in $A_0(H')$ and T_φ denoting the compactly supported Radon measure associate to φ, we have

$$\operatorname{Tr} U(T_\varphi) = \sum_{h \in \Lambda} \sum_{j_h} \operatorname{Vol}(A_h^{j_h}) \frac{\varphi(h) \operatorname{Tr}(T_h^{j_h})}{|\det(1 - \mathrm{d}T_n | N_{j_n})|}.$$

Let $\mathfrak{k} + \mathfrak{a} + \mathfrak{n}^+$ be an Iwasawa decomposition of $\mathfrak{g}$. As usual $G = KAN^+, M = C_K(A)$ and $B = MAN^+$. Then for any vector space E we set $C^q(\mathfrak{n}^+, E) = \mathrm{Hom}_{\mathbb{R}}(\Lambda^q \mathfrak{n}^+, y)$ – i.e. the space of alternating multilinear maps of $\mathfrak{n}^+ \times \ldots \times \mathfrak{n}^+ \to E$. $C^q(\mathfrak{n}^+, E)$ inherits its topology from E, since $\mathfrak{n}^+$ is finite dimensional. Let U be a representation of G on V and let V_∞ be the Garding space of V. Then $d_g : C(n^+, V_\infty) \to C^{q+1}(n^+, V_\infty)$ is defined by

$$\mathrm{d}f(X_1 \wedge \cdots \wedge X_{q+1}) = \sum (-1)^{j+1} U(X_j) f(X_1 \wedge \cdots \wedge \hat{X}_j \wedge \cdots \wedge X_{q+1}).$$

Clearly $d_{q+1} d_q = 0$, so we set $H^q(\mathfrak{n}^+, V_\infty) = \ker d_q / \mathrm{Im}(d_{q-1})$.

For g in MA and f in $C^q(\mathfrak{n}^+, X)$ we set

$$T_{g,q}(f)(X_1 \wedge \cdots \wedge X_q) = U(f) f(\mathrm{Ad}\, g^{-1}) X_1 \wedge \cdots \wedge \mathrm{Ad}(g^{-1}) X_q).$$

Then $T_{g,q}$ is continuous $T_{g_1,q}T_{g_2,q} = T_{g_1g_2,q}$ and $T_{g,q+1}d_q = d_qT_{g,q}$. Let $T^*_{g,q}$ denote the associated action of $T_{g,q}$ on $H^q(\mathfrak{n}^+, V_\infty)$.

THEOREM (Osborne) 18.2.6. Let (U, V) denote a principal series representation (not necessarily unitary) or an irreducible unitary representation of $G = SL(2, \mathbb{R})$; then d_q have closed range, $H^q(\mathfrak{n}^+, V_\infty)$ are finite dimensional (only $H^0(\mathfrak{n}^+, V_\infty)$ and $H^1(\mathfrak{n}^+, V_\infty)$ are nontrivial in this case), Tr U is defined and analytic on G' and

$$\sum(-1)^q \operatorname{Tr}(T_{g,q}) = \sum(-1)^q \operatorname{Tr}(T^*_{g,q})$$

for g in $\{MA \cap G' | \det(1 - \operatorname{Ad}(g)|_{\mathfrak{n}^+}) > 0\}$.

We note that Guillemin has outlined a similar nonelliptic Lefschetz version of the Selberg trace formula for $SL(2, R)$ using the ideas of geometric quantization on contact manifolds.

18.3. NONCOMPACT SPACES WITH FINITE VOLUME

The results of the last section have been extended to the noncompact case and historically it was the noncompact case that interested the number theorists. This transition has recently been successfully studied by applying scattering theory to this problem. This work was led by Gel'fand, Fadeev, Phillips, and Lax. Space does not permit a thorough investigation of the connections with scattering theory. We only point out that a key ingredient of the theory is an application of the Stone–vonNeumann theorem made by Sinai.

The classic example is the case $\Gamma = SL(2, Z) \subset SL(2, R)$ which produces a manifold $M = \Gamma\backslash G/K$ which is noncompact but does have finite volume, $\pi/3$. It was this case that Selberg treated originally. Only recently has the general case been treated. For the classical case we refer the reader to the beautiful exposition of Kubota, Lax, and Phillips, and the papers Fadeev and coworkers.

The harmonic analysis in this noncompact, finite volume situation becomes much more interesting. The left regular representation of G on $L^2(G/\Gamma)$ now decomposes into the sum of three mutually orthogonal closed subspace

$$L^2(G/\Gamma) = L^2_{\text{cus}} \oplus L^2_{\text{Eis}} \oplus L^2_{\text{res}}.$$

These subspaces are formed by taking the closures of the subspaces spanned by the cusp forms, wave packets of Eisenstein series and square integrable residues of the Eisenstein series. On $L^2_{\text{cus}} \oplus L^2_{\text{res}}$, U is a discrete direct sum of irreducible unitary representations of G with finite multiplicities while $U|L^2_{\text{Eis}}$ is a direct integral of unitary principal series representations.

The spectral theory of $\Gamma\backslash\mathscr{P}$ is fairly direct. The eigenfunctions of the continuous portion of the spectrum of $\Delta_{\mathscr{P}}$ arise as follows. Let

$$\Gamma_\infty = \begin{pmatrix} 1 & \mathbb{Z} \\ 0 & 1 \end{pmatrix};$$

for $z = x + iy$ let $a_s(z) = y^s$. Then a_s is an eigenfunction of Δ, and a_s is invariant by Γ_∞. We make a_s invariant by Γ by summing over $\Gamma_\infty\backslash\Gamma$:

$$E(z;s) = \sum_{\gamma\in\Gamma_\infty\backslash\Gamma} a_s(\gamma z).$$

Note that if

$$\gamma = \begin{pmatrix} k & l \\ m & n \end{pmatrix}, \quad \text{then} \quad a_s(\gamma z) = \frac{y^s}{|ms+n|^{2s}}; \quad \text{so} \quad E(z,s)$$
$$= \sum \frac{y^s}{|mz+n|^{2s}},$$

where the greatest common divisor of m and n is one.

Here $\Delta_{\mathscr{P}}\, E(z,s) = s(s-1)E(z,s)$. $E(z,s)$ is called the Eisenstein series. We generalize this concept as follows. For any real number $x \in R \cup \infty$, x is called a *cusp* if $\Gamma_x = \{\gamma\in\Gamma \mid \gamma x = x\}$ is generated by a parabolic element; i.e. there is a parabolic element fixing x. We assume that there are only a finite number of inequivalent cusps, $s_1 = \infty, s_2, \ldots, s_h$. We set $\Gamma_i = \Gamma_{s_i}$ and let $g_i \in G$ be such that

$$g_i\infty = s_i \quad \text{and} \quad g_i^{-1}\Gamma_i g_i = \Gamma_0 = \begin{pmatrix} 1 & Z \\ 0 & 1 \end{pmatrix}.$$

We assume that $\Gamma_\infty = \Gamma_0$. Then the *Eisenstein series for cusp* s_i is

$$E_i(z,s) = \sum_{\Gamma_i\backslash\Gamma} y(g_i^{-1}\gamma z)^s \quad \text{for } z \text{ in } \mathfrak{p} \text{ and } s \text{ in } C.$$

Again it follows that $\Delta_{\mathscr{P}} E_i = s(s-1)E_i$. E_i admits a Fourier series expansion at the cusp s_j

$$E_i(g_j z, s) = \sum_{m\in\mathbb{Z}} a_{ij,m}(y,s)\exp(2\pi i m x),$$

where $a_{ij,m}(y,s) = \int_0^1 E_i(g_j z, s)\exp(-2\pi i m x)\,dx$. The general formula for

$a_{ij,m}$ has been developed in Kubota and elsewhere. In particular we note that at s_1 we have $a_{11,0}(y,s) = y^s + \varphi(s)y^{1-s}$ where

$$\varphi(s) = \pi^{1/2}\Gamma\frac{(s-\frac{1}{2})}{\Gamma(s)}\zeta\frac{(2s-1)}{\zeta(2s)}.$$

And in general $a_{ij,0}(y,s) = \delta_{ij}y^s + \varphi_{ij}(s)y^{1-s}$. $\Phi(s) = (\varphi_{ij}(s))$ is called the constant term matrix. It is easy to check that $\Phi(s)$ is a symmetric matrix and that $y^2 d^2 a_{ij,0}/dy^2 = s(s-1)a_{ij,0}$.

THEOREM 18.3.1. $\varphi_{ij}(s)$ is holomorphic in $\mathrm{Re}\, s > \frac{1}{2}$ except for a finite number of simple poles on $(\frac{1}{2}, 1]$. $\Phi(s)$ is meromorphic on the whole s-plane and satisfies $\Phi(s)\Phi(s-1) = I$.

THEOREM 18.3.2. $E_i(z,s)$ are holomorphic in $\mathrm{Re}\, s > \frac{1}{2}$ except at a finite number of poles of $\varphi_{ii}(s)$; $E_i(z,s)$ are meromorphic on the whole s-plane; and if $\mathscr{E}(z,s) = {}^t(E_1, \ldots, E_h)$ then $\mathscr{E}(z,s) = \Phi(s)\mathscr{E}(z, 1-s)$

Recall that a *modular form of weight* k ($k \in Z$) is a holomorphic function on $\mathscr{P}$ which satisfied (i) $f(\gamma z) = J(\gamma, z)^{-k} f(z)$ and (ii) f is bounded at infinity. Here $J(\gamma, z) = (cz+d)^{-2}$. If a modular form f vanishes at ∞ (the cusp), then f is called a cusp form. The C-vector space of cusp forms is denoted M_k^0.

The *Maass wave-forms* are nonholomorphic cusp forms i.c. bounded smooth K-invariant functions φ on G which satisfy $\mathfrak{C}\varphi = [(1-s^2)/4]\varphi$. Let $W_s(\Gamma)$ denote this space of wave forms. Setting $f_\varphi(z) = \varphi(g)$ where $g(i) = z$ we see that $W_s(\Gamma)$ is isomorphic to the space of smooth C-valued functions f on $\mathscr{P}$ which satisfy (i) $f(\gamma z) = f(z)$ (ii) $\Delta_{\mathscr{P}} f = [(1-s^2)/4] f$ and (iii) f is bounded.

THEOREM 18.3.3. In the case $\Gamma = SL(2, Z)$ we have $L^2(\Gamma\backslash G) = L_0^2(\Gamma\backslash G) \oplus C \oplus \int_0^\infty H(s)\,ds$ where $(V^{0,s}, H(s))$ are class one principal series. The discrete representation T^{2k} (resp. the principal series $V^{0,(2s-1)/4}$) occur in $R(g)|L_0^2(\Gamma\backslash G)$ with multiplicity equal to the dimension of cusp forms M_k^0 and (resp. dimension of the space of Maass wave forms $W_s(\Gamma)$).

Let Θ be the space of incomplete theta series (v. Kubota [K5]), and let $\hat{\Theta}$ denote the subspace of Θ spanned by eigenfunctions of Δ in Θ. Then the spectrum of Δ_M is discrete on $\hat{\Theta}$ and L_0 and is purely continuous on $\hat{\Theta}_0$. The spectral theorem of Δ_M is then:

THEOREM 18.3.4. $L_2(\Gamma\backslash G/K) = \hat{\Theta}_0 \oplus \hat{\Theta} \oplus L_0^2(\Gamma\backslash G/K)$ where $\hat{\Theta}_0$ is the orthogonal complement of $\hat{\Theta}$ in Θ. Let F_k be an ONB for $\hat{\Theta} \oplus L_0$ where

$\Delta F_k = \lambda_k F_k$. Then any f in $L_2(\Gamma\backslash H)$ can be decomposed as

$$f(z) = \sum_k (f, F_k)F_k(z) + \frac{1}{4\pi}\int_{-\infty}^{\infty}(f, E(z,s))E(z,s)\,\mathrm{d}r,$$

where $s = \frac{1}{2} + ir$.

The relation to scattering theory is given as follows. Let $\mathscr{S}(z)$ denote the scattering matrix for the automorphic wave equation $u_{tt} = Lu = \Delta_M u + \frac{1}{4}u$. Then $\mathscr{S}(z)$ is regular on the lower half-plane except for poles at $\{-i\lambda_j\}$ for 'relevant' λ_js. Here L has eigenvalues λ_j^2; A has eigenvalues $\pm\lambda_j$, where

$$A = \begin{pmatrix} 0 & I \\ L & 0 \end{pmatrix}.$$

We set $U(t) = e^{At}$. The poles of $\mathscr{S}$ coincide with $-i$ times the eigenvalues of the infinitesimal generator B of semigroup $Z(t) = P_+U(t)P_-$, where P_+, P_- are the orthogonal projections on the outgoing, incoming subspaces.

THEOREM 18.3.5. (Fadeev–Pavlov).

$$\mathscr{S}(z) = -a^{-2iz}\frac{\Gamma(\frac{1}{2})\Gamma(iz)\zeta(2iz)}{\Gamma(\frac{1}{2}+iz)\zeta(1+2iz)}.$$

$\mathscr{S}(z)$ is related to quantum mechanics as follows. The Schrödinger wave equation $u_t = iLu$ has the scattering matrix $\mathscr{S}^s(z) = \mathscr{S}(\sqrt{z})$. The Schrödinger scattering matrix has, under suitable conditions, an analytic extension which is holomorphic in the physical plane except for poles at the bound state energies $\{-\lambda_j^2\}$ and meromorphic in the nonphysical plane.

We return now to the general case of a symmetric space G/K of rank one with $\Gamma\backslash G/K$ of finite volume.

Let d be the number of equivalence classes of Γ-cuspidal minimal parabolic subgroups, where the equivalence relation is given by conjugacy with an element of Γ. Let $P^1 = P, P^2, \ldots, P^d$ represent these classes. Each is conjugate by a k_i in K to P – i.e. $P^i = k_i P k_i^{-1}$, $1 \leq i \leq d$. When referring to P^i, all elements viz. ρ^i, v^i etc. will have a superscript i.

DEFINITION 18.3.6. The *Eisenstein series* is given by

$$E(P^i, v^i, x) = \sum_{\gamma\in\Gamma\backslash\Gamma\cap P^i} \exp(v^i + \rho^i)(H^i(x\gamma)).$$

DEFINITION 18.3.7. The constant term of E is

$$E_{P^j}(P^i, v^i, x) = \mathrm{Vol}(N^j/\Gamma\cap N^j)^{-1}\int_{N^j/\Gamma\cap N^j} E(P^i, v^i, xn^j)\,\mathrm{d}n^j.$$

We have then

$$E_{P^i}(P^i, v^i, x) = \sum_{w \in W_{ij}} M_{ij}(w, v^i) \exp(wv^i + \rho^j)(H^j(x)),$$

where W_{ij} is the set of bijections $A^i \to A^j$ induced by inner automorphisms of G. We define the matrice $M(v)$ by $M(v) = (M_{ij}(v))$ where $M_{ij}(v) = M_{ij}(k_j w k_i^{-1}, k_i v)$. And we define the vector $E(v,x)$ with entries $E(P^i, v^i, x)$ where $v^i = k_i v$.

THEOREM 18.3.8.

(i) $E(v, x) = M(v)E(-v, x)$ for v in Λ.

(ii) M has no poles in $\mathrm{Re}(v) > \delta$; its poles in $\mathrm{Re}(v) > 0$ are finite in number, simple and lie on the line segment $\{v \in \Lambda \mid 0 < v \leqq \delta\}$.

(iii) $M(v)M(-v) = I$.

(iv) $M^*(v) = M(\bar{v})$ where $*$ is adjoint with respect to the Hermitian structure on $\Lambda^{\mathbb{C}}$ given by $\langle .,\vartheta . \rangle$.

Using these results we are able to write down the partition function on $\Gamma \backslash G/K$.

THEOREM (Gangolli–Warner–Venkov) 18.3.9. The partition function on $\Gamma \backslash G/K$ is

$$\vartheta(\beta) = \sum n_j \exp(-(r_j^+)^2 + \delta_0^2)\beta) = |Z(\Gamma)| \mathrm{Vol}(G/\Gamma) \frac{1}{4\pi} \times$$

$$\times \int_{-\infty}^{\infty} e^{-(r^2 + \delta_0^2)\beta} |c(r)|^2 \, dr + (4\pi\beta)^{-1/2} |u_\gamma| j(\gamma) C(h(\gamma)) \times$$

$$\times \exp(\delta_0^2 \beta + u_\gamma^2/4\beta) + \frac{1}{4\pi} \int_{-\infty}^{\infty} \exp - (r^2 + \delta_0^2)\beta \psi(ir)\psi(ir)^{-1} \, dr +$$

$$+ \tfrac{1}{4} \exp(-\delta_0^2 \beta)(d - \mathrm{Tr}(M(0)) \times$$

$$\times \frac{-d}{2\pi} \int_{-\infty}^{\infty} \exp - (r^2 + \delta_0^2)\beta \Gamma'(1 + ir)/\Gamma(1 + ir) \, dr +$$

$$+ K_5 \int \exp(-(r^2 + \delta_0^2)\beta) \, dr + K_6 \int \exp(-(r^2 - \delta_0^2)\beta) J(r) \, dr,$$

where K_5 and K_6 are constants and $J(r)$ is in [G7] and $\Psi(r)$ is $\det M(r)$.

For a proof see Gangolli–Warner G7 or Venkov V5.

Similarly to the compact case we would like to have

$$\lim_{\beta \to 0} \beta^{n/2} \vartheta(\beta) = C_G \mathrm{Vol}(G/\Gamma) \quad \text{so that} \quad N(r) \sim C_G r^{n/2}.$$

However, to date this estimate is not available. It can be shown that there is a k sufficiently large so the $N(r) \sim C_G r^{k/2}$. Or in other words

THEOREM 18.3.5. The Weyl function $N(r) = \sum_{\{j|(r_j^+)^2 \leq r} n_j$ is tempered.

PROBLEMS

EXERCISE 18.1. Show that for the case $M = \Gamma \backslash SL(2,R)/SO(2)$, compact, that the partition function determines the Length spectrum with multiplicity. (We note that in general this exercise is an open problem.)

EXERCISE 18.2. For $s = |k| - m$ with m in Z and $0 \leqq m < |k|$ let $A(s, L)$ denote the space of analytic automorphic forms of weight s and multiplier L; i.e. L is a unitary representation of Γ in V and we are considering functions $f: \mathscr{P} \to V$ which satisfy $J(\gamma, z, s)^{-1} f(\gamma z) = L(\gamma) f(z)$ for all γ in Γ. Here $J(g, z, s) = ((cz+d)/|cz+d|)^s$ for g in G, z in $\mathscr{P}$ and s in C. Use the STF with the function

$$f(t) = \begin{cases} 1 \text{ if } t = s \text{ or } 1 - s \\ 0 \text{ otherwise} \end{cases}$$

to show that $\dim A(s, L) = (2s-1) \operatorname{Vol}(\Gamma \backslash \mathscr{P}) \dim(V)/4$. Extend this to $0 \leqq k \leqq 1$ by noting that the spectrum of Δ_k with multiplier L is the same as the spectrum of Δ_{-k} with multiplier $\bar{L}$ and checking that the spectrum of $-\Delta_k$ and $-\Delta_{1-k}$ are the same outside the point $k(1-k)$, show that $\dim A(k, L) - \dim A(1-k, \bar{L}) = (2k-1) \operatorname{Vol}(\Gamma \backslash \mathscr{P}) \dim(V)/4 = (2k-1)(g-1) \dim V/2$ since $\operatorname{Vol}(\Gamma \backslash \mathscr{P}) = 2g - 2$.

EXERCISE 18.3. Consider the case of solid state physics where $G =$ Euclidean motion group and Γ is a discrete subgroup such that $\Gamma \backslash G$ is compact. Γ in this case is called a *space group*. If R is the right regular representation of G on E^3 then if $H = (-\hbar^2/2m)\Delta$ is the Hamiltonian we have $R(g)H = HR(g)$ for g in G. Thus G is a symmetry group. Let L be an irreducible unitary representation of Γ and let U be the induced representation from L of G. Then $U = \int_{\hat{G}} U(\lambda)\, d\lambda$ since G is type I. Thus by Schur's lemma we have $H = \int_{\hat{G}} h(\lambda) I \, d\lambda$. In fact $h(\lambda) = a\lambda^2$. Note that we also have $H = \int_\Gamma H(L)\, dL$. Thus show that for each irreducible unitary representation of Γ the induced representation is a discrete sum of irreducible unitary representations with $\dim \operatorname{Hom}_G(U(\lambda), U) =$ multiplicity of $h(\lambda)$ in $H(L)$ divided by $\dim(L)$. Thus we see that the spectrum is in one–one correspondence with the irreducible unitary representations of G.

Chapter 19

Quantum Field Theory

19.1. APPLICATIONS TO QUANTUM FIELD THEORY

The results of geometric quantization and representation theory are playing an ever increasing role in quantum theory. We review a few examples of this interrelationship in this chapter. Much of quantum field theory involves what must be best described as purely formal relationships, or at least that is the way that we will treat the formalism. We will show that many of these formal expressions have already appeared in quantum statistical mechanics.

Quantum field theory studies objects similar to those in statistical mechanics – viz. generalized density matrices where the partition function is given by $Z = \mathrm{Tr}(\rho) = \int \mathrm{d}g\,\mathrm{d}\varphi \exp(iI(g,\varphi))$ where $\mathrm{d}g$ is a measure on the 'space of metrics' and $\mathrm{d}\varphi$ is a measure on the 'space of fields'. Expanding the action $I(g,\varphi)$ about the background field and metric (g_0, φ_0) we have $I(g,\varphi) = I + I_1(\tilde{g}) + \bar{I}_1(\tilde{\varphi})$ to first order where $\tilde{g} = g - g_0$ and $\tilde{\varphi} = \varphi - \varphi_0$. Thus to first order we have

$$\log Z = iI_0 + \log \int \mathrm{d}g \exp(iI_1(\tilde{g})) + \log \int \mathrm{d}\varphi \exp(iI_1(\tilde{\varphi})).$$

The second and third terms are said to describe the contributions of thermal gravitons and matter waves on the background. Nominally the term $I_1(\tilde{\varphi})$ has the form $I_1(\tilde{\varphi}) = \frac{1}{2}\int \tilde{\varphi} A \tilde{\varphi}(-g_0)^{1/2}\,\mathrm{d}^4x$ where A is a second order differential operator, e.g. $A = -\nabla_\alpha \nabla^\alpha + m^2 + \xi R$. Here for $\xi = \frac{1}{6}$ and $m^2 = 0$, A is the conformally invariant wave equation.

Hawking in his germinal study in this area assumed that A is a normal operator with a complete set of eigenfunctions say $A\varphi_n = \lambda_n \varphi_n$. Expanding $\tilde{\varphi}$ as $\tilde{\varphi} = \sum a_n \varphi_n$ and setting $\mathrm{d}\varphi = \mu \prod_n \mathrm{d}a_n$ we have formally

$$Z(\tilde{\varphi}) = \int \mathrm{d}\tilde{\varphi} \exp iI_1(\tilde{\varphi})) = \frac{\mu}{2} \prod_n \int \mathrm{d}a_n \exp(-\lambda_n a_n^2) =$$
$$= \frac{\mu}{2} \pi^{1/2} \prod_n \lambda_n.$$

DEFINITION 19.1.1. The *zeta function* $\zeta_A(s)$ of A is the function $\eta_A(s) = \sum \lambda_n^{-s}$.

Since $\zeta'_A(s) = -\sum \ln(\lambda_n)\lambda_n^{-s}$ we have formally

$$\zeta'_A(0) = -\sum \ln(\lambda_n) = -\ln(\prod \lambda_n) = -\ln(\det A).$$

This suggests the definition

DEFINITION 19.1.2. The *determinant* of operator A is given by $\det(A) = \exp(-\zeta'_A(0))$.

Accepting this one is able to show that

THEOREM 19.1.3. $\log Z(\tilde{\varphi}) = \frac{1}{2}\zeta'_A(0) + \frac{1}{2}\log((\frac{1}{4}\pi\mu^2)\zeta_A(0)$ or $\det(A) = (2\pi\mu^2)^{-\zeta_A(0)}\exp(-\zeta'_A(0))$.

To make this description somewhat more concrete as well as more historical we review the fifth parameter formalism. Consider the scalar field φ with Lagrangian $\mathscr{L} = \frac{1}{2}g^{\mu\nu}\partial_\mu\partial_\nu\varphi - \frac{1}{2}\xi R\varphi^2 - \frac{1}{2}m^2\varphi^2$, where R is the scalar curvature of space-time manifold M, ξ is an arbitrary real number, m^2 is the mass squared and $\hbar = c = 1$. The associated field equation is $H(x)\varphi = 0$ where $H(x) = -\nabla^\mu\nabla_\mu + \xi R + m^2$. The Green's function G of $H(x)\varphi = 0$ is formally given by

$$G(x, x') = \int_0^\infty i\,\mathrm{d}s\langle x|\exp(-isH)|x'\rangle$$
$$= i\int_0^\infty \mathrm{d}s\langle x, s|x', 0\rangle$$

in terms of a basis $|x\rangle$ where $\langle x|x'\rangle = [-g(x)]^{-1/2}\,\delta(x - x')$. Here we have adopted the physicists notation to make him more comfortable.

DEFINITION 19.1.4. The *action* of φ is defined as

$$S = \int \mathrm{d}^4x\,\mathscr{L} = -\tfrac{1}{2}\int \mathrm{d}^4x\,\sqrt{-g}\,\varphi^*(x)H(x)\varphi(x).$$

DEFINITION 19.1.5. The *energy momentum tensor* is defined by the functional derivative

$$T^{\mu\nu}(x) = \frac{2}{\sqrt{-g(x)}}\delta S/\delta g_{\mu\nu}(x)$$

which in our case is

$$T^{\mu\nu}(x) = \partial^\mu\varphi\partial^\nu\varphi - \tfrac{1}{2}g^{\mu\nu}\partial_\alpha\varphi\partial^\alpha\varphi - \tfrac{1}{2}g^{\mu\nu}m^2\varphi^2 + $$
$$+ \xi[g^{\mu\nu}\nabla_\alpha\nabla^\alpha(\varphi^2) - \nabla^\mu\nabla^\nu(\varphi^2) + G^{\mu\nu}\varphi^2],$$

where $G^{\mu\nu} = R^{\mu\nu} - \frac{1}{2}g^{\mu\nu}R$.

THEOREM 19.1.6. If $\xi = \frac{1}{6}$ and $m = 0$ (i.e. the conformally invariant case) then $T^\mu_\mu = 0$.

Proof. This is left as an exercise for the reader.

DEFINITION 19.1.7. We define the *expected value* of $T^{\mu\nu}$ by

$$\langle T^{\mu\nu} \rangle = \frac{2}{\sqrt{-g(x)}} \int d\varphi \delta S/\delta g_{\mu\nu} \exp(iS) / \int d\varphi \, e^{is}$$
$$= \frac{2}{\sqrt{-g(x)}} \delta W / \delta g_{\mu\nu}(x),$$

where $W = i \ln \int d\varphi \exp(iS)$.

Thus the partition function is $Z = \int d\varphi \exp(iS) = \exp(iW)$.

Writing $\varphi = \sum a_n \varphi_n(x)$ (for physicists $\varphi_n(x) = \langle x | \varphi_n \rangle$) we have formally

$$S = -\tfrac{1}{2} \int \varphi^* H \varphi \sqrt{-g} \, d^4x = -\tfrac{1}{2} \sum a_n^2 \lambda_n.$$

Setting $d\varphi = \prod_n \mu \, da_n$ (where μ is a normalization constant) we have formally

$$Z = \prod_n \int \mu \, da_n \exp(-\tfrac{1}{2} i a_n^2 \lambda_n) = \left\{ \prod_n \frac{2\pi\mu^2}{i\lambda_n} \right\}^{1/2}$$
$$= \left\{ \det\left(\frac{iH}{2\pi\mu^2} \right) \right\}^{-1/2}.$$

Putting these results together we have

THEOREM 19.1.8. $W = i \ln \det(iH/2\pi\mu^2) = -l\zeta'(0) - \frac{1}{2}\zeta(0) \ln(-2\pi i \mu^2)$.

Using the parametrix formalism of Chapter 16 for compact manifolds we can show that $\langle x, s | x', 0 \rangle$ can be written as

THEOREM 19.1.9.

$$\langle x, s | x', 0 \rangle = \exp(-im^2 s) \frac{i}{(4\pi i s)^2} \exp(i\tau^2/4s) P(s, s'; is)$$

where

$$i\partial P/\partial s = -\nabla^\mu \nabla_\mu P + \xi R P - \frac{i\tau}{s} \frac{\partial P}{\partial \tau}.$$

COROLLARY 19.1.9. If we define the zeta function $\zeta(\nu)$ by $\zeta(\nu) = \mathrm{Tr}\, G^\nu$ then we have

$$\zeta(v) = \frac{1}{\Gamma(v)(4\pi)^2} \int d^4x \sqrt{-g}\, i\, ds (is)^{v-3} P(x, x; is) e^{-ism^2}.$$

Using the parametrix expansion

$$P(x, x'; \beta) = a_0 + \beta a_1(x, x') + \beta^2 a_2(x, x') + \cdots$$

we leave it to the reader to check that formally

THEOREM 19.1.10.

$$\zeta(0) = \frac{i}{(4\pi)^2} \int d^4x \sqrt{-g}\, [a_2(x, x) - m^2 a_1(x, x) + \tfrac{1}{2} a_0 m^4]$$

and

$$\begin{aligned}\zeta'(0) = \frac{i}{32\pi^2} \bigg\{ &(\gamma - 3/2) \int d^4x \sqrt{-g} \,\times \\ &\times [a_2(x, x) - m^2 a_1(x, x) + \tfrac{1}{2} m] - \\ &- \int d^4x \sqrt{-g} \int i\, ds \ln is \frac{\partial^3}{\partial (is)^3} \,\times \\ &\times P(x, x; is) \exp(-ism^2).\end{aligned}$$

The propagator $\langle x, s | x', 0 \rangle$ is interesting since formally it satisfies the Schrödinger equation

$$i \frac{\partial}{\partial s} \langle x, s | x', 0 \rangle = H(x) \langle x, s | x', 0 \rangle$$

with

$$\lim_{s \to 0} \langle x, s | x', 0 \rangle = (-g)^{-1/2} \delta(x - x').$$

It is for this reason that the name fifth parameter formalism is coined.

We leave it to the reader to check that in the case $(R^3, g_{\alpha\beta})$ the propagator has the form

$$\langle x, s | x', 0 \rangle = \frac{i}{(4\pi i s^2)} \exp(-im^2 s) \exp(i\tau^2/4s),$$

where

$$\tau = \int_0^s ds' \left(g_{\alpha\beta} \frac{dx^\alpha}{ds'} \frac{dx^\beta}{ds'} \right)^{1/2}.$$

DEFINITION 19.1.11. Space time M is said to be an *Einstein space* if $M = S^3(a) \times R$.

THEOREM 19.1.12. If M is an Einstein space then

$$\langle x, s \mid s', 0 \rangle = (4\pi i s)^{-1/2} \exp\left(\tfrac{1}{4} i(\tau - \tau')^2/s\right) K(q, q'; s),$$

where $K(q,q';s)$ is the propagator on $S^3(a)$ satisfying

$$\left(i\frac{\partial}{\partial s} + \Delta_{S^3} - \frac{R}{6} \right) K(q, q'; s) = i\delta(s)\delta(q, q')$$

for q, q' in $S^3(a)$. Note that we have selected $\xi = \frac{1}{6}$.

Using representation theory this equation can be immediately solved as we saw in Chapter 17 to give

$$K(q, q'; s) = (4\pi i s)^{-3/2} \left[a \sin(s/a)^{-1} \sum (s + 2\pi n a) \times \right.$$
$$\left. \times \exp\left(\frac{i}{4s}(s + 2\pi n a)^2 \right) \right].$$

We leave it to the reader to write out a closed expression for the Green's function $G(x, x')$ in this case.

DEFINITION 19.1.13. The *effective Lagrangian* is defined by the coincidence limit

$$L_{\text{eff}} = -\frac{i}{2} \int_0^\infty \frac{ds}{s} \langle x, s \mid x, 0 \rangle \exp(-ism^2).$$

As we have seen in Chapter 16 the propagator $\langle x, s \mid x, 0 \rangle$ has an expansion of the form

$$\langle x, s \mid x', 0 \rangle = \exp(-i\lambda_n s)\, \varphi_n(x)\, \varphi_n^*(x'),$$

where $H\varphi_n = E_n\varphi_n$. If the space time is compact we have already seen many examples where H is given by the Laplace–Beltrami operator Δ_M and is thus selfadjoint and with discrete eigenvalues. The coincidence limit and subsequent integration is equivalent to finding the partition function

$$\begin{aligned} \vartheta(s) &= \operatorname{Tr} \langle x, s \mid x, 0 \rangle \\ &= \sum d(n) \exp(-iE_n s), \end{aligned}$$

where $d(n)$ is the degeneracy of eigenvalue E_n. Thus the effective action is

$$W_{\text{eff}} = \frac{1}{2\mathrm{Vol}(M)} \int \frac{\mathrm{d}s}{s} \vartheta(is) \exp(-ism^2).$$

EXAMPLE 19.1.14. Let M deSitter space time i.e. $M = S_4^1(a) = SO(1,4)/SO(1,3)$. The reader can check that the effective action is given as above with $d(n) = (n/6)(n+1)(2n+1)$ and $\mathrm{Vol}(M) = 8\pi^2 a^4/3$.

Using the Hadamard expansion for the parametrix we write

$$\langle x, s | x, 0 \rangle = i(4\pi is)^{-d/2} \sum_{n=0}^{\infty} a_n(x)\,(is)^n.$$

THEOREM 19.1.15.

$$\begin{aligned} L_{\text{eff}} = {} & \frac{1}{32\pi^2} \lim_{\nu \to 1} \left\{ \frac{\frac{1}{2}a_0 m^4 - a_1 m^2 + a_2}{\nu - 1} \right\} + \\ & + \sum_{n=0}^{2} \frac{(-m^2)^{2-n}}{(2-n)!} a_n(\psi(3-n) + \gamma) + \\ & + \sum_{n=3}^{\infty} a_3 (m^2)^{2-u} \Gamma(n-2), \end{aligned}$$

where $\psi(z) = \Gamma'(z)/\Gamma(z)$ and $\gamma = -\psi(1)$.

DeWitt's philosophy of renormalization is to subtract all a_0, a_1, and a_2 terms leaving

$$L_{\text{finite}} = L_{\text{eff}} - \frac{1}{32\pi^2} \left[\tfrac{1}{2} a_0 m^4 - a_1 m^2 + a_2 + \sum_{n=0}^{2} \{\ldots\} \right].$$

Using the following formal identity

THEOREM 19.1.16.

(i) $\langle T^\mu_\mu \rangle = -2\partial L_{\text{eff}}/\partial \ln m^2$,

(ii) $\int \mathrm{d}^4x \sqrt{-g}\, \langle T^\mu_\mu \rangle = -i\zeta(0) = (1/4\pi^2) \int \mathrm{d}^4x \sqrt{-g}\, a_2(x)$.

We have

COROLLARY 19.1.17. $\mathrm{Lim}_{m^2 \to 0} \langle T^\mu_\mu \rangle = -(1/16\pi^2) a_2$ where $a_2(x) = \frac{1}{6}(\frac{1}{5} - \xi) \times R^{;\alpha}_{;\alpha} + \frac{1}{2}(\frac{1}{6} - \xi)^2 R^2 + \frac{1}{180} R_{\alpha\beta\gamma\delta} R^{\alpha\beta\gamma\delta} - \frac{1}{180} R_{\alpha\beta} R^{\alpha\beta}$.

The fact that classically $T^\mu_\mu = 0$ but that $\langle T^\mu_\mu \rangle \neq 0$ is called the *trace anomally*. We note that prior to renormalization $\langle T^\mu_\mu \rangle \to 0$ as $m^2 \to 0$. Thus the trace anomally is an artefact of the renormalization about which much has been written.

19.2. Static Space Times and Periodization

Static space times are described by products $M = M^3 \times R$; clearly Einstein space time is static. Fundamental groups arose in the study of static space times in the work of Hawking, Gibbons and others on black holes. Looking at the Schwarzschild metric in the form.

$$-\mathrm{d}s^2 = \frac{32M^3}{r}\exp(-r/2M)(\mathrm{d}|V|^2 + |V|^2\,\mathrm{d}(t/2m)^2 + \\ + r^2[\mathrm{d}\vartheta^2 + \sin^2\vartheta\,\mathrm{d}\varphi^2],$$

where $|V| = \exp(r/4M)(r/4M - 1)$, we see there will be a singularity (i.e. an infinite curvature) at the origin $|V| = 0$ (i.e. $r = 2M$) unless we identify t and $t + 8\pi M$. This led Gibbons to reason that the blackhole is endowed with a temperature T gives by $\beta_0 = 1/kT_0 = 8\pi M$, i.e. the blackhole is in equilibrium with a heat bath at a temperature proportional to β^{-1} with partition function $Z = \int \mathrm{d}\varphi \exp(iS)$.

The general philosophy of the blackhole analysis is just an example of what we considered in the Selberg trace formalism. That is, to view the Schrödinger equation on a manifold M_∞. If the propagator $\rho_\infty(x,x',s)$ is then summed over all classical paths we have a propagator on $M_\beta = M_\infty/\Gamma$ where $\Gamma = \mathbb{Z}(\beta)$, the infinite cyclic group with period β. And the propagator is given as in Chapter 18 by

$$\rho_\beta(x, x', s) = \sum_{\gamma \in \Gamma} \rho_\infty(\gamma x, x', s)$$

or more generally by

$$\rho_\beta(x, x', s) = \sum_{\gamma \in \Gamma} \mathfrak{X}(\gamma)\rho_\infty(\gamma x, x', s).$$

Example (Sommerfeld–Carslaw) 19.2.1. Consider the Schrödinger equation on the cone with the metric $\mathrm{d}s^2 = \mathrm{d}r^2 + r^2\,\mathrm{d}\varphi^2$; it is given by

$$\left(i\frac{\partial}{\partial\tau} + \nabla^2\right)\rho_\beta(r, \varphi, r', \varphi', s) = i\delta(\tau)\delta(\varphi - \varphi')\frac{\delta(r - r')}{(rr')^{1/2}}$$

for $|\varphi|, |\varphi'| < \beta/2$. If φ has period $\varphi + \beta, \beta = \infty$, Carslaw in 1909 showed that

$$\rho_\infty(r, \varphi r', \varphi') = \frac{-i}{4\pi\tau} \exp(i(r^2 + r'^2)/\tau \times$$

$$\times \int_{-\infty}^{\infty} d\mu \exp(i\mu(\varphi - \varphi')) \exp(-\tfrac{1}{2} i\,|\mu|\,\pi) J_{|\mu|}\left(\frac{rr'}{2\tau}\right). \qquad (*)$$

To obtain the propagator on the cone we form

$$\rho_\beta(r, r', s) = \sum_{m \in Z} \exp(2\pi i m \delta) \rho_\infty(\mathbf{r}_m, \mathbf{r}', s),$$

where $\mathbf{r}_m = (\mathbf{r}, \varphi + m\beta)$. Thus ρ_β has period β in φ. Using $(*)$ we have the Fourier series

$$\rho_\beta(r, r', s) = -\frac{-i}{2\pi s} \exp(i(r + r')^2/4s) \sum_n \exp\left(2\pi \frac{i(n+\delta)}{\beta} |\varphi - \varphi'|\right) \times$$

$$\times \exp\left(i \frac{(n+\delta)}{\beta} \pi^2\right) J_{2n/\beta}\left(\frac{rr'}{2s}\right).$$

We note that if $\beta = 2\pi$ and $\delta = 0$ then we have

$$\rho_{2\pi}(r, r', s) = \frac{-1}{4\pi s} \exp(i(r - r')^2/4s).$$

The case $\beta = 2\pi$ and $\delta \neq 0$ is just the Aharonov–Bohm set up where δ is the electromagnetic flux through the axis.

The Gibbons–Hawking and Perry set up is just the Carslaw–Sommerfeld model of the cone extended to include variable ϑ.

Consider now the general situation. Let $\rho(x, x', s)$ satisfy the Schrödinger equation

$$\left(i\frac{\partial}{\partial\tau} - \Delta + \frac{R}{6}\right) \rho(x, x', \tau) = i g^{-1/2} \delta(x - x') \delta(\tau).$$

Assume that ρ is periodic of period β_0 in imaginary time. We set

$$\zeta(s, \beta_0) = \frac{\exp(i\pi s/2)}{(s)} \int_0^\infty d\tau\, \tau^{s-1} \int \rho_3(x, x', \tau)\, d^4x \sqrt{-g}$$

$$= \frac{i}{\Gamma(s)} \exp(i\pi s/2) \int \frac{d\tau\, \tau^{s-1}}{(4\pi i \tau)^{1/2}} \sum_m \exp(i m^2 \beta_0^2 / 4\tau) \rho_3(\tau), \qquad (+)$$

where

$$\rho_3(\tau) = \int d^4x\, g^{1/2} \rho_3(x, x, \tau).$$

However,

$$\vartheta_3\left(0, \frac{\beta_0^2}{4\pi\tau}\right) = \sum_m \exp(im^2\beta_0^2/4\tau)$$

and if we use the Poisson summation formula we have

$$\frac{1}{(4\pi i\tau)^{1/2}} \vartheta_3\left(0, \frac{\beta_0^2}{4\pi\tau}\right) = \frac{1}{\beta_0} \vartheta_3\left(0, \frac{4\pi\tau}{\beta_0^2}\right).$$

Thus we have

$$\zeta(s, \beta_0) = \frac{i}{\beta_0} \frac{\exp}{\Gamma(s)} (i\pi s/2) \int_0^\infty d\tau\, \tau^{s-1} \sum_n \exp(i4\pi n^2 \tau/\beta_0^2).$$

Here $\rho_3(x, x', s)$ satisfies the equation

$$\left(\frac{\partial}{\partial\tau} - \Delta_2 - \frac{R}{6}\right) \rho_3(x, x', \tau) = ig^{-1/2} \delta(x - x')\delta(\tau).$$

If

$$\left(\Delta_2 + \frac{R}{6}\right) \varphi_k = \omega_k^2 \varphi_k$$

we have

$$\rho_3(x, x, \tau) = \sum \exp(-i\omega_k^2 \tau) | \varphi_k \rangle \langle \varphi_k |$$

(using physicist's notation). Thus we have

THEOREM 19.2.2.

$$\zeta(s, \beta_0) = \frac{i}{\beta_0} \sum_{n, \omega_k} d(k)/[\omega_k^2 + (4n^2\pi^2/\beta_0^2)]^s,$$

where $d(k)$ is the degeneracy of ω_k.

THEOREM 19.2.3. The regularized one-loop effective Lagrangian can be given in terms of zeta functions by

$$L_{\text{eff}} = -\frac{i}{2} \frac{\zeta(0, \beta_0)}{s-1} + \zeta'(0, \beta_0).$$

The zero temperature term is in the term given by $m = 0$ in $(+)$. Call this term $\bar{L}$. Then we have

COROLLARY 19.2.4.

$$\bar{L}_{\beta_0} = -\beta_0^{-1} \sum_{\omega_k} d(k) \ln(1 - \exp(-\beta\omega_k) + \bar{L}_\infty .$$

Using the expansion from Chapter 16

$$\rho(x, x, s) = (4\pi i s)^{-3/2} \sum_{l=0,1/2,1,3/2\ldots} a_l (is)^l + W$$

we have

THEOREM 19.2.5. Asymptotically

$$\zeta(s, \beta_0) = \frac{i}{\beta_0}\zeta_{M_3}(s) - \frac{i\pi^{3/2-2s}}{8\Gamma(s)} a_l \left(\frac{\beta_0}{4}\right)^{2s+l-2} \Gamma(s+l-3)\zeta_R(2s+2l-3)$$

as $\beta_0 \to 0$.

Here $\zeta_R(s)$ is the standard Riemann zeta function.

These results allow us to calculate the free energy, entropy, energy, etc. in the high temperature limit.

Finally we note that using Theorem 18.3.9 we have the explicit form of the effective Lagrangian on a noncompact, but finite, volume space time manifold. It is left to the reader to develop the physics of this example.

19.3. EXAMPLES OF ZETA FUNCTIONS IN QUANTUM FIELD THEORY

In this section we consider scalar fields on static space times of the form $T \otimes M^3$ where the spatial section M^3 is a Clifford–Klein space form of the flat or spherical type – R^3/Γ or S^3/Γ. As we have seen earlier quantum mechanics and quantum field theory has been considered on multiply connected Riemannian spaces $M = N/\Gamma$. The kernels on M and N are related by $K(q', q'', \tau) = \sum_{\gamma\in\Gamma} a(\gamma)K(q', q''\gamma, \tau)$ as we have seen. The multiplier $a(\gamma)$ has entered this expression somewhat formally to this point in our presentation. We now want to explore its physical meaning in some greater depth and show how it is related somewhat to the concept of 'twisted fields' which we mention in Exercise 19.3.

EXAMPLE 19.3.1. Consider the following simple example of $M = S^1 = R/Z$. If we require the periodicity condition $\psi(q\gamma) = a(\gamma)\psi(q)$, then in this example this requires that $\psi(\vartheta + 2\pi n) = a(\gamma_n)\psi(\vartheta)$ where $\vartheta\gamma_n = \vartheta + 2\pi n$ and $a(\gamma_n) = \exp(2\pi i n\alpha) 0 \leqq \alpha \leqq \frac{1}{2}$. The eigenvalues of the Laplacian $d^2/d\vartheta^2$ are $-k^2$ where $k = n - \alpha, n \in Z$. Thus the fields $\varphi^{(\alpha)}$ on S^1 are parametrized by α.

As we have seen above the vacuum average of the Hamiltonian in this case is given by $E = \langle H \rangle = i \int_{(d-1)} \zeta'_d(0, \ldots)$ where $\zeta_d(s)$ is the equal time zeta function on d-dimensional space time and the integration is over the spatial coordinates. We recall that

$$\zeta_d(s, \ldots) = \frac{1}{(4\pi)^{1/2}} \frac{(s - \frac{1}{4})}{\Gamma(s)} \zeta_{d-1}(s - \tfrac{1}{2}, \ldots).$$

Thus we have $E = \zeta_{d-1}(-\frac{1}{2})$. In the present example $d = 2$ and thus

$$\zeta_1^{(\alpha)}(s) = \sum_k k^{-2s} = \sum_{n=-\infty}^{\infty} (n - \alpha)^{-2s}.$$

If we use the Hurwitz–Lerch–Hermite zeta function

$$\zeta_{\mathrm{HLH}}(s, w) = \sum_{n=b}^{\infty} (n + w)^{-s}$$

we have

$$\zeta_1^{(\alpha)}(s) = \zeta_{\mathrm{HLH}}(2s, \alpha) + \zeta_{\mathrm{HLH}}(2s, 1 - \alpha).$$

If we use the relation

$$\zeta_{\mathrm{HLH}}(-p, w) = \frac{-1}{p+1} \varphi_{p+1}(w),$$

where $\varphi_p(w)$ are given by Bernoulli polynomials with the property $\varphi_{2k}(1 - w) = \varphi_{2k}(w)$, then we have

$$E = -\pi^2 \sum_{n=1}^{\infty} n^{-2} \cos(2\pi n\alpha) = \alpha - \alpha^2 - 1/6.$$

In particular for $\alpha = 0$, $E = -1/6$ (which was derived by Ford for real fields); this corresponds to the trivial representation $a(\gamma) = 1$ of $\Gamma = Z$. And for $\alpha = \frac{1}{2}$, $E = 1/12$ which corresponds to the nontrivial representation $a(\gamma) = (-1)^n$.

EXAMPLE 19.3.2. Consider now the case of real fields on flat space time where M^3 is the Klein bottle wave guide, $R \otimes K_2$. If we represent K_2 as R^2 with $(x, y) \sim (x + pa, (-1)^p y + 2mb)$ $p, m \in Z$, then the Feynman–Green function $G(x, x') = \zeta(1, x, x')$ for this space time is given by

$$G(x, x') = \frac{1}{4\pi^2} \sum_{m,n=-\infty}^{\infty} [\{(x - x' - 2na)^2 + (y - y' - 2mb)^2 + (z - z')^2 - \\ - (t - t')^2]^{-1} + \{(x - x' - (2n + 1)a)^2 + (y + y' - 2mb)^2 + \\ + (z - z')^2 - (t - t')^2\}^{-1}].$$

Using the results from the last section the reader can check that the energy density of the first term in this expression is

$$\langle T_{00} \rangle_1 = -(32\pi^2)^{-1} \sum_{-\infty}^{\infty}{}' (n^2a^2 + m^2b^2)^{-2},$$

while the second term gives

$$\langle T_{00} \rangle_2 = \pi^{-2} \sum_{-\infty}^{\infty} \left\{ \frac{2a^2(4\xi - 1)(2n + 1)^2}{[(2n + 1)^2 a^2 + 4(y - mb)^2]^3} - \right. \\ \left. - \frac{6\xi - 1}{[(2n + 1]^2 a^2 + 4(y - mb)^2]^2} \right\}.$$

The reader can check that by integrating over the Klein bottle the total energy for a unit distance in the z direction is

$$E(a, b) = -7(16\pi a^2)^{-1} \zeta_{\text{HLH}}(3) - ab(16\pi^2)^{-1} \sum_{m,n=-\infty}^{\infty} (n^2a^2 + m^2b^2)^{-2}.$$

The infinite Möbius strip model is obtained by letting $b \to \infty$; thus in the conformally coupled case ($\xi = 1/6$) we have

$$\langle T_{00} \rangle_{m=b} = -(16\pi^2 a^4)^{-1} \zeta_{\text{HLH}}(4) - \frac{2a^2}{3\pi^2} \sum_{-\infty}^{\infty} \frac{(2n + 1)^2}{[(2n + 1)^2 a^2 + 4y^2]^3}.$$

The two dimensional Epstein zeta functions allow some simplification in these formulae. These zeta functions are given by

$$Z\begin{pmatrix} g^t \\ h^t \end{pmatrix}(s, A) = \sum_{m_1, m_2=-\infty}^{\infty} \exp(2\pi i h^t m) [(m + g)A(m + g)^t]^{-s},$$

where m, g, h are column matrices,

$$m = \begin{pmatrix} m_1 \\ m_2 \end{pmatrix}$$

and A is a 2×2 matrix. Z satisfies the functional equation

$$Z\begin{pmatrix} g^t \\ h^t \end{pmatrix}(s, A) = (\det A)^{-1/2}\pi^{2s-1}\frac{\Gamma(1-s)}{\Gamma(s)} \times$$

$$\times \exp(-2\pi i g^t h) Z\begin{pmatrix} h^t \\ -g^t \end{pmatrix}(1-s, A^{-1}).$$

The reader can check that

$$\langle T_{00}\rangle_1 + \langle T_{00}\rangle_2 = -(32\pi^2)^{-1} Z\begin{pmatrix} 0 & 0 \\ 0 & 0 \end{pmatrix}(2, A) -$$

$$-(16\pi^2)^{-1}((4\xi - 1)a^2\frac{\partial}{\partial a^2} + 6\xi - 1)Z\begin{pmatrix} \frac{1}{2} & y/b \\ 0 & 0 \end{pmatrix}(2, A)$$

where

$$A = \begin{pmatrix} a^2 & 0 \\ 0 & b^2 \end{pmatrix}.$$

And similarly

$$E(a, b) = \frac{-7}{16\pi a^2}\zeta_{\mathrm{HLH}}(3) - \frac{ab}{16\pi^2}Z\begin{pmatrix} 0 & 0 \\ 0 & 0 \end{pmatrix}(2, A).$$

Using Hardy's expression

$$Z\begin{pmatrix} 0 & 0 \\ 0 & 0 \end{pmatrix}(s, 1) = 4\zeta_{\mathrm{HLH}}(s)\beta(s)$$

we have in the case $a = b$

$$E(a, a) = -\frac{1}{4\pi^2 a^2}\left[\frac{7\pi}{4}\zeta_{\mathrm{HLH}}(3) - \zeta_{\mathrm{HLH}}(2)\beta(2)\right].$$

EXAMPLE 19.3.3. The Casimir geometry of infinite parallel plates has been examined by several authors. Using the Dirichlet boundary conditions the eigenfunctions are $(2/a)\sin(m\pi x/a), m = 1, 2, \ldots$ and the associated zeta function is

$$\zeta_1(s, x, x') = \frac{2}{a}\sum_{m=1}^{\infty}\left(\frac{m\pi}{a}\right)^{-2s} \sin(m\pi x/a)\sin(m\pi x'/a).$$

Using the Epstein zeta function

$$Z\begin{pmatrix} g \\ h \end{pmatrix}(s) = \sum_{m=-\infty}^{\infty} |m + g|^{-s}\exp(2\pi i m h)$$

and the identity

$$Z\begin{pmatrix} g \\ h \end{pmatrix}(2s) \pm \pi^{2s-1/2}\frac{\Gamma(\frac{1}{2}-s)}{\Gamma(s)}\exp(-2\pi igh)Z\begin{pmatrix} h \\ -g \end{pmatrix}(1-2s)$$

we have

$$\zeta_1(s,x,x') = \frac{a^{2s-1}}{2\pi^{1/2}}\frac{\Gamma(\frac{1}{2}-s)}{\Gamma(s)}Z\begin{pmatrix} (x-y')/2a \\ 0 \end{pmatrix}(1-2s) -$$

$$-Z\begin{pmatrix} (x+x')/2a \\ 0 \end{pmatrix}(1-2s).$$

If we let $\mathbf{x} = (t, x, y, z)$ denote the coordinates in this space time, then the coincidence limit of the 4-dimensional zeta function is

$$\lim_{\substack{t'\to t \\ z'\to z \\ y'\to y}} \zeta_4(s,\mathbf{x},\mathbf{x}') = \frac{i}{(4\pi)^{3/2}}\frac{\Gamma(s-3/2)}{\Gamma(s)}\zeta_1(s-3/2,x,x').$$

We leave it to the reader to show that in this case we have

$$\langle T_{00}\rangle = \frac{-\pi^2}{12a^4}\left[\zeta_{\mathrm{HLH}}(-3) + (1-6\xi)Z\begin{pmatrix} 0 \\ x/a \end{pmatrix}(-3)\right]. \qquad (*)$$

The minimal coupling case of this example was treated by DeWitt (i.e. the case $\xi = 0$) which we check by using

$$Z\begin{pmatrix} 0 \\ x/a \end{pmatrix}(-3) = \frac{3}{4\pi^4}Z\begin{pmatrix} x/a \\ 0 \end{pmatrix}(4)$$

and

$$\sum(\vartheta - n)^{-4} = \pi^4(\operatorname{cosec}^4\pi\vartheta - \tfrac{2}{3}\operatorname{cosec}^2\pi\vartheta).$$

We note that

$$Z\begin{pmatrix} x/a \\ 0 \end{pmatrix}(4)$$

diverges as the plates are approached – i.e. $x \to 0$ or $x \to a$. Using the results of the last section the reader can check that the total energy per unit plate area is given by

$$E = \frac{i}{2}\lim_{s\to 1}\frac{\zeta_4(s-1)}{s-1} = \frac{-\pi^2}{12a^3}\zeta_{\mathrm{HLH}}(-3)$$

which is the integral of the first term in (*).

EXAMPLE 19.3.4. The rectangular wave guide with Dirichlet boundary conditions has zeta function

$$\zeta_2(s) = \tfrac{1}{2}\pi^{-2s}\left[\tfrac{1}{2}Z\begin{pmatrix}0 & 0\\ 0 & 0\end{pmatrix}(s, A) - b^{2s}\zeta_{\mathrm{HLH}}(2s) - a^{2s}\zeta_{\mathrm{HLH}}(2s)\right],$$

where

$$A = \begin{pmatrix}a^2 & 0\\ 0 & b^2\end{pmatrix}.$$

Using $E = (1/8\pi)\zeta_2'(-1)$ the reader can check that

$$E(a, b) = \frac{1}{32\pi}\left[(a^{-2} + b^{-2})\zeta_{\mathrm{HLH}}(3) - \frac{abZ}{\pi}\begin{pmatrix}0 & 0\\ 0 & 0\end{pmatrix}(2, A)\right].$$

EXAMPLE 19.3.5. 'Twisted fields' on the Klein bottle space time are given by taking $\alpha(\gamma_{pm}) = \exp(2\pi ip\alpha)$ or $a(\gamma_{pm}) = \exp(2\pi ip\alpha)(-1)^m, 0 \leqq \alpha \leqq \frac{1}{2}$. The total energy for twisted fields is easily seen to be

$$E(a, b, \alpha) = -(ab/16\pi^2)Z\begin{pmatrix}0 & 0\\ 0 & 0\end{pmatrix}(2, A)\frac{-4}{a^2}\times$$

$$\times\sum_{n=1}^{\infty}(\cos(2\pi\alpha n) - \tfrac{1}{8}\cos(4\pi\alpha n))n^{-3}.$$

EXAMPLE 19.3.6. We now consider space times of the form $T \otimes S^3/\Gamma$. We have already considered the conformally coupled scalar field on the Einstein universe $T \otimes S^2$. Using $\zeta_{S^3}(s) = \sum n^2(n^2/a^2)^{-s} = a^{2s}\zeta_{\mathrm{HLH}}(2s-2)$ we have $E = \frac{1}{2}\zeta_{S^3}(-\frac{1}{2}) = 1/240a$.

The case $T \otimes RP(3)$, $RP(3) = S^3/Z_2$ has two one dimensional representations of Z_2, $a(\pm 1) = 1$ and $a(\pm 1) = \pm 1$. In the first case the zeta function is $a^{2s}\sum_{j=0,1,2\ldots}^{\infty}(2j+1)^{-2(s+1)}$ with $E = -7/20a$ while in the second case $\zeta(s) = a^{2s}\sum_{j=1/2,3/2,\ldots}(2j+1)^{-2(s+1)}$ with $E = 1/30a$.

The case of the lens space time $T \otimes L(3)$, $L(3) = S^3/Z_m$ is similarly treated. The reader can check that

$$\langle T_{00}\rangle = -(1440\pi^2a^4)^{-1}(m^4 + 10m^2 - 14)$$

and

$$E = -(720a)^{-1}(m^3 + 10m - 14m^{-1}).$$

The general case of space times of the form $T \otimes S^3/\Gamma$ has been treated by Dowker and Banach in D29. In particular they have found that the general expression for $\langle T_{ab}\rangle$ is not of the same form as $R_{ab} - \frac{1}{2}g_{ab}R$ since it contains additional geometrical structure. This leads to difficulties in the 'back reaction' problem where one would like to use $\langle T_{\mu\nu} \rangle$ on the right-hand side of Einstein's field equations in hopes that this represents the back reaction of the field on the geometry. Furthermore we have found that 'twisted fields' have a very different back reaction as compared to untwisted fields.

Problems

EXERCISE 19.1. The Proca field equation was introduced in Chapter 16. In this case show

$$Z(0) = (\det(\Delta_1 + m^2))^{-1/2}(\det(\Delta_0 + m))^{1/2},$$

where Δ_p is the Laplacian on p-forms.

EXERCISE 19.2. Show that if M_3 is an infinite wave guide that the zero point energy $E = -\bar{L}_\infty$ is given by

$$E = -8\pi a^2 \zeta_R(2)\beta(2) - \tfrac{1}{2}\pi\zeta_R(3),$$

where $\beta(s)$ is given by $\beta(s) = \sum_{n=0}^{\infty}(-1)^n(2n+1)^{-s} s > 2$.

EXERCISE 19.3. Consider the case of massless fields on $S^1 \times R^3$ where x^μ, $\mu = 0, 1, 2, 3$ with periodicity of S^1 represented in the x^3 direction. Assume the period is a. The Green's function for the field equation satisfies $\Box G(x, x') = -\delta(x, x')$. If $G_0(x, x')$ is the Minkowski Green's function, show that

$$G(x, x') = \sum_{n=-\infty}^{\infty} G_0(x, x' + nae_3).$$

Set $G_{\text{ren}} = \sum_{n \neq 0} G_0(x, x' + nae_3)$. Show that

$$\langle T^{\mu\nu} \rangle = i[\partial^2/\partial x_\mu \partial x_\nu G_{\text{ren}}(x, x')]|_{x=x'}$$

$$= \frac{1}{\pi^2}\sum_{n=1}^{\infty}\left(\frac{1}{na}\right)^4 \begin{pmatrix} -1 & & & 0 \\ & 1 & & \\ & & 1 & \\ 0 & & & -3 \end{pmatrix} =$$

$$= \frac{\pi^2}{90a^2} \begin{pmatrix} -1 & & & 0 \\ & 1 & & \\ & & 1 & \\ 0 & & & -3 \end{pmatrix}.$$

The 'twisted' field is defined by

$$G_{\text{ren}}^{\text{twist}}(x,x') = \sum_{n\neq 0} (-1)^n G_0(x, x' + nae).$$

Show in this case that

$$\langle T^{\mu\nu}_{\text{twist}}\rangle = \frac{1}{\pi^2}\sum_{n=1}^{\infty}\frac{(-1)^n}{(na)^4}\begin{pmatrix}-1 & & & 0\\ & 1 & & \\ & & 1 & \\ 0 & & & -3\end{pmatrix} =$$

$$= -\frac{7}{8}\frac{\pi^2}{90a^4}\begin{pmatrix}-1 & & & 0\\ & 1 & & \\ & & 1 & \\ 0 & & & -3\end{pmatrix}.$$

Note that the energy density $\langle T^{00}\rangle$ is negative whereas $\langle T^{00}_{\text{twist}}\rangle$ is positive. Thus the twist has raised the energy of the vacuum.

Treat the other cases $\sum \times R$ where $\sum = R^3/\Gamma$ is one of the 18 Clifford–Klein space forms. E.g. show that for $\sum = S^1 \times S^1 \times R$ with identical periods for the S^1 factors has

$$\langle T^{\mu\nu}\rangle = \frac{1}{2\pi^2 a^4}\sum_{\substack{m_1 n=-\infty\\ (m_1 n)\neq(0,0)}}^{\infty}\frac{1}{(m^2+n^2)^2}\begin{pmatrix}-1 & & & 0\\ & 1 & & \\ 0 & & -1 & \\ & & & -1\end{pmatrix} =$$

$$= \frac{305}{a^4}\begin{pmatrix}-1 & & & \\ & 1 & & 0\\ 0 & & -1 & \\ & & & -1\end{pmatrix}$$

Consider the twisted fields in this case and treat the cases where $\sum =$ $= K^2 \times R$ where K^2 is the Klein bottle and $\sum = M^2 \times R$ where M^2 is the Mobius strip of length a and infinite width.

Chapter 20

Coherent States and Automorphic Forms

20.1. Coherent States and Automorphic Forms

Perelomov was the first to consider the concept of 'coherent states' for an IUR[(2)] of any Lie group G. We will examine the case $G = SU(1,1)/Z_2$, the group of motions of the unit disc D, and we will outline the general theory in some exercises. As we know the group of 2×2 matrices

$$\begin{pmatrix} \alpha & \beta \\ \bar{\beta} & \bar{\alpha} \end{pmatrix}$$

in $SL(2,\mathbb{C})$ with $|\alpha|^2 - |\beta|^2 = 1$ acts on D. For a demi-integer $k \geqq 1, k = 1, 3/2, 2, \ldots$ let H^k denote the Hilbert space of square-integrable, for the measure $d\mu_k(z) = (1 - |z|^2)^{2k-2}\, dx\, dy$ functions f on D. H has the inner product

$$\langle f_1, f_2 \rangle = \frac{2k-1}{\pi} \int_D f_1(z)\, \overline{f_2(z)}\, d\mu_k(z).$$

We let G act on H^k by

$$(T^k(g)\, f)(z) = (\beta z + \bar{\alpha})^{-2k}\, f(\alpha z + \bar{\beta}/\beta z + \bar{\alpha}).$$

It is easy to see that $g \to T^k(g)$ is a UR[(1)] of G in H^k. The holomorphic functions on D in H^k form a Hilbert subspace $\mathscr{H}^k$ of H^k. And T^k leaves $\mathscr{H}^k$ stable. Thus $T^k | \mathscr{H}^k$ is an IUR[(2)] of G. It is easy to check that the functions

$$f_p(z) = \left[\frac{\Gamma(2k+p)}{\Gamma(2k)\Gamma(p+1)}\right]^{1/2} z^p, \qquad p = 0, 1, 2 \ldots$$

form an ONB[(3)] for $\mathscr{H}^k$ and

$$T^k(u_\nu) f_p^k = x_{-k-p}(u_\varphi) f_p^k$$

[(1)] **UR = Unitary Representation.**
[(2)] **IUR = Irreducible Unitary Representation.**
[(3)] **ONB = Orthonomal Basis.**

where

$$u_\varphi = \begin{pmatrix} e^{i\varphi/2} & 0 \\ 0 & e^{-i\varphi/2} \end{pmatrix} \quad \text{and} \quad x_k(u_\varphi) = e^{ik\varphi}.$$

$T^k | \mathscr{H}^k$ is one of two discrete series of G. The other is given by the anti-holomorphic functions on D and demi-integers $\leqq -1$. If ψ_0 is the vector of least weight $\psi_0 = f_0 = 1$, then

$$T^k(g)\psi_0 = (\beta z + \bar{\alpha})^{-2k} = e^{i\Phi}\psi_\zeta,$$

where

$$\zeta = -\beta/\bar{\alpha} \in D, \quad \Phi = 2k \arg \alpha$$

and

$$\psi_\zeta(z) = (1 - |\zeta|^2)^k (1 - \zeta z)^{-2k}$$

Given an IUR $(T, \mathbf{H})$ of a Lie group G, let

$$H = \{h \in G \mid T(h)\psi_0 = e^{i\Phi(h)}\psi_0, \Phi(h) \in R, \psi_0 \in \mathbf{H}\}.$$

For each class $x \in M = G/H$, choose a representative $g(x) \in G$. Let $\psi_x = T(g(x))\psi_0$. Then the set $\{\psi_x\}$ is called a system of *coherent states* of type (T, ψ_0).

In our example above $M = G/H = D$ and to each point ζ of D there corresponds a coherent state ψ_ζ.

THEOREM 20.1.1. We have the following:

(a) $$\psi_\zeta(z) = (1 - |\zeta|^2)^k \sum_{n=0}^{\infty} f_n(\zeta) f_n(z),$$

(b) $$\frac{2k-1}{\pi} \int d\mu(\zeta) P_\zeta = I,$$

where

$$d\mu(\zeta) = (1 - |\zeta|^2)^{-2} d\xi \, d\eta, \qquad \zeta = \xi + i\eta$$
$$P_\zeta = |\zeta\rangle\langle\zeta|$$

is the projection operator on state ψ_ζ, and I is the unit operator.

(c) If $$\mathfrak{C} = \frac{K_+K_- + K_-K_+}{2} - K_0^2.$$

is the Casimir operator for (T^k, G), where we have

$$[K_0, K_\pm] = \pm K_\pm, [K-, K_+] = 2K_0,$$

then $|\zeta\rangle = \psi_\zeta = (1 - |\zeta|^2)^k e^{\zeta K_+} |0\rangle$.

We leave it to the reader to

(d) Define the ONB $\{|n\rangle\}$ for $(T^k, \mathcal{H}^k)$ and relate it to $\{|\zeta\rangle\}$, where

$$K_{\pm}|n\rangle = (n+1)(n+2k)|n+1\rangle$$
$$K_{-}|n\rangle = n(n+2k-1)|n-1\rangle$$
$$K_0|n\rangle = (k+n)|n\rangle.$$

(e) $\|\psi\|_k^2 = \int \mathrm{d}\mu_k(\zeta)\,|\psi(\zeta)|^2.$

(f) $\langle\eta|T(g)|\zeta\rangle = e^{i\Phi}\langle\eta|\zeta'\rangle$

$$\Phi = 2k\,\mathrm{Arg}(\alpha - \zeta\beta) \quad \text{and} \quad \zeta' = (\alpha\zeta - \beta)(-\bar{\beta}\zeta + \bar{\alpha}).$$

THEOREM (Onofri) 20.1.2. Let G be a compact semi-simple Lie group with Cartan subgroup H. Let

$$\mathfrak{g}_{\mathbb{C}} = \mathfrak{h}_{\mathbb{C}} \oplus \sum_{\alpha\in\Delta} \mathfrak{g}_\alpha$$

be the Cartan decomposition, Δ_+ the set of positive roots, $\delta = \frac{1}{2}\sum_{\alpha\in\Delta_+}\alpha$, and $(U, \mathbf{H})$ a UIR, with highest weight λ. Then there exists a vector $|0\rangle \in \mathbf{H}$ such that $U(h)|0\rangle = e^{\lambda}(h)|0\rangle, h\in H$.

$$X|0\rangle = 0, \qquad X \in \mathfrak{n}_+ = \sum_{\alpha\in\Delta_+} \mathfrak{g}_\alpha.$$

$U(g)$ extends to a holomorphic representation T of $G_{\mathbb{C}}$ with stability subgroup of $\langle 0|$ being the Borel subgroup corresponding to $\mathfrak{b} = \mathfrak{h}_{\mathbb{C}} + \mathfrak{n}$. Let $\pi(b)$ be the holomorphic character of B defined by $\langle 0|T(b) = = \pi(b)\langle 0|$.

(a) $T(g^{-1})^+|0\rangle$ is a system of coherent states of type $(T, \mathbf{H})$; and $\Psi(g) = \langle 0|T(g^{-1})|\Psi\rangle$ is holomorphic on $G_{\mathbb{C}}$ and defines a holomorphic section of the homogeneous line bundle $E_\pi(G/H, \mathbb{C})$ associated to $B \to G_{\mathbb{C}} \to G_{\mathbb{C}}/B$ by the holomorphic character π.

(b)

$$|z\rangle = |g\cdot 0\rangle = U(g)|0\rangle / \langle 0|U(g)|0\rangle$$

(c) $$\langle\Psi_1|\Psi_2\rangle = \mathrm{d}_\lambda \int_{G/H} \bar{\Psi}_1(z)\,\Psi_2(z) \times \exp[-f(z,\bar{z})]\,\mathrm{d}\mu(z),$$

where

$$z = g\cdot 0,\ f(z,\bar{z}) = \log|\langle 0|U(g)|0\rangle|^{-2}, \text{ and}$$

$$\mathrm{d}_\lambda = \prod_{\alpha\in\Delta_+} \langle\lambda+\rho,\alpha\rangle/\langle\rho,\alpha\rangle = \dim \mathbf{H}.$$

Thus **H** is isomorphic to a Hilbert space of holomorphic functions bounded by $\exp\frac{1}{2}f(z,\bar{z})$.

(d) For details on the classical generating function and the expectation value of the corresponding self-adjoint operator on the coherent states, v. Onofri [O2].

For the extension of the above to $SU(N,N)$, resp. $SO(n,2)$ see Perelomov [P8], resp. Onofri [O2].

Returning to our example, $D=G/H$, the basis $\{\psi_\zeta\}$ is *over* or *super complete*. The question arises: is there a complete subsystem. Let Γ be a discrete subgroup of G such that $\mathrm{Vol}(\Gamma\backslash D)$ is finite. If $\Gamma=\{\gamma_n\}$, we consider the set of states $\{\psi_n\}$ where $\psi_n=\psi_{\gamma_n\zeta}$ for $\zeta\in D$.

We considered in Section 18.1 automorphic forms on the upper half-plane $\mathscr{P}$. We need a bit more generality which we include now. If $\mathscr{P}$ is the Poincaré half-plane then associated to $M=\Gamma\backslash\mathscr{P}$ is a compact Riemann surface $\hat{M}$. For $p\in M$ we set $n_p=|\Gamma_p|$, 1, ∞ as p is elliptic, regular or parabolic.

$\mathrm{Vol}(\Gamma\backslash\mathscr{P})=(2n-2)\pi-\sum_i 2\pi/n_{p_i}$ and Euler–Poincaré characteristic

$$x(\hat{M})=\sum_i 1-n+1=2g-2,$$

where $g=$ genus of M, and $c=\sum_i 1=$ the complete number of cycles.

So

$$\mathrm{Vol}(M)=\mathrm{Vol}(\hat{M})=2\pi(2g-2)+2\pi\sum_{p\in\hat{M}}\left(1-\frac{1}{n_p}\right).$$

The Riemann–Roch formula for modular forms becomes

$$d_k(\Gamma)=\begin{cases}0 & k<0\\ g & k=1, \quad \text{if } \Gamma\backslash\mathscr{P} \text{ is compact}\\ g-1+\mathrm{Card}(\hat{M}\backslash M) & k=1, \quad \text{if } M \text{ is non-compact}\\ (2k-1)(g-1)+\sum\limits_{p\in M}\left[k\left(1-\dfrac{1}{n_p}\right)\right], & k\geqq 2.\end{cases}$$

The numbers $(g,c,n_1,\dots,n_c)$ are called the signature of Γ, E.g. the modular group has signature $(0,3;2,3,\infty)$.

Now as on the upper-half plane we say that a function $f_m(z)$ on D is a *automorphic form* of weight $m \in Z$ if $f(z)$ is analytic on D and is regular on $D \cup \mathscr{P}, \mathscr{P} =$ set of parabolic vertices, and satisfies

$$f\left(\frac{\alpha_n z + \bar{\beta}_n}{\beta_n z + \bar{\alpha}_n}\right) = (\beta_n z + \bar{\alpha}_n)^{2m} f_m(z),$$

where

$$\gamma_n = \begin{pmatrix} \alpha_n & \beta_n \\ \overline{\beta_n} & \overline{\alpha_n} \end{pmatrix} \in \Gamma.$$

An automorphic form is called *parabolic* if it vanishes at all parabolic vertices. Let $d_m(\Gamma)$ (resp. $d_m^+(\Gamma)$) denote the dimension of the space of automorphic (resp., parabolic) forms. Let m_0(resp. m_0^+) denote the least m for which $d_m(\Gamma) \geqq 2$, (resp. $d_m^+(\Gamma) \geqq 2$). If $\Gamma \backslash D$ is compact, then

$$d_m(\Gamma) = d_m^+(\Gamma), \quad m_0 = m_0^+.$$

THEOREM (Perelomov) 20.1.3. Let $\{\psi_\zeta\}$ be a system of coherent states of type (T^k, ψ_0). Then the subsystem $\{\psi_n\}$, $\psi_n = \psi_{\zeta_n}$ associated with $\Gamma \subset G$ is incomplete if $k > m_0^+ + 1/2$.

Proof. One needs only the fact that there exists a parabolic form $f_{m_0^+}(\zeta)$ of weight m_0^+ having a zero at an arbitrary point ζ_0 of D; and so it vanishes at all points ζ_n. The theorem follows since

EXERCISE 20.1.4. $\{\psi_{\zeta_n}\}$ is incomplete iff ζ_n is the set of zeros of a function $f(\zeta)$ in $\mathscr{H}^k$.

EXERCISE 20.1.5. Show that $f_{m_0^+}(\zeta) \in \mathscr{H}^k$.

Let S_Γ denote the area of the fundamental domain of $\Gamma \backslash D$ for the measure $d\mu(\zeta)$: viz.

$$S_\Gamma = \pi\left[g - 1 + \tfrac{1}{2} \sum_{i=1}^{c} \left(1 - \frac{1}{n_i}\right)\right].$$

E.g. for the modular group $S_\Gamma = \pi/12$.

(Note that this volume is $\frac{1}{4}$ the volume on the upper half-plane in measure $d\mu = y^{-2}\, dx\, dy$.)

THEOREM (Perelomov) 20.1.6. If $S_\Gamma < \infty$, then the system $\{\psi_n\}$ of type (T^k, ψ_0) is complete when $S_\Gamma < \pi/(2k-1)$.

Using the Riemann–Roch formula for $d_m(\Gamma)$ we have

THEOREM (Perelomov) 20.1.7. If Γ has signature $(0, c; l_1, \ldots, l_{c_1}, \infty, \ldots, \infty)$ and admits an automorphic form with a zero in the fundamental domain, then there are only 21 such discrete subgroups Γ and they are listed in [P9].

PROBLEMS

EXERCISE 20.1.1. An area of recent study is the transition from strong to weak coupling in lattice gauge theories. In two dimensions, lattice gauge theory considers a plane square lattice $L = \{x = n_1 e_1 + n_2 e_2 | (n_1, n_2) \in Z^2\}$ and gauge variables $U_{ij} \in G$ where (i, j) denotes any pair of nearest neighbor points in L and G is any compact group. Nominally $G = U(N)$ or $SU(N)$. The U_{ij} are random variables (such that $U_{ij} = U_{ji}^{-1}$) whose probability distribution is given in terms of a positive symmetric class function Φ on G. The *multiple loop variables* are defined by

$$W_n(g, N) = \langle (1/N)\operatorname{Tr}(U^n) \rangle = \int_N \frac{\mathrm{d}U\, \Phi(U) \operatorname{Tr}(U^n)}{\int \mathrm{d}U\, \Phi(U)},$$

where $\mathrm{d}U$ denotes Haar measure on G, $U \in G$. Lattice gauge theorists are interested in the limits of W_n as $N \to \infty$.

A specific example of Φ has been used which is based on the so called *Vallain action*

$$e^{-A(\vartheta)} = \sum_{n \in Z} \exp\left(\frac{-N}{\lambda}(\vartheta + 2n\pi)^2\right) = \left(\frac{\lambda}{2\pi N}\right) \sum_{n \in Z} \exp\left(\frac{-\lambda n^2}{4N}\right) \exp\,(\mathrm{i}n\vartheta)$$

which we recognize as Jacobi's ϑ_3-function. We already noted that ϑ_3 is a solution of the heat equation or Bloch's equation on S^1. The higher dimensional cases can be built similarly on the heat equation on G as we see below.

Taking $\Phi(U) = \det(\vartheta_3(U|q))$ where $\vartheta_3(z) = \sum_{n \in Z} q^{n^2} z^n$, then in terms of polar coordinates on $U(N)$ – i.e. $U = S \operatorname{diag}(e^{i\vartheta_1}, \ldots, e^{i\vartheta_N}) S^T$, show that

$$W_n(q, N) = \frac{Z^{-1}}{N} \int \frac{\mathrm{d}\vartheta_1}{2\pi} \cdots \int \frac{\mathrm{d}\vartheta_N}{2\pi} |\Delta(e^{i\vartheta})| \prod_{j=1}^{2N} \vartheta_3(e^{i\vartheta_j}|q) \sum_{m=1}^{N} e^{in\vartheta_m},$$

where

$$Z(\lambda, N) = \frac{1}{(2\pi)^N N!} \int_{-\pi}^{\pi} \ldots \int d\vartheta_1 \ldots d\vartheta_N |\Delta(e^{i\vartheta})|^2 \prod_j e^{-A(\vartheta_j)}.$$

Here $\Delta(e^{i\vartheta}) = \prod_{1 \le k < j \le N} (e^{i\vartheta_j} - e^{i\vartheta_k})$ is Vandermonde's determinant and $q = \exp(-\lambda/4N)$.

Let $\varphi_j(z)$ denote the polynomials which satisfy

$$\int_{-\pi}^{\pi} \frac{d\vartheta}{2\pi} \sum q^{n^2} e^{in\vartheta} \overline{\varphi_j(e^{i\vartheta})} \varphi_k(e^{i\vartheta}) = \delta_{jk}.$$

We leave it to the reader to check that in terms of φ_j we can write

$$W_n(q, N) = \frac{1}{N} \sum_{j=0}^{N-1} \int_{-\pi}^{\pi} d\vartheta\, \vartheta_3(e^{i\vartheta}|q) e^{in\vartheta} |\varphi_j(e^{i\vartheta})|^2$$

$$= \int_{-\pi}^{\pi} d\vartheta\, e^{in\vartheta} \rho_N(\vartheta, q),$$

where ρ_N represents the probability density for eigenvalues of U:

$$\rho_N(\vartheta, q) = \frac{1}{N} \sum_{j=0}^{N-1} |\varphi_j(e^{i\vartheta})|^2 \vartheta_3(e^{i\vartheta}|q).$$

Onofri has shown that the $1/N$ expansions of Z and W_n is convergent for sufficiently small $|\lambda/N|$ and its coefficients are analytic in λ near the real axis (i.e. no 'Gross–Witten' singularity to all orders in $1/N$). Onofri shows that it is still not possible to commute the strong coupling limit with the planar limit (i.e. $\lambda \to \infty$ and $N \to \infty$).

EXERCISE 20.1.2. Show that W_n is formally identical to the ground state expectation value of the observable $w_n = \sum_{j=1}^{N} e^{in\varphi_j}$ for a quantum system consisting of N noninteracting fermions whose one particle eigenfunctions are the polynomials $\varphi_j(z)$. Thus show that

$$\langle \mathrm{Tr}(U^n) \rangle = \sum_{j=0}^{N} \langle \varphi_j | z^n | \varphi_j \rangle.$$

Show that $\langle (1/N) \mathrm{Tr}\, U \rangle = (1/N) q((1 - q^{2N})/(1 - q^2))$; etc.

EXERCISE 20.1.3. Show

$$Z(\lambda, N) = \left(\frac{\lambda}{4\pi N}\right)^{N/2} \prod_{k=1}^{N-1} (1 - e^{-\frac{1}{2}\lambda(1 - k/N)})^k.$$

EXERCISE 20.1.4. Let $\mathfrak{X}_{(l_1,\ldots,l_N)}$ be the fundamental characters of $U(N)$ given by Weyl's formula. Show that for $G = U(N)$

$$\frac{e^{-A}}{Z} = \sum_{l_1 < l_2 < \cdots < l_N} \mathfrak{X}_{(l_1,\ldots,l_N)} q^{\Sigma_j (l_j - j + 1)^2} \prod_{j>k} \left[\frac{1 - q^{2(l_j - l_k)}}{1 - q^{2(j-k)}}\right].$$

EXERCISE 20.1.5. Let $K_{U(N)}(\vartheta_j, \lambda)$ and $K_{SU(N)}(\vartheta_j, \lambda)$ denote the kernels of the heat equations $\partial K/\partial\lambda = \Delta_G K$. Use the relationship $K_{U(N)}(\vartheta_j + \alpha, \lambda) = \sum_{v \in Z^N} K_{SU(N)}(\vartheta_j^v, \lambda) K_{U(N)}(\alpha^v, \lambda/N)$ where $\vartheta_j^v = \vartheta_j - 2\pi v/N + 2\pi v \delta_{jN}$ and $\alpha^v = \alpha + 2\pi v/N$ to derive a character expansion for $(e^{-A}/Z)_{SU(N)}$.

EXERCISE (Cushman) 20.1.6. Let $(R^4, \Omega = dx_1 \wedge dy_1 + dx_2 \wedge dy_2)$ be a symplectic vector space with Hamiltonian $H(x_1,x_2,y_1,y_2) = \frac{1}{2}(x_1^2 + x_2^2 + y_1^2 + y_2^2) + V = T + V$. Let $X_T = \Omega^{\#}(dT)$. Assume $\mathscr{L}_{X_T}(H) = 0$. Let $z_1 = 2(x_1x_2 + y_1y_2)$, $z_2 = 2(x_2y_1 - x_1y_2)$, $z_3 = (x_1^2 + y_1^2 - x_2^2 - y_2^2)$, $z_4 = (x_1^2 + x_2^2 + y_1^2 + y_2^2)$.

The flow of the linear Hamiltonian vector field X_T defines an S^1-action Φ_t on R^4. Show that $S^{-1}(h) = \{x \in R^4 | x_1y_2 - x_2y_1 = h\}$, $h \neq 0$, is a smooth manifold which is invariant under Φ_t. Show that Φ_t restricts to a free and proper S^1 action on $S^{-1}(h)$ with orbit manifold $M(h)$; $\pi_h: S^{-1}(h) \to M(h) = S^{-1}(h)/S^1$. Show that $M(h)$ is a symplectic manifold with symplectic form Ω_h where $\pi_h^* \Omega_h = i_h^* \Omega$ where $i_h: S^{-1}(h) \to R^4$ is the inclusion map.

EXERCISE 20.1.7. Let

$$SU(1,1) = \left\{ \begin{pmatrix} a & 0 & b & 0 \\ 0 & a & 0 & b \\ c & 0 & d & 0 \\ 0 & c & 0 & c \end{pmatrix} \in \mathrm{Sp}(4,R) \middle| ad - bc = 1 \right\}$$

with Lie algebra

$$su(1,1) = \left\{ \begin{pmatrix} a & 0 & b & 0 \\ 0 & a & 0 & b \\ c & 0 & -a & 0 \\ 0 & c & 0 & -a \end{pmatrix} \in sp(4,R) \right\};$$

$su(1,1)$ has basis

$$e_1 = \left(\begin{array}{cc|cc} 1 & 0 & \multicolumn{2}{c}{0} \\ 0 & 1 & & \\ \hline \multicolumn{2}{c|}{0} & -1 & 0 \\ & & 0 & 1 \end{array}\right), \quad e_2 = \left(\begin{array}{cc|cc} \multicolumn{2}{c|}{0} & \multicolumn{2}{c}{0} \\ \hline 1 & 0 & & \\ 0 & 1 & \multicolumn{2}{c}{0} \end{array}\right),$$

$$e_3 = \left(\begin{array}{cc|cc} & & 1 & 0 \\ \multicolumn{2}{c|}{0} & 0 & 1 \\ \hline \multicolumn{2}{c|}{0} & \multicolumn{2}{c}{0} \end{array}\right).$$

Let $\{e_i^*\}$ denote the dual basis to $\{e_i\}$. Show that the linear symplectic action $SU(1,1) \times R^4 \to R^4 : (V, x) \to Vx$ has momentum mapping $J : R^4 \to \to su(1,1)^*$ given by $J(x)v = \frac{1}{2}\Omega(vx, x)$, $v = a_1 e_1 + a_2 e_2 + a_3 e_3$. Viz. $J(x) = = z_1 e_1^* + \frac{1}{2} z_2 e_2^* + \frac{1}{2} z_3 e_3^*$. Show that $J(Vx)v = (\mathrm{Ad}^t_{V^{-1}} J(x))V = J(x)V^{-1}vV$ for v in $su(1,1)$, V in $SU(1,1)$ and x in R^4; i.e. J is Ad^* invariant. Show that $SU(1,1)$ acts transitively on $S^{-1}(h)$. Thus $J(S^{-1}(h))$ is the $SU(1,1)$) coadjoint orbit $\mathcal{O}_h^*$ through $J((0,h,1,0))$ for $h \neq 0$.

Let $B(v, v') = -\frac{1}{4}\mathrm{Tr}(vv')$ be the Killing form on $su(1,1)$. B induces a linear map $b : su(1,1) \to su(1,1)^*$ by $b(v)v' = B(v, v')$. Show that $\mathcal{O}_h = b^{-1} \circ J(S^{-1}(h))$ is an orbit of the adjoint action of $SU(1,1)$ with symplectic form Ω given by $\Omega(w)([v,w],[v',w]) = B(w,[v,v'])$ for w, v, v' in $su(1,1)$. Note that we can also map $su(1,1)$ into R^3 by $z_1 e_1 + z_2 e_2 + z_3 e_3 \xrightarrow{\lambda} (z_1, z_2, z_3)$.

Since H is invariant under the S^1 action Φ_t in 20.2.4 there is an induced Hamiltonian K on $M(h)$ which satisfies $K \circ \pi_h = H \circ i_h$. Let $\tilde{J}(x) = = \lambda \circ b^{-1} \circ J(x)$. Show that $\tilde{J}(S^{-1}(h)) = \{h^2 = -z_2 z_3 - z_1^2, z_2 \leqq 0, z_3 \geqq 0\}$ can be identified with $M(h)$. Make R^3 into a Lie algebra and describe the Euler equations $\dot{z} = X_K(z)$.

EXERCISE 20.1.8. Following the notation of 20.1.6 set $S = z_4$, $N = \frac{1}{2}z_3$, $Q = \frac{1}{2}z_2$, $P = z_1$. Show that

$$\mathscr{L}_P = \mathscr{L}_{X_P} = x_1 \frac{\partial}{\partial x_1} + x_2 \frac{\partial}{\partial x_2} - y_1 \frac{\partial}{\partial y_1} - y_2 \frac{\partial}{\partial y_2},$$

etc. Show that $\{P, N\} = \mathscr{L}_N P = -2N$, $\{P, Q\} = \mathscr{L}_Q(P)$, $\{Q, N\} = \mathscr{L}_N Q = = P$. – i.e. P, Q, N span a simple Lie algebra. Check that $\{S, P\} = \{S, Q\} = = \{S, N\} = 0$.

References and Historical Comments

Short versions of the history of harmonic analysis have been presented by Mackey in *Bull. Amer. Math. Soc.* **69** (1963), 628–686 and in Kirillov K8. We direct the reader to these articles for a more thorough historical discussion. We high-light the topics presented in this text.

The major emphasis on harmonic analysis in this volume has been directed toward an understanding of geometric realizations of various irreducible unitary representations of Lie groups. This approach arose in the work of Borel, Weil and Tits for the case of compact semisimple Lie groups. This work has many connections to other areas of mathematics – e.g. the work of Borel, Hirzebruch, Bott, Kostant, etc. Unfortunately physicists were not attracted by these developments.

Simultaneous with the development of the Borel–Weil theorem many other currents were amove in harmonic analysis. Mackey had generalized the work of Wigner into what is now called the Mackey machine (see the little group method). The Mackey machine has been studied at length by physicists. In addition to extending the induction technique from finite and compact groups, Mackey had abstracted the Stone–von Neumann theorem; furthermore he had begun the study of the space $\hat{G}$ as an abstract entity.

Gelfand, Naimark, and Bargmann also at this time were beginning to describe the representation theory for noncompact semisimple Lie groups. This theory matured in the major works of Harish–Chandra. Finally the seminal work of Selberg appeared which would provide the cornerstone to connect these developments.

Following Langlands' study of Selberg's work, Langlands formulated his conjectures regarding the geometric realization of the discrete series, which conjectures become an analogue of the Borel–Weil theorem. The Langlands' conjectures were immediately demonstrated at various levels of completeness – culminating in the work of Schmid and others which combines the work of Harish–Chandra and the geometric approach of Borel–Weil.

At the time of the Langlands' conjectures another major approach to describing the geometric realizations of a wider class of Lie groups was being developed. This view is sometimes referred to as Kostant's 'big picture'. The 'big picture' in various aspects was developed simultaneously by Kirillov, Kostant, and Souriau. Kirillov sketched the 'big picture' via his trace formula in the nilpotent and the compact semisimple cases. As we have shown in the text, Kirillov's results in the compact semisimple case are a natural outgrowth of the Borel–Weil–Hirzebruch work. Kirillov developed the 'big picture' for nilpotent Lie groups by building on Mackey's generalized Stone–von Neumann theory.

Kostant's influences in this period are inadequately represented in his papers. His unpublished MIT lectures from this period are far more visionary. Kostant and Auslander immediately extended Kirillov's result to type I solvable Lie groups. Pukansky's work in connection with Kirillov and Kostant must be mentioned.

The influence of the Kostant and Kirillov approach was also felt on the continent where Dixmier, Duflo, Bernat, Vergne, and others were expanding on these results. Also on the continent and independent of all others Souriau was developing his approach to geometric quantization. Geometrically Souriau's approach coincided with the 'big picture'. Souriau provided many beautiful examples from physics in his development of geometric quantization. Slowly physicists began to apply bits and pieces of the representation theoretic aspects of geometric quantization to problems in physics. We hope the reader now sees that the major results for many elementary quantum systems lie in the Borel-Weil theorem and its related results.

The geometry per se of geometric quantization provides the physicist with a deeper understanding of the differential geometry of mechanics. The geometric structures of mechanics arise ever so naturally when one studies the geometric realizations of the underlying symmetry groups of physics.

The work of Maass, Selberg, Langlands, and others provided an emphasis in a related area of study – automorphic forms. Here the work of Roelke, Tamagawa, and Maurin stands out. Gelfand provided the crucial impetus in this area leading to subsequent work by Gangolli on asymptotics and Fadeev *et al.* and Lax–Philips on scattering theory and automorphic forms. (Of course a major result in Lax–Phillips goes back to the Stone–von Neumann theorem.)

We have not dealt explicitly with Feyman path integrals in the text. However, there is an intimate connection between many of our results and the development of these results historically. E.g. the reader should note

K1, K2, D6, and so on. We cite also D33. A survey of related results is presented in S13a. For connections to geometric quantization see also S25. Work of S. Albeverio and R. Høegh-Krohn in *Mathematical Theory of Feynman Path Integrals*, Lecture Notes in Math. No. 523 (1976); *Invent. math.* 40 (1977), 59–106; and Poisson formula and zeta-function of Schrödinger operator (preprint) must be mentioned in this connection.

We turn now to a chapter by chapter literature survey. In Chapter 0 we review the history of high temperature asymptotics. The general reference is Fowler F5 and Berger B8. The cited results are in M39, V6a, K3a.

The study of geometry of orbit spaces in Lie algebras (0.9) must go back to Ehresmann (*Ann. Math.* **35** (1934), 396–443). However, prior to Kostant and Kirillov the major geometric ideas were developed by Bott (v. Bott and Samelson B18a).

For further details regarding scattering theory (0.10) the reader is referred to L5, C12a, C12b. The results due to Lieb are in L11b and Cwikel in C14a.

The references on quantum field theory (0.11) are discussed below in connection to Chapter 19. Appropriate background reading is provided by DeWitt in D5 and D2.

The references for Chapter 1 include any standard book on representations of Lie groups and algebras. E.g. Vilenkin V8, Helgason H11, Mackey M4, Kirillov K8, Maurin M21, Warner W5. For the results on spectral theory see Maurin M15–M22 and Brunig–Heintze B21. The papers by Roelcke R12, Tamagawa T3 and Elstrodt E3 are also relevant in this connection.

The references for Chapter 2 for semidirect products are Mackey M4, Kirillov K8, Bernat B11. The reader is also directed to Wolf, W20. The basic reference for Fock space is Bargmann B5.

The basic reference for Chapter 3 is Cartan C6. For background and review see N2a. For 3.2 see Atiyah A8. Regarding the extension of Schrödinger's equation to arbitrary Riemannian manifolds see Dowker D16 and D18 and references cited therein. The results in 3.4 are due to Liberman L7 and Lichnerowicz L8-L10. The philosophy of 3.6 is due to Kostant K16 and Souriau S26; v. also Kirillov K8.

As in Chapter 3 the basic reference for Chapter 4 is Cartan C6. Much work has been done on contact manifolds by Sasaki and his school. Example 4.1.20 is mentioned in Gel'fand G9 or G10 although I presume it is a classical result. The results on nonisomorphism of contact structures are due to Abe A1, Chern C11, Lutz L13, and Martinet M8. For 4.3 Reeb's work R6 is the classic in this area. For Palais' result see P1. The other basic reference is Boothby–Wang B14. For generalized Brieskorn manifolds see Abe A1. Results on slice diagrams are in Janich JS1. The basic references for 4.4 and 4.5 are Cartan C6, Reeb R6 and Gray G20. See also Sasaki S3. For 4.6 see B14. Example 4.6.4 is due to Wolf W14. For 4.7 see Gray G20. For 4.8 see Boothby B12-13.

The reference for Chapter 5 is van Hove H29. Other work prior to van Hove's includes Groenwold G20a. Several books relating to quantization have been written recently; these include Souriau, Simms, Sniatycki, Weinstein, Wallach, Guillemin-Sternberg, Maslov, Leray, and Abraham-Marsden (2nd edition). The Stone–von Neumann theorem is discussed in several places – Cartier C7, Wallach W3, Mackey M4, and Riefel R11. The approach taken in Example 5.2.4 follows Streater's paper S28 and Onofri O11. Kobayashi's result is in K10. For 5.3 see K16 and S26.

For Chapter 6, 6.1 and 6.3 the reader is referred to Bernat B11. Section 6.2 follows Fischer and Williams F3 and Rawnsley R2. For 6.3 the references are Atiyah A12, Hess H18 and Kostant K17.

For Chapter 7, 7.1, the standard references are Kostant K16, Bernat B11, Vergne V6, Kirillov K8, etc. The references for 7.2 are Miscenko-Fomin M32, Miscenko M33, Dikii D9, Manakov M7, and Arnol'd A5. The references for Example 7.2.7 are Adler A3 and Moser M37; for example 7.2.9 see Olshanetsky–Perelomov O3–O6. For more details on Exercise 7.2 see Wolf W19. For further details on Morse theory in orbit spaces (7.3) the reader is directed to B18a, ROa-ROc and T2b.

Chapter 8, 8.1, follows closely Sataka's work S5–S8. For Murakami's results see M40. The references for 8.2 are Kirillov K5 (for a summary see Wallach W3), Bernat B11, Pukansky P16, Streater S29 and Auslander Kostant A13.

The basic references for Chapter 9 are Cartan's collected works, any book on representation theory of Lie algebras – e.g. Warner W5. For Kostant's work see K16. The results on complex structures are contained in Borel–Hirzebruch B17. The standard references for Borel–Weil are B16 and Bott B18. The reader is also directed to Tits T11. For related geometry see Griffiths G22. Kirillov's character formula is in K6. See also his book K8 and Pukansky's paper P16. The basic reference for 9.3 is Lipsman L12; see also Pukansky P18.

The basic references for Chapter 10 are Wang W4, Bott B18, Ise I1, Wolf W17, Borel-Hirzebruch B17, and Griffiths G21-25. For 10.2 see S22 and G25. For further details on *R*-spaces (10.3) and the Schubert cell decomposition the reader is directed to D31a, H24a, K18b, T2a, T2b and

T2c. For Exercise 10.1 see Ise I1. For Exercise 10.2 and 10.3 see Wolf W18.

Chapter 11 is based on Czyz's work C15-18. The Kodaira–Spencer result is in K11. For 11.2 the reader should also see Simms S23. The Kodaira-Hirzebruch result is in H23. The standard references for 11.3 are Souriau S27 and E1. Maslov's work has been reviewed in many places and related to Morse theory.

The basic reference for Chapter 12 is Wolf W16. The approach to Example 12.1.6 is due to Takahashi T2. For Kirillov's result see K6. The results in 12.2 have been discussed in several lectures by Kostant; see K18a.

Chapter 13 is based on Rawnsley's thesis R1 and later paper R4. The reader is also referred to Souriau S27, Robertson-Noonan and Hannabuss H1.

The references for Chapter 14 are Warner W5, Bargmann B4, Harish-Chandra H3-5. Example 14.1.6 follows T2. The generalization of the Borel-Weil theorem was suggested in Langlands L4; it was proven in various steps by Schmid, Okamoto, Hotta, Parthasarathy, Narasimhan-Okamoto, etc.

The basic references for Chapter 15, 15.1, are Kodaira K12, Ise I2, Matsushima M11, Hirzebruch H20-21, Matsushima–Murakami M13, Selberg S20. A more recent version of the stated vanishing theorems is given by F. Williams in *Osaka J. Math.* **18** (1981), 147–160. For 15.2 the reader is referred to Wolf–Tirao T9, Harish–Chandra H3, and Bargmann B4.

For Chapter 16 the classic references are von Neumann N3 and Dirac D10. The basic reference on limiting theorems is Fowler F5. For 16.2 see Epstein E4, Minakshisundaram M27-31, and for background reading see Kac K3. As has been brought to our attention in Guillemin–Sternberg G26 p.xi the original concept of the inverse spectral problem or the problem of spectral geometry was suggested by Sir Arthur Schuster in 1882 in his report on spectroscopy or 'spectrum analysis'. We hope in this chapter that the reader begins to appreciate that Schuster was right and that the backbone of quantum statistical mechanics is spectral geometry; and, furthermore, that the basic results in spectral geometry were due to the statistical mechanicians Fowler, his student Mulholland, Kirkwood and others.

For 16.3 the basic references are Molchanov M34 and Varadhan V2. For the Hadamard solution the reader is directed to Combet C13, DeWitt D5, Yosida Y1, Parker P2, etc. As we just noted the work on spectral geometry goes back to Schuster, the famous lecture at Göttingen by a physicist (see K3) and work by Sommerfeld and Jeans leading to Weyl's papers W9–10 (see also W11), then the more recent work of McKean and Singer, Seeley, Gilkey, Berger, etc. For 16.4 see M31 and Eskin E5. The references for 16.5 are L11a, S24a and M23. There are many excellent treatments of the Ising model; e.g. v. C. Gruber, *Group Analysis of Classical Lattice System*, Lecture Notes in Physics No. 66 (1977). The reader is also directed to M. Sato *et al.*, in *Publ. RIMS* (*Kyoto*) **16** (1980), 531–584. For Exercise 16.2 see Perelemov P12. For Exercise 16.3 see Dikii D8 and M26. For Exercise 16.4 see Husimi H31, Dirac D11 and Copson C14.

The basic references for Chapter 17 are Eskin, Benabdallah and, of course, Cartan. For 17.2 see Mulholland M39 and Cahn-Wolf C1. The C. DeWitt reference is D6 and the Bourbaki volume is B19. For 17.2 see Eskin E5.

The references for Chapter 18 are Selberg S20, McKean M24, Olshanskii O7, Cartier C7, Gel'fand *et al.*, G10, Ehrenpreis–Mautner ES1, Langlands L2-4, Wallach W1, Gangolli-Warner G6, Gangolli G3-5, Molchanov M34, Hotta H27, Brunig B21, Kolk K13, Duistermaat D31, Osborne O12, and Guillemin G28. For background reading the works by Hejhal H8-9 are especially recommended. For a more recent work regarding 1.18.23, the reader is referred to F. Williams, 'Discrete Series Multiplicities in $L^2(\Gamma\backslash G)$, in *Am. J. Math.* (to appear). Williams gives the best possible result on the multiplicity of the discrete series representations in $L^2(\Gamma\backslash G)$. Also Williams clarifies the Hotta–Parthasarathy result (our Theorem 14.1.8); viz. both assumptions (i) and (ii) may be dropped. The proof is due to Wallach. We wish to thank Williams for his helpful comments. For the noncompact but finite volume case the references are Selberg S20, Faddeev V4, Lax-Phillips L5, Kubota K19, Venkov V3 and V5, etc.

The basic references for Chapter 19 are DeWitt D1-5, Hawking H7, Dowker D14-30, Isham I3, Parker P2, and Carslaw C5.

The references for Chapter 20 are Perelomov P9-11 and Onofri O8; see also R3.

Bibliography

A1. K. Abe, 'On a Generalization of the Hopf Fibration', *Tohoku Math. J.* **29** (1977), 335–374; **30** (1978), 177–210.

A2. R. Abraham and J. Marsden, *Foundations of Mechanics*, second edition, (Benjamin, Reading, 1978).

A3. M. Adler, 'On a Trace Functional for Formal Pseudo Differential Operators and the Symplectic Structure for KdV Type Equations', *Invent. math.* **50** (1979), 219–248.

A4. A. Andreotti and E. Vesentini, 'Carleman Estimates for the Laplace Beltrami Operator on Complex Manifolds', *Publ. IHES* **25** (1965), 81–130.

A5. V. I. Arnol'd, 'Sur la geometrie differentielle des groupes de Lie...', *Ann. Inst. Fourier* **16** (1966), 319–361.

A6. V. I. Arnol'd, 'On a Characteristic Class Entering into Conditions of Quantization', *Func. Anal. Appl.* **1** (1967), 1–13.

A7. V. I. Arnol'd, *Mathematical Methods in Classical Mechanics*, (Springer-Verlag, New York, 1978).

A8. M. F. Atiyah, 'Complex Connections in Fibre Bundles', *Trans. Amer. Math. Soc.* **85** (1957), 181–207.

A9. M. F. Atiyah and R. Bott, 'A Lefschetz Fixed Point Formula for Elliptic Complexes', *Ann. Math.* **86** (1967), 374–407; **88** (1968) 451.

A10. M. F. Atiyah *et al.*, 'On the Heat Equation and the Index Theorem', *Invent. math.* **19** (1973), 279–330.

A11. M. F. Atiyah and W. Schmid, 'A Geometric Construction of the Discrete Series', *Invent. math.* **42** (1977), 1–62.

A12. M. F. Atiyah, 'Riemann Surfaces and Spin Structures', *Ann. sc. Ecole Norm. Sup.* **4** (1971), 47–62.

A13. L. Auslander and B. Kostant, 'Polarization and Unitary Representations of Solvable Lie Groups', *Invent. math.* **14** (1971), 255–354.

B1. R. Banach and J. S. Dowker, 'Automorphic Field Theory', *J. Phys.* **A12** (1979), 2527–2543.

B2. R. Banach and J. S. Dowker, 'The Vacuum Stress Tensor for Automorphic Fields on Some Flat Spacetimes', ibid. 2545–2562.

B3. M. Bander and C. Itzykson, 'Group Theory and the Hydrogen Atom', *Rev. Mod. Phys.* **38** (1966), 330–358.

B4. V. Bargmann, 'Irreducible Unitary Representations of the Lorentz Group', *Ann. Math.* **48** (1947), 568–640.

B5. V. Bargmann, 'On a Hilbert Space of Analytic Functions and an Associated Integral Transform, I', *Comm. Pure Appl. Math.* **14** (1961), 187–214.

B6. A.-I. Benabdallah, 'Noyau de diffusion sur les espaces homogenes compacts', *Bull. Soc. math. Fr.* **101** (1973), 265–283.

B7. L. Berard-Bergery, 'Laplacien et geodesiques fermes sur les formes d'espace hyperbolique compact', *Sem. Bourbaki exp.* **406** (1971/72).

B7a. F. A. Berezin, 'General Concept of Quantization', *Comm. math. Phys.* **40** (1975), 153–174.

B8. M. Berger *et al.*, *Le Spectre d'un Variete Riemannienne*, Lecture Notes in Math. No. 194 (1971).

B9. M. Berger, 'Geometry of the Spectrum', *Proc. Symp. Pure Math.* **27** (1975), 129–152.

B10. P. Bernat, 'Sur les representations unitaires des groupes de Lie resolubles', *Ann. Ec. Norm. Sup.* **82** (1965), 37–99.

B11. P. Bernat, *et al.*, *Representations des Groupes de Lie Resolubles*, (Dunod, Paris, 1972).

B11a. A. Bohm, *Rigged Hilbert Spaces and Quantum Mechanics*, Lecture Notes in Phys. No. 78 (1978).

B12. W. M. Boothby, 'Homogeneous Complex Contact Manifolds', *Proc. Symp. Pure Math.* **3** (1961), 144–154.

B13. W. M. Boothby, 'A Note on Homogeneous Complex Contact Manifolds', *Proc. Amer. Math. Soc.* **13** (1962), 276–280.

B14. W. M. Boothby, and H. C. Wang, 'On Contact Manifolds', *Ann. Math.* **68** (1958), 721–734.

B15. A. Borel, 'Kählerian Coset Spaces of Semisimple Lie Groups', *Proc. Nat. Acad. Sci.* **40** (1954), 1140–1151.

B16. A. Borel and A. Weil, *Representations lineares et espaces homogenes kähleriens des groupes de Lie compacts*, Sem. Bourbaki (1954) exp. 100 by J. P. Serre.

B17. A. Borel and F. Hirzebruch, 'Characteristic Classes and Homogeneous Spaces', *Amer. J. Math.* **80** (1958), 459–538; **81** (1959), 315–382.

B18. R. Bott, 'Homogeneous Vector Bundles', *Ann. Math.* **66** (1957), 203–248.

B18a. R. Bott and H. Samelson, 'Applications of Morse Theory to Symmetric Spaces', *Amer. J. Math.* **70** (1958), 964–1028.

B19. N. Bourbaki, *Elements de Mathematique, Groupes et Algebres de Lie*, Chapters 4–6 (Hermann, Paris, 1968).

B20. F. Bruhat, 'Travaux de Harish-Chandra', *Sem. Bourbaki Exp.* **143** (1957).

B21. J. Bruning and E. Heintze, 'Representations of Compact Lie Groups and Elliptic Operators', *Invent. math.* **50** (1979), 169–203.

C1. R. S. Cahn and J. A. Wolf, 'Zeta Functions and their Asymptotic Expansions for Compact Locally Symmetric Spaces of Negative Curvature', *Comment. Math. Helv.* **51** (1976), 1–21.

C2. R. S. Cahn, P. Gilkey and J. A. Wolf, 'Heat Equation, Proportionality Principle and Volume of Fundamental Domains', in *Differential Geometry and Relativity*, (D. Reidel Publ. Co., Dordrecht, Holland, 1976), pp. 43–54.

C2a. R. S. Cahn, 'Asymptotic Expansion of Zeta Function on Compact Semisimple Lie Groups', *Proc. AMS* **54** (1976), 459–462.

C3. P. Candelas and D. J. Raine, 'General Relativistic Quantum Field Theory: An Exactly Soluble Model', (preprint).

C4. T. Carleman, 'Über die asymptotische verteilund der eigenwerte partieller differential gleichungen', *Ber. Ver. Akad. Leipzig* **88** (1936), 119–132.

C5. H. S. Carslaw, *Proc. London Math Soc.* **20** (1898), 121; **8** (1910), 365; **18** (1919), 291.

C6. E. Cartan, *Lecons sur les invariants integraux*, (Hermann, Paris, 1922).

CS1. E. Cartan, 'Sur la determination d'un systeme orthogonal complete dans un espace de Riemann symetrique clos', *Rend. del. Cir. Mat. Palermo* **53** (1929), 217–252.

C7. P. Cartier, 'Quantum Mechanical Commutation Relations and Theta Functions', in *Algebraic Groups and Discontinuous Subgroups*, (Amer. Math. Soc., Providence, 1965), pp. 361–383.

C8. P. Cartier, *Theorie des groupes, fonctions theta et modules des varietes abeliennes*, Sem. Bourbaki 1967/68 Exp., p. 338.

C9. P. Cartier, *Some Numerical Computations Relating to Automorphic Functions, in Computers in Number Theory*, (Academic Press, New York, 1971), pp. 37–48.

C10. J. Chazarain, 'Formule de Poisson pour les varietes Riemanniennes', *Invent. meth.* **24** (1974), 65–82.

C11. S. Chern, *Pseudogroupes infinis*, Colloq Int. SNRS Strasbourg, 1953, p. 119.

C12. Y. Colin de Verdiere, 'Spectre du laplacien et longueurs des geodesiques periodiques', *Compos. Math* 27 (1973), 83–106, 159–164.

C12a. Y. Colin de Verdiere, *La matrice de scattering pour l'operateur de Schrödinger sur la droite réelle*, Sem. Bourbaki (1979/80), 557.

C12b. Y. Colin de Verdiere, 'Une formule de traces pour l'opérateur de Schrödinger dans R^3', (preprint).

C12c. Y. Colin de Verdiere, 'Spectre conjoint d'operateurs pseudo-differentiels qui commutent', *Duke, Math, J.* **46** (1979), 169–182.

C13. E. Combet, 'Parametrix et invariants sur les varietes compactes', *Ann. Sci. Ecole Norm. Sup.* **3** (1970), 247–271.

C14. E. T. Copson, 'On an Elementary Solution of a Partial Differential Equation of Parabolic Type', *Proc. Roy. Soc. Edin.* **61A** (1941), 37–60.

C14a. M. Cwikel, 'Weak Type Estimates for Singular Values and the Number of Bound States of Schrödinger Operators', (preprint).

C15. J. Czyz, *Bull. Polon Acad. Sci.* **26** (1978), 129–138.

C16. J. Czyz, *On Some Approach to Geometric Quantization*, Lecture Notes in Math. 676 (1978).

C17. J. Czyz, 'On Geometric Quantization and its Connection with Maslov Theory', *Rep. Math. Phys.* **15** (1979), 57–97.

C18. J. Czyz, 'On Geometric Quantization of Compact and Complex Manifolds', (preprint, 1979).

D1. B. S. DeWitt, 'Dynamical Theory in Curved Spaces', *Rev. Mod. Phys.* **29** (1957), 377.

D2. B. S. DeWitt, 'Quantum Field Theory in Curved Spacetime', *Phys. Reports* **19C** (1975), 295–357.

D3. B. S. DeWitt, 'Quantum Theory: The New Synthesis', (preprint Univ. Texas).

D4. B. S. DeWitt, C. F. Hart and C. J. Isham, 'Topology and Quantum Field Theory', (preprint, Univ. Texas).

D5. B. S. DeWitt, *Dynamical Theory of Groups and Fields*, (Gordon and Breach, New York, 1965).

D6. C. M. DeWitt, 'L'integrale fonctionelle de Feynman', *Ann. Inst. H. Poincaré* **A11** (1969), 153–206.

D7. C. M. DeWitt, and M. G. G. Laidlaw, *Phys. Rev.* **D3** (1971), 1375.

D8. L. A. Dikii, 'Trace Formulas for Sturm-Liouville Differential Operators', *Uspehi Mat. Nauk.* (*N.S.*) **13** (1958), 111–143.

D9. L. A. Dikii, 'Hamiltonian Systems Connected with the Rotation Group', *Func. Anal. Appl.* **6** (1972), 326–327.

D10. P. A. M. Dirac, *The Principles of Quantum Mechanics*, (Oxford University Press, London, 1958).

D11. P. A. M. Dirac, *Proc. Camb. Phil. Soc.* **30** (1934), 150.

D12. J. Dixmier, 'Representations integrables du groupe de DeSitter', *Bull. Soc. math. Fr.* **89** (1961), 9–41.

D13. J. Dixmier, 'Polarisations dans les algebres de Lie', *Ann. sc. Ecole Norm. Sup.* **4** (1971), 321–335.

D14. J. S. Dowker, 'Quantum Mechanics on Group Space and Huygen's Principle', *Ann. Phys.* **62** (1971), 361–382.

D15. J. S. Dowker, 'On the Degeneracies of Quantum Systems', *Inter. J. Theo. Phys.* **8** (1973), 319–320.

D16. J. S. Dowker, 'Covariant Feynman Derivation of Schrödinger Equation in Riemann Space', *J. Phys.* **A7** (1974), 1256.

D17. J. S. Dowker and D. F. Pettengill, 'The Quantum Mechanics of Ideal Asymmetric Top with Spin', ibid. **A7** (1974), 1527–1536.

D18. J. S. Dowker, 'Covariant Schrödinger Equations', in *Proc. of Conf. on Functional Integration and Applications*, (Clarendon Press, Oxford, 1975).

D19. J. S. Dowker, 'Quantum Mechanics and Field Theory on Multiply Connected and on Homogeneous Spaces', *J. Phys.* **A5** (1972), 936–943.

D20. J. S. Dowker and R. Critchley, 'Covariant Casimir Calculations', *J. Phys.* **A9** (1976), 535–540.

D21. J. S. Dowker and R. Critchley, 'Scalar Effective Lagrangian in DeSitter Space', *Phys. Rev.* **D13** (1976), 224–234.

D22. J. S. Dowker and R. Critchley, 'Effective Lagrangian and Energy Momentum Tensor in Desitter Space', *Phys. Rev.* **D13** (1976), 3224.

D23. J. S. Dowker and R. Critchley, 'The Stress Tensor Conformal Anomaly for Scalar and Spinor Fields', (preprint).

D24. J. S. Dowker and M. B. Altaire, 'Spinor Fields in an Einstein Universe', *Phys. Rev.* **D17** (1978), 417–422.

D25. J. S. Dowker, 'Quantum Field Theory on a Cone', *J. Phys.* **A10** (1977), 115–124.

D26. J. S. Dowker, 'Conformal Anomalies in Quantum Field Theory', (preprint).

D27. J. S. Dowker, 'Single Loop Divergences in Six Dimensions', *J. Phys.* **A10** (1978), L63.

D28. J. S. Dowker and R. Critchley, 'Vacuum Stress Tensor in an Einstein Universe, Finite Temperature Effects', *Phys. Rev.* **D15** (1977), 1484.

D29. J. S. Dowker and R. Banach, 'Quantum Field Theory on Clifford Klein Space Times', (preprint).

D30. J. S. Dowker and G. Kennedy, 'Finite Temperature and Boundary Effects in Static Space Time', *J. Phys.* **A11** (1978), 895–920.

D30a. J. J. Duistermaat and V. W. Guillemin, 'The Spectrum of Positive Elliptic Operators and Periodic Bicharacteristics', *Invent. math.* **29** (1975), 39–79.

D31. J. J. Duistermaat, J.A.C. Kolk, and V. S. Varadarjan, 'Spectra of Compact Locally Symmetric Manifolds of Negative Curvature', *Invent. math.* **52** (1979), 27–94.

D31a. J. J. Duistermaat, J.A.C. Kolk, and V. S. Varadarjan, 'Functions, Flows and Oscillatory Integrals on Flag Manifolds', (preprint).

D32. T. E. Duncan, 'Brownian Motion, the Heat Equation and Some Index Theorems', (preprint).

DS1. S. A. Dunne, 'Application of the Kirillov Theory to the Representations of 0(2, 1)', *J. Math. Phys.* **10** (1969), 860–868.

D33. E. B. Dynkin, 'Brownian Motion in Certain Symmetric Space and Nonnegative Eigenfunctions of the Laplace Beltrami Operator', *Amer. Math. Soc. Transl.* (1968), 203–228.

E1. J. Elhadad, 'Sur l'interpretation en geometrie symplectique des etats quantiques de l'atome d'hydrogen', *Conv. Geom. simp. e Fis. mat. INDAM, Rome*, (1973), 259–291.

ES1. L. Ehrenpreis and F. I. Mautner, 'Some Properties of the Fourier Transform on Semisimple Lie Groups II, III', *Trans. Amer. Math. Soc.* **84** (1957), 1–55; **90** (1959), 431–484.

E2. L. Ehrenpreis, 'An Eigenvalue Problem for Riemann Surfaces', *Advances in Theory of Riemann Surfaces* (1971), 131–140.

E3. J. Elstrodt, 'Die Resolvente zum Eigenwertproblem der automorphen Formen in der hyperbolischen Ebene', *Math. Ann.* **203** (1973), 295–330; *Math. Z.* **132** (1973), 99–134; *Math. Ann.* **208** (1974), 99–132.

E4. P. Epstein, 'Zur theorie allgemeiner Zetafunktionen I, II', *Math. Ann.* **56** (1903), 615–644; **63** (1907), 205–246.

E5. L. D. Eskin, *The Heat Equation on Lie Groups, In Memoriam: N. G. Cebotarev*, Izdat. Kazan Univ. (1964), pp. 113–132.

E6. L. D. Eskin, 'The Heat Equation and the Weierstrass Transform on Certain Symmetric Spaces', *Amer. Math. Soc. Transl.* **75** (1968), 239–254.

F1. L. Faddeev, 'Expansion in Eigenfunctions of the Laplace Operator on the Fundamental Domain of a Discrete Group on the Lobacevskii Plane', *Trans. Moscow Math. Soc.* **17** (1967), 357–386.

F2. R. P. Feynman, *Statistical Mechanics*, (Benjamin, Reading, 1972).

F3. H. R. Fischer and F. L. Williams, 'Complex Foliated Structures', *Trans. Amer. Math. Soc.* **252** (1979), 163–195.

F4. M. Forger and H. Hess, 'Universal Metaplectic Structures and Geometric Quantization', *Comm. math. Phys.* **64** (1979), 269–278.

F5. R. H. Fowler, *Statistical Mechanics*, (Cambridge University Press, Cambridge, 1936).

F6. W. H. Furry, 'Isotropic Rotational Brownian Motion', *Phys. Rev.* **107** (1957), 7.

G1. M.P. Gaffney, 'Asymptotic Distributions Associated with the Laplacian for Forms', *Comm. Pure Appl. Math.* **11** (1958), 535–545.

G2. R. Gangolli, 'Positive Definite Kernels on Homogeneous Spaces', *Ann. Inst. H. Poincaré* **3** (1967), B121–225.

G3. R. Gangolli, 'Asymptotic Behavior of Spectra of Compact Quotients of Certain Symmetric Spaces', *Acta Math.* **121** (1968), 151–192.

G4. R. Gangolli, 'On the Length Spectra of Certain Compact Manifolds of Negative Curvature', *J. Diff. Geom.* **12** (1977), 403–424.

G5. R. Gangolli, 'Zeta Functions of Selberg's Type for Compact Space Forms of Symmetric Spaces of Rank One', *Ill. J. Math.* **21** (1977), 1–41.

G6. R. Gangolli and G. Warner, 'On Selberg's Trace Formula', *J. Math. Soc. Japan* **27** (1975), 328–343.

G7. R. Gangolli and G. Warner, 'Zeta Functions of Selberg's Type for Some Noncompact Quotients of Symmetric Spaces of Rank One', (preprint, Univ. Washington).

G8. S. S. Gelbart, *Automorphic Forms on Adele Groups*, (Princeton University Press, Princeton, 1977).

G9. I. M. Gel'fand and S. V. Fomin, 'Geodesic Flows on Manifolds of Constant Negative Curvature', *Amer. Math. Transl.* **1** (1965), 49–65.

G10. I. M. Gel'fand *et al.*, *Representation Theory and Automorphic Functions*, (W. B. Sauders Co., Philadelphia, 1969).

G11. P. B. Gilkey, 'The Spectral Geometry of Real and Complex Manifolds', *Proc. Symp. Pure Math.* **27** (1975), 265–280.

G12. P. B. Gilkey, 'The Spectral Geometry of Riemannian Manifold', *J. Diff. Geom.* **10** (1975), 601–618.

G13. P. B. Gilkey, *The Index Theorem and the Heat Equation*, (Publish or Perish, Inc., Boston, 1974).

G14. R. Godemont, *Introduction aux travaux de Selberg*, Sem. Bourbaki, Exp. 144 (1956/57).

G15. R. Godemont, 'The Decomposition of $L^2(G/\Gamma)$ for $\Gamma = SL_2(Z)$', *Proc. Symp. Pure Math.* **9** (1966), 211–224.

G16. R. Godemont, 'The Spectral Decomposition of Cusp Forms', ibid. 225–234.

G17. R. Godemont, *Fonctions holomorphes de carre sommable dans le demi plan de Siegel*, Sem. H. Cartan 1957/58 Exp. 6.

G18. M. J. Gotay and J. A. Isenberg, 'Geometric Quantization and Gravitational Collapse', (preprint).

G19. M. J. Gotay, 'Functional Geometric Quantization and van Hove's Theorem', (preprint).

G20. J. W. Gray, 'Some Global Properties of Contact Structure', *Ann. Math.* **69** (1959), 421–450.

G20a. H. J. Groenwold, 'On the Principles of Elementary Quantum Mechanics', *Physica* **12** (1946), 405–460.

G20b. G. W. Gibbons, 'Quantized Fields Propagating in Plane Wave Space Times', *Comm. Math. Phys.* **45** (1975), 191–202.

G20c. M. L. Glasser, 'The Evaluation of Lattice Sums', *J. Math. Phys.* **14** (1973), 409; 701.

G21. P. A. Griffiths, 'The Differential Geometry of Homogeneous Vector Bundles', *Trans. Amer. Math. Soc.* **109** (1963), 1–34.

G22. P. A. Griffiths, 'Some Geometric and Analytic Properties of Homogeneous Complex Manifolds', *Acta Math.* **110** (1963), 115–155.

G23. P. A. Griffiths, 'Hermitian Differential Geometry and the Theory of Positive and Ample Holomorphic Vector Bundles', *J. Math. Mech.* **14** (1966), 117–140.

G24. P. A. Griffiths, 'Curvature and Cohomology of Locally Homogeneous Vector Bundles', (preprint).

G25. P. A. Griffiths and W. Schmid, 'Locally Homogeneous Complex Manifolds', *Acta Math.* **123** (1969), 253–302.

G26. V. Guillemin and S. Sternberg, *Geometric Asymptotics*, (Amer. Math. Soc., Providence, 1977).

G27. V. Guillemin, 'Lectures on Spectral Theory of Elliptic Operators', *Duke Math. J.* **44** (1977), 485–517.

H1. K. C. Hannabuss, 'The Localizability of Particles in deSitter Space', *Proc. Camb. Phil. Soc.* **70** (1971), 283–302.

H2. J.-I. Hano and S. Kobayashi, 'A Fibering of a Class of Homogeneous Complex Manifolds', *Trans. Amer. Math. Soc.* **94** (1960), 233–243.

H3. Harish-Chandra, 'Representations of Semisimple Lie Groups IV-VI', *Amer. J. Math.* **77** (1955), 743–777; **78** (1956), 1–41; 564–628.

H4. Harish-Chandra, 'Discrete Series for Semisimple Lie Groups', *Acta Math.* **113** (1965), 241–318; **116** (1966), 1–111.

H5. Harish-Chandra, 'Harmonic Analysis on Semisimple Lie Groups', *Bull. Amer. Math. Soc.* **76** (1970), 529–587.

H6. Harish-Chandra, 'Automorphic Forms on Semisimple Lie Groups', *Lecture Notes in Math.* **62** (1968).

H7. S. W. Hawking, 'Zeta Function Regularization of Path Integrals in Curved Spacetime', *Comm. math. Phys.* **55** (1977), 133–148.

H8. D. A. Hejhal, 'The Selberg Trace Formula and the Riemann Zeta Function', *Duke Math. J.* **43** (1976), 441–482.

H9. D. A. Hejhal, 'The Selberg Trace Formula for PSL(2, R)', Vol. I, *Lecture Notes in Math.* **548** (1976).

H10. S. Helgason, 'Fundamental Solutions of Invariant Differential Operators on Symmetric Spaces', *Am. J. Math.* **84** (1963), 565–601.

H11. S. Helgason, *Differential Geometry on Symmetric Spaces*, (Academic Press, New York, 1962).

H12. R. Hermann, *Algebraic Topics in Systems Theory*, III of Interdisc. Math., (Math Sci Press, Brookline, mass., 1973).

H13. R. Hermann, *Associative Algebras, Spinors, Clifford, and Cayley Algebra*, VII, ibid., (1974).

H14. R. Hermann, *Geometry of Nonlinear Differential Equations, Backlund Transformations and Solitons* Part A, XII, ibid., (1976).

H15. R. Hermann, *Lie Algebras and Quantum Mechanics*, (Benjamin, New York, 1970).

H16. R. Hermann, *Vector Bundles in Mathematical Physics*, ibid., (1970).

H17. R. Hermann, *Physical Aspects of Lie Group Theory*, (University of Montreal Press, Montreal, 1974).

H18. H. Hess and D. Krausser, 'Lifting Classes of Principle Bundles', (preprint).

H19. H. Hess, 'On a Geometric Quantization Scheme Generalizing Those of Kostant-Souriau and Czyz', (preprint).

H20. F. Hirzebruch, *Automorphic formen und der satz von Riemann-Roch*, Symposium internacional de Topologia Algebraica, Mexico (1958), pp. 129–144.

H21. F. Hirzebruch, *Characteristic numbers of homogeneous domains, Seminars on Analytic Functions*, (Princeton, 1957), II, pp. 92–104.

H22. F. Hirzebruch, *Topological Methods in Algebraic Geometry*, (Springer–Verlag, New York, 1966).

H23. F. Hirzebruch and K. Kodaira, 'On the Complex Projective Spaces', *J. Math.* **36** (1957), 201–216.

H24. R. Hotta, 'Elliptic Complexes on Certain Homogeneous Spaces', *Osaka J. Math.* **7** (1970), 117–160.

H24a. R. Hotta, 'The Generalized Schubert Cycles and the Poincaré Duality', *Osaka J. Math.* **4** (1967), 271–278.

H25. R. Hotta, 'On a Realization of the Discrete Series for Semisimple Lie Groups', *J. Math Soc. Japan* **23** (1971), 384–407.

H26. R. Hotta and R. Parthasarathy, 'A Geometric Meaning of the Multiplicity of Integrable Discrete Classes in $L^2(\Gamma\backslash G)$', *Osaka J. Math.* **10** (1973), 211–234.

H27. R. Hotta and R. Parthasarathy, 'Multiplicity Formulae for Discrete Series', *Invent. math.* **26** (1974), 133–178.

H28. L. Von Hove, 'Sur la probleme des relations entre les transformations unitaires de la mecanique et les transformations canonique', *Acad. Roy. Belgique Bull. Cl. Sci.* **37** (1951), 610–620.

H29. L. Von Hove, 'Sur certains representations unitaires d'un groupe infini de transformations', *Mem. Acad. Roy. Belg. Cl. Sc.* **26** (1952), 61–102.

H30. H. Huber, 'Zur analytischen Theorie hyperbolischer Raumformen und Bewegungsgruppen', *Math. Ann.* **138** (1959), 1–26; **142** (1961), 385–398; **143** (1961), 463–464.

H31. K. Husimi, 'Some Formal Properties of the Density Matrix', *Proc. Phys.-Math. Soc. Japan* **22** (1940), 264–314.

I1. M. Ise, 'Some Properties of Complex Analytic Vector Bundles over Compact Homogeneous Spaces', *Osaka Math. J.* **12** (1960), 217–252.

I2. M. Ise, 'Generalized Automorphic Forms and Certain Holomorphic Vector Bundles', *Amer. J. Math.* **86** (1964), 70–108.

I3. C. J. Isham, 'Twisted Quantum Fields in Curved Space-Time', *Proc. Roy. Soc. London* **A362** (1978), 383–404.

J1. A. Joseph, 'Derivations of Lie Brackets and Canonical Quantization', *Comm. math. Phys.* **17** (1970), 210.

JS1. K. Janich, 'Differenzielbare *G*-Mannigfaltigkeiten', *Lecture Notes in Math.* **59** (1968).

K1. M. Kac, 'On Some Connections between Probability Theory and Differential and Integral Equations', *Proc. of the Second Berkeley Symposium on Mathematical Statistics and Probability*, (1951), pp. 189–215.

K2. M. Kac, *Probability and Related Topics in Physical Sciences*, (Interscience, New York, 1959).

K3. M. Kac, 'Can One Hear the Shape of a Drum?', *Amer. Math. Mon.* **73** (1966), 1–23.

K3a. L. S. Kassel, 'Mathematical Methods for Computing Thermodynamic Functions from Spectroscopic Data', *J. Chem. Phys.* **1** (1933), 576–585.

K4. J. B. Keller, 'Corrected Bohr-Sommerfeld Quantum Conditions for Nonseparable Systems', *Ann. Phys.* **4** (1958), 180–188.

K5. A. A. Kirillov, 'Unitary Representations of Nilpotent Lie Groups', *Usp. Mat. Nauk* **17** (1962), 4, 57–110.

K6. A. A. Kirillov, 'Characters of Unitary Representations of Lie Groups', *Func. Anal. Appl.* **2** (1969), 2, 40–55; **3**, 1, 36–47.

K7. A. A. Kirillov, 'Constructions of Unitary Representations of Lie Groups', *Vestnik Moskov. Univ. Ser. I, Meh.* **25** (1970), 2, 41–51.

K8. A. A. Kirillov, *Elements of the Theory of Representations*, (Springer–Verlag, New York, 1976).

K8a. J. G. Kirkwood, *Phys. Rev.* **44** (1933), 31; *J. Chem. Phys.* **1** (1933), 597.

K9. S. Kobayashi, 'Remarks on Complex Contact Manifolds', *Proc. Amer. Math. Soc.* **10** (1959), 164–167.

K10. S. Kobayashi, 'Irreducibility of Certain Unitary Representations', *J. Math. Soc. Japan* **20** (1968), 638–642.

K11. K. Kodaira and D. C. Spencer, 'Groups of Complex Line Bundles over Compact Kähler Manifolds', *Proc. Nat. Acad. Sci.* **39** (1953), 1273–1278.

K12. K. Kodaira, 'On Kähler Varieties of Restricted Type', *Ann. Math.* **60** (1954), 20–48.

K13. J. A. C. Kolk, 'The Selberg Trace Formula and Asymptotic Behavior of Spectra', (thesis, Rijksuniverseteit, 1974).

K14. Y. Konno, 'On Vanishing Theorems of Square Integrable $\bar{\partial}$-Cohomology Spaces', *Osaka J. Math.* **9** (1972), 183–216.

K15. B. Kostant, *Quantization and Unitary Representations*, in Lecture Notes in Math. No. 170 (1970), pp. 87–208.

K16. B. Kostant, *On Certain Unitary Representations which Arise from a Quantization Theory*, in Lecture Notes in Physics No. 6 (1970), pp. 237–253.

K17. B. Kostant, 'Symplectic Spinors', in *Conv. Geom. simpl. e Fis. mat. INDAM (Rome)* (1973), pp. 139–152.

K18. B. Kostant, 'On Whittaker Vector and Representation Theory', *Invent. math.* **48** (1978), 101–184.

K18a. B. Kostant, 'The Solution to a Generalized Toda Lattice and Representation Theory', *Ad. Math.* **34** (1979), 195–338.

K18b. B. Kostant, 'Lie Algebra Cohomology and Generalized Schubert Cells', *Ann. Math.* **77** (1963), 72–144.

K19. T. Kubota, *Introduction to Eisenstein Series*, (Halsted Press, New York, 1973).

K20. M. Kuga, *Fiber Varieties over a Symmetric Space whose Fibers are Abelian Varieties*, (The Univer. of Chicago, 1963/64).

L1. S. Lang, $SL_2(R)$, (Addison-Wesley, Reading, mass., 1975).

L2. R. P. Langlands, 'The Dimension of Spaces of Automorphic Forms', *Am. J. Math.* **85** (1963), 99–125.

L3. R. P. Langlands, 'Eisenstein Series', *Proc. Symp. Pure Math.* **9** (1966), 235–252.

L4. R. P. Langlands, 'Dimensions of Spaces of Automorphic Forms', *Proc. Symp. Pure Math.* **9** (1966), 253–257.

LS1. R. P. Langlands, 'On the Functional Equation Satisfied by Eisenstein Series', *Lecture Notes in Math.* **544** (1976).

L5. P. D. Lax and R. S. Phillips, *Scattering Theory for Automorphic Functions*, (Princeton University Press, Princeton, 1977).

L6. J. Leray, *Analyse lagrangienne et mecanique quantique*, (RCP 25 IRMA Strasbourg).

L7. P. Libermann, *Sur les automorphismes infinitesmaux des structures contact*, Coll. Geom. Diff. Global Bruxelles (1958), pp. 37–59.

L8. A. Lichnerowicz, 'Theoremes de reductivite sur des algebres d'automorphes', *Rend. di Mat.* **22** (1963), 197–244.

L9. A. Lichnerowicz, 'L'algebre de Lie des automorphes infinitesimaux sympletique', *Symposia Math.* **XIV** (INDAM, 1973), 11–24.

L10. A. Lichnerowicz, *Derivations et cohomologies des algebres de Lie attaches d'une variete symplectique et une variete contact Geometrie symplectique et physique mathematique*, (Coll. Inter. CNRS, 1974), pp. 29–44.

L11. A. Lichnerowicz, 'Champs spinoriels et propagateurs en relativite generale', *Bull. Soc. math. Fr.* **92** (1964), 11–100.

L11a. E. Lieb, *Comm. math. Phys.* **31** (1973), 327.

L11b. E. Lieb, 'Estimates for the Eigenvalues of the Laplacian and Schrödinger Operators', *Bull. AMS* **82** (1976), 751–753.

L12. R. L. Lipsman, 'Orbit Theory and Harmonic Analysis on Lie Groups with Cocompact Nilradical', (preprint, Univ. Maryland).

L13. R. Lutz, 'Sur quelques properties des formes differentielles en dimension trois', (thesis, Univ. Louis-Pasteur-Strasbourg, 1971).

M1. H. Maass, 'Uber eine neue Art von nichtanalytischen automorphen Funktionen...', *Math. Ann.* **121** (1949), 141–183.

M2 H. Maass, *Lectures on Modular Functions of One Complex Variable*, (Tata Lecture Notes, Bombay, 1964).

M3. G. W. Mackey, *Mathematical Foundations of Quantum Mechanics*, (Benjamin, New York, 1963).

M4. G. W. Mackey, *Induced Representations of Groups and Quantum Mechanics*, (Benjamin, New York, 1968).

M5. G. W. Mackey, 'Induced Representations of Locally Compact Groups and Applications', in *Functional Analysis and Related Fields*, (Springer-Verlag, New York, 1968), pp. 132–166.

M6. G. W. Mackey, 'Ergodic Theory and its Significance for Statistical Mechanics and Probability Theory', *Ad. Math.* **12** (1974), 178–268.

MS1. G. W. Mackey, *The Theory of Group Representations*, (The University of Chicago Lecture Notes, 1955).

MS2. G. W. Mackey, *Unitary Groups Representations*, (Benjamin, Reading, 1978).

M6a. G. W. Mackey, 'A Theorem of Stone and von Neumann', *Duke Math. J.* **16** (1949), 313–326.

M7. S. V. Manakov, 'A Remark on the Integration of Eulerian Equations of the Dynamics of an *n*-Dimensional Rigid Body', *Func. Anal. Appl.* **10** (1976), 93–95.

M8. J. Martinet, 'Formes de contact sur les varietes de dimension 3', *Lecture Notes in Math.* **209** (1972), 142–163.

M9. V. P. Maslov, *Theorie des perturbations et methode asymptotiques*, (Dunod, Paris, 1972).

M10. Y. Matsushima, 'On the First Betti Number of Compact Quotient Spaces of Higher Dimensional Symmetric Spaces', *Ann. Math.* **75** (1962), 312–330.

M11. Y. Matsushima, 'On Betti Number of Compact Locally Symmetric Riemann Manifold', *Osaka Math. J.* **14** (1962), 1–20.

M12. Y. Matsushima, 'A Formula for the Betti Numbers of Compact Locally Symmetric Riemannian Manifolds', *J. Diff. Geom.* **1** (1967), 99–109.

M13. Y. Matsushima and S. Murakami, 'On Vector Bundle Valued Harmonic Forms and Automorphic Forms on Symmetric Riemannian Manifolds', *Ann. Math.* **78** (1963), 365–416.

M14. Y. Matsushima, 'On Certain Cohomology Groups Attached to Hermitian Symmetric Spaces', *Osaka J. Math.* **2** (1965), 1–35; **5** (1968), 223–241.

M15. K. Maurin, 'Automorphic Functions and Spectral Theory Non Compact Case', *Bull. Acad. Polon. Sci., Ser. sci. math., ast., phys.* **12** (1964) 385–390.
M16. K. Maurin, 'Dualitatsatz fur automorphe funktionen auf einer lokalkompakten unimodularen gruppe', ibid. **14** (1966), 493–496; **15** (1967), 465–472.
M17. K. Maurin and L. Maurin, 'Enveloping Algebra of a Locally Compact Group and Its Self Adjoint Representations', ibid. **11** (1963), 525–529.
M18. L. Maurin, 'Verschiedene definitionen der automorphen funktionen und ihre aquivalenz', ibid. **15** (1967), 473–478.
M19. K. and L. Maurin, 'Automorphic Forms, Spectral Theory and Group Representations', ibid. **13** (1965), 199–203.
M20. K. Maurin, 'Allgemeine eigenfunktionsentwicklungen, unitaire darstellungen lokalkompakter gruppen and automorphe funktion', *Math. Ann.* **165** (1966), 204–222.
M21. K. Maurin, *General Eigenfunction Expensions and Unitary Representations of Topological Groups*, (PWN, Warsaw, 1968).
M22. K. Maurin and L. Maurin, 'A Generalization of the Duality Theorem of Gelfand-Piateckii-Sapiro and Tamagawa Automorphic Forms', *J. Fac. Sci. Univ. Tokyo* **17** (1970), 331–339.
M23. H. P. McKean, 'Kramers-Wannier Duality for the 2-Dimensional Ising Model as an Instance of the Poisson Summation Formula', *J. Math. Phys.* **5** (1964), 775–776.
M24. H. P. McKean, 'Selberg's Trace Formula as Applied to a Compact Riemann Surface', *CPAM* **25** (1972), 225–246.
M25. H. P. McKean and I. Singer, 'Curvature and the Eigenvalues of the Laplacian', *J. Diff. Geom.* **1** (1967), 43–69.
M26. H. P. McKean and P. von Moerbeke, 'The Spectrum of Hill's Equation', *Invent. Math.* **30** (1975), 217–274.
M27. S. Minakshisundaram, 'Zeta Functions on the Sphere', *J. Indian Math. Soc.* **13** (1949), 41–48.
M28. S. Minakshisundaram, 'On Epstein Zeta Functions', *Can. J. Math.* **1** (1949), 320–326.
M29. S. Minakshisundaram, 'Zeta Functions on the Unitary Sphere', ibid. **4** (1952), 26–30.
M30. S. Minakshisundaram, 'Eigenfunctions on Riemann Manifolds', *J. Indian Math. Soc.* **17** (1953), 158–165.
M31. S. Minakshisundaram and A. Pleijel, 'Some Properties of the Eigenfunctions of the Laplace Operator on Riemannian Manifolds', *Can. J. Math.* **1** (1949), 242–256.
M32. A. S. Mischenko and A. T. Fomenko, 'Euler Equations on Finite Dimensional Lie Groups', *Math. USSR Izv.* **12** (1978), 371–390.
M33. A. S. Mischenko, 'Integral Geodesics of Flow on Lie Groups', *Func. Anal. Appl.* **4** (1970), 232–235.
M33a. J. Mickelsson and J. Niederle, 'On Non-Linear Realizations of the Group $SU(2)$', *Comm. math. Phys.* **16** (1970), 191–206.
M34. S. A. Molchanov, 'Diffusion Processes and Riemannian Geometry', *Russ. Math. Sur.* **30** (1975), 1–64.
M35. A. Morimoto, 'On Normal Almost Contact Structure', *J. Math. Soc. Japan* **15** (1963), 420–436.
M36. A. Morimoto, 'On Normal Almost Contact Structure with a Regularity', *Tohoku Math. J.* **16** (1964), 90–104.
M37. J. Moser, 'Three Integrable Hamiltonian Systems Connected with Isospectral Deformation', *Adv. Math.* **16** (1975), 197–220.

M38. H. Moscovici, 'A reciprocity Theorem for Unitary Representations of Lie Groups', *Israel J. Math.* **15** (1973), 230–236.

M39. H. P. Mulholland, 'An Asymptotic Expansion', *Proc. Camb. Phil. Soc.* **24** (1928), 280–289.

M40. S. Murkami, *Cohomology Groups of Vector Valued Forms on Symmetric Spaces*, (Univ. Chicago Lecture Notes, 1966).

N0. M. Nakamura and H. Umegaki, 'Heisenberg's Commutation Relations and the Plancherel Theorem', *Proc. Japan Acad.* **37** (1961), 239–242.

N1. M. S. Narasimhan and K. Okamoto, 'Analogue of the Borel Weil Bott Theorem for Symmetric Pairs of Noncompact Type', Ann. Math. **91** (1970), 486–511.

N2. M. Nichanian, 'Les Transformees de Fourier des Distributions de Type Positif sur *SL* (2, *R*) et la Formule des Traces', *Lecture Notes in Math.* **497** (1975), 349–367.

N2a. E. Nelson, *Tensor Analysis*, (Princeton University Press, Princeton, 1967).

N3. J. von Neumann, *Mathematical Foundations of Quantum Mechanics*, (Princeton University Press, Princeton, 1955).

N4. L. Nirnberg, A complex Frobenius theorem, Seminars on Analytic Functions (Princeton, 1957) I, 172–189.

O1. K. Okamoto, 'On Induced Representations', *Osaka J. Math.* **4** (1967), 85–94.

O2. K. Okamoto and H. Ozeki, 'On Square Integrable $\bar{\partial}$-Cohomology Spaces Attached to Hermitian Symmetric Spaces', ibid. **4** (1967), 95–110.

O3. M. A. Ol'shanetskii and A. M. Perelomov, 'Geodesic Flows on Symmetric Spaces of Zero Curvature and Explicit Solution of Generalized Calogero Model, *Func. Anal. Appl.* **10** (1976), 237–239.

O4. M. A. Ol'shanetskii and A. M. Perelomov, 'Explicit Solutions of Classical Generalized Toda Models', *Invent. Math.* **54** (1979), 261–269.

O5. M. A. Ol'shanetskii and A. M. Perelomov, 'Explicit Solution of the Calogero Model', *Lett. Nuovo Cim.* **16** (1976), 333–339.

O6. M. A. Ol'shanetskii and A. M. Perelomov, 'Explicit Solution of Some Completely Integrable Systems', ibid. **17** (1976), 97–101.

O7. G. I. Olsankii, 'On the Frobenius Duality Theorem', *Funct. Anal. i Priloz* **3** (1969), 49–58.

O8. E. Onofri, 'A Note on Coherent State Representations of Lie Groups', *J. Math. Phys.* **16** (1975), 1087–1089.

O9. E. Onofri and M. Pauri, 'Analyticity and Quantization', *Lett. Nuovo Cim.* **3** (1972), 35–42.

O10. E. Onofri and M. Pauri, 'Dynamical Quantization', *J. Math. Phys.* **13** (1972), 533–543.

O11. E. Onofri, 'Quantization Theory for Homogeneous Kähler Manifolds', (preprint).

O12. M. S. Osborne, 'Lefschetz Formulas on Nonelliptic Complexes', (thesis, Yale, 1972).

O13. H. Ozeki and M. Wakimoto, 'On Polarizations of Certain Homogeneous Spaces', *Hiroshima Math. J.* **2** (1972), 445–482.

P1. P. S. Palais, 'A Global Formulation of the Lie Theory of Transformation Groups', *Mem. Amer. Math. Soc.* **22** (1957).

P2. L. Parker, *Aspects of Quantum Field Theory in Curved Spacetime: Effective Action and Energy Momentum Tensor*, Cargese Lectures, (ed. Levy-Deser, Plenum, 1978).

P3. R. Parthasarathy, 'A Note on the Vanishing of Certain "L^2 Cohomologies"', *J. Math. Soc. Japan* **23** (1971), 676–691.

P4. R. Parthasarathy, 'Dirac Operators and the Discrete Series', *Ann. Math.* **96** (1972), 1–30.

P5. V. K. Patodi, 'Curvature and the Fundamental Solution of the Heat Operator', *J. Indian Math. Soc.* **34** (1970), 269–285.

P6. V. K. Patodi, 'Curvature and the Eigenforms of the Laplace Operator', *J. Diff. Geom.* **5** (1971), 233–249.

P7. S. J. Patterson, 'The Laplacian Operator on a Riemann Surface', *Compositio Math.* **31** (1975), 83–107; **32** (1976), 71–112.

P8. S. J. Patterson, 'On the Cohomology of Fuchsian Groups', *Glasgow Math. J.* **16** (1975), 123–140.

P8a. L. Pauling and E. B. Wilson, *Introduction to Quantum Mechanics*, (McGraw-Hill, New York, 1935).

P8b. B. S. Pavlov and L. D. Fadeev, 'Scattering Theory and Automorphic Functions', *J. Sov. Math.* **3** (1975), 522–548.

P9. A. M. Perelomov, 'Coherent States for Arbitrary Lie Groups', *CMP* **26** (1972), 222–236.

P10. A. M. Perelomov, 'Note on Completeness of Systems of Coherent States', *Teori. i. Matem. tuz.* **6** (1971), 213–224.

P11. A. M. Perelomov, 'Coherent States for the Lobachevskian Plane', *Func. Anal. Appl.* **7** (1973), 215–222.

P12. A. M. Perelomov, 'Schrödinger Equation Spectrum and KdV Type Invariants', *Ann. Inst. H. Poincaré* **A24** (1976), 161–164.

P12a. A. M. Perelomov, *Comm. math. Phys.* **64** (1978), 237.

P13. F. Perrin, *Ann. Ec. norm. Sup.* **45** (1928), 1.

P14. A. Pleijel, 'A Study of Certain Green's Functions with Applications in Theory of Vibrating Membranes', *Ark. f. Mat.* **2** (1954), 553–569.

P15. A Preissmann, Quelques Proprietes globales des espaces de Riemann', *Comm. Math. Helv.* **15** (1943), 175–216.

P16. L. Pukansky, 'On the Characters and the Plancherel Formula of Nilpotent Groups', *J. Func. Anal.* **2** (1967), 255–280.

P17. L. Pukansky, 'Characters of Algebraic Solvable Groups', ibid. **3** (1969), 435–494.

P18. L. Pukansky, 'Unitary Representations of Lie Groups with Co-compact Radical and Applications'. *Trans. Amer. Math. Soc.* **236** (1978), 1–50.

R0a. S. Ramanujan, 'An Application of Morse Theory to Certain Symmetric Spaces', *J. Indian Math. Soc.* **42** (1968), 243–275.

R0b. S. Ramanujan, 'Application of Morse Theory to Some Homogeneous Spaces', *Tohoku Math. J.* **21** (1969), 343–353.

R0c. S. Ramanujan, 'On Stiefel Manifolds', *J. Math. Soc. Japan* **2** (1969), 543–548.

R1. J. H. Rawnsley, 'Some Applications of Quantization', (thesis, Oxford, 1972).

R2. J. H. Rawnsley, 'On the Cohomology Groups of a Polarization and Diagonal Quantization', *Trans. Amer. Math. Soc.* **230** (1977), 235–255.

R3. J.H. Rawnsley, 'Coherent State and Kähler Manifolds', *Quart. J: Math.* **28** (1977), 403–415.

R4. J. H. Rawnsley, 'DeSitter Symplectic Spaces and their Quantizations', *Proc. Camb. Phil. Soc.* **76** (1974), 473–480.

R5. D. Ray, 'On Spectra of Second Order Differential Operators', *Trans. Amer. Math. Soc.* **77** (1954), 299–321.

R6. G. Reeb, 'Sur certaines properties topologique des trajectoires des systemes dynamique', *Mem. Acad. R. Belgique* **27** (1952).

R7. G. Reeb, 'Varietes de Riemann dont toutes les geodesiques sont fermees', *Bull. Cl. Sci. Bruxelles* **36** (1950), 324–329.

R8. G. Reeb, 'Trois problemes de la theorie des systemes dynamique', *Colloq. Geom. Diff. Global CBRM* (1958), 89–94.

R9. P. Renouard, 'Varietes symplectiques et quantification', (thesis, Orsay, 1969).

R10. A. G. Reyman and M. A. Semenov-Tian-Shansky, 'Reduction of Hamiltonian Systems and Affine Lie Algebras', *Invent. Math.* **53** (1979), 81–100.

R11. M. A. Rieffel, 'On the Uniqueness of the Heisenberg Commutation Relations', *Duke Math. J.* **39** (1972), 745–752.

R11a. H. P. Robertson and T. W. Noonan, *Relativity and Cosmology*, (Saunders, Philadelphia, 1968).

R12. W. Roelcke, 'Das Eigenwerten Problem der Automorphen Formen in der Hyperbolischen Ebene', *Math. Ann.* **167** (1966), 292–337; **168** (1967), 261–324.

S0a. S. Sankaran, 'Heisenberg, Commutation Relations and Pontryagin's Duality Theorem', *Math. Z.* **98** (1967), 387–390.

S0b. S. Sankaran, 'Imprimitivity and Duality', *Bull. Unione Mat. Ital.* **7** (1973), 251–259.

S1. S. Sasaki, 'On the Differential Geometry of Tangent Bundles of Riemann Manifolds', *Tohoku Math. J.* **10** (1958), 338–354; **14** (1962), 146–155.

S2. S. Sasaki, 'On Differential Manifolds with Certain Structures which are Closely Related to Almost Contact Structure', ibid. **12** (1960), 459–470.

S3. S. Sasaki, *Almost Contact Manifolds*, (Lecture Notes, Tohoku University, 1965).

S4. S. Sasaki, and Y. Hatakeyama, 'On Differentiable Manifolds with Contact Metric Structure', **14** (1962), 249–271.

S5. I. Satake, 'On Unitary Representations of a Certain Group Extension', *Sugaku, Math. Soc. Japan* **21** (1969), 241–253.

S6. I. Satake, 'Factors of Automorphy and Fock Representations', *Adv. Math.* **7** (1971), 83–110.

S7. I. Satake, 'Fock Representations and Theta Functions', in *Advances in Theory of Riemann Surfaces*, (Princeton University Press, Princeton, 1971), pp. 393–405.

S8. I. Satake, 'Unitary Representations of a Semidirect Product of Lie Groups on $\bar{\partial}$-Cohomology Spaces', *Math. Ann.* **190** (1971), 177–202.

S9. W. Schmid, 'Homogeneous Complex Manifolds and Representations of Semisimple Lie Groups', (thesis, University of California, Berkeley, 1967).

S10. W. Schmid, 'On a Conjecture of Langlands', *Ann. Math.* **93** (1971), 1–42.

S11. W. Schmid, 'Some Properties of Square Integrable Representations of Semisimple Lie Groups', *Ann. Math.* **102** (1975), 535–564.

S12. I. S. Schulman, 'A Path Integral for Spin', *Phys. Rev.* **176** (1968), 1558–1569; *Phys. Rev.* **188** (1969), 1139–1142.

S13. I. S. Schulman, *JMP* **12** (1971), 304–308.

S13a. I. S. Schulman, *Techniques and Applications of Path Integration*, (Wiley, New York, 1981).

S14. R. L. E. Schwarzenberger, 'Crystallography in Spaces of Arbitrary Dimension', *Proc. Camb. Phil. Soc.* **76** (1974), 23–32.

S15. J. Schwinger, 'On Gauge Invariance and Vacuum Polarization', *Phys. Rev.* **83** (1951), 664–679.

S16. J. Schwinger, 'The Theory of Quantized Fields V', *Phys. Rev.* **93** (1954), 615–628.

S17. J. Schwinger, 'On the Euclidean Structure of Relativistic Field Theory', *Proc. Nat. Acad. Sci.* **44** (1958), 956–965.

S18. J. Schwinger, 'Euclidean Quantum Electrodynamics', *Phys. Rev.* **115** (1959), 721–731.

S19. R. T. Seeley, 'Complex Powers of an Elliptic Operator', *Proc. Symp. Pure Math.* **10** (1967), 288–307.

S20. A. Selberg, 'Harmonic Analysis and Discontinuous Groups in Weakly Symmetric Riemannian Spaces with Applications to Dirichlet Series', *J. Indian Math. Soc.* **20** (1956), 47–87.

S21. A. Selberg, *Automorphic Functions and Integral Operators, Seminars in Analytic Functions*, Vol. 2, (1957), pp. 152–161.

S22. M. A. Semenov-Tian-Shansky, 'On a Property of Kirillov Integral', *Mat. Ind. Steklov* (*LOMI*) **37** (1973), 53–65.

S22. M. A. Semenov-Tian-Shansky, *Izv. Akad. Nauk* **40** (1976), 562–592.

S23. D. J. Simms, '*Geometric Quantization of Energy Levels in the Kepler Problem*, Conv. di Geom. Simp. e Fis. Mat. INDAM, Rome (1973).

S24. D. J. Simms, and N. M. J. Woodhouse, 'Lectures on Geometric Quantization', *Lecture Notes in Physics* **53** (1976).

S24a. B. Simon, 'The Classical Limit of Quantum Partition Functions', *Comm. math. Phys.* **71** (1980), 247–276.

S25. J. Sniatycki, *Geometric Quantization and Quantum Mechanics*, (Springer-Verlag, New York, 1980).

S26. J. -M. Souriau, *Structure des Systemes dynamiques*, (Dunod, Paris, 1970).

S27. J. -M. Souriau, *Sur la variete de Kepler*, Conv. di Geom. Simp. e Fis. Mat. INDAM, Rome, (1973).

S28. R. F. Streater, 'Canonical Quantization', *Comm. math. Phys.* **2** (1966), 354–374.

S29. R. F. Streater, 'The Representations of the Oscillator Group', ibid. **4** (1967), 217–236.

S30. K. F. Stripp and J. C. Kirkwood, 'Asymptotic Expansion of the Partition Function of the Asymmetric Top', *J. Chem. Phys.* **19** (1951), 1131–1133.

T1. E. A. Tagirov, 'Consequences of Field Quantization in DeSitter Type Cosmological Models', *Ann. Phys.* **76** (1973), 561–579.

T2. R. Takahashi, 'Sur les representations unitaires des groupes de Lorentz generalises', *Bull. Soc. math. Fr.* **91** (1963), 289–433.

T2a. M. Takeuchi, 'Cell Decompositions and Morse Equalities on Certain Symmetric Spaces', *J. Fac. Sci., Univ. Tokyo* **12** (1965), 81–192.

T2b. M. Takeuchi, 'Nice Functions on Symmetric Spaces', *Osaka J. Math.* **6** (1969), 283–288.

T2c. M. Takeuchi, and S. Kobayashi, 'Minimal Imbeddings of *R*-Spaces', *J. Diff. Geom.* **2** (1968), 203–215.

T3. T. Tamagawa, 'On Selberg Trace Formula', *J. Fac. Sci. Univ. Tokyo* **8** (1960), 363–386.

T4. S. Tanno, 'A Theorem of Regular Vector Fields', *Tohoku Math. J.* **17** (1965), 235–238.

T5. S. Tanno, 'The Topology of Contact Riemannian Manifolds', *I11. J. Math.* **12** (1968), 700–717.

T6. S. Tanno, 'Sasakian Manifolds with Constant ϕ-holomorphic Sectional Curvature', *Tohoku Math. J.* **21** (1969), 501–507.

T7. S. Takizawa, 'On the Contact Structures of Real and Complex Manifolds', *Tohoku Math. J.* **15** (1963), 227–252.

T8. S. Takizawa, 'On Soudures of Differentiable Fibre Bundles', *J. Math. Kyoto Univ.* **2** (1963), 237–276.

T9. J. A. Tirao and J. A. Wolf, 'Homogeneous Holomorphic Vector Bundles', *Indiana Univ. Math. J.* **20** (1970), 15–31.

T10. J. A. Tirao, 'Square Integrable Representations of Semi-Simple Lie Groups', *Trans. Amer. Math. Soc.* **190** (1974), 57–75.

T11. J. Tits, 'Sur certaines classes d'espaces homogenes de groupes de Lie', *Mem. de l'Acad. Royale Belgique* **29** (1955).

V1. V. S. Varadarajan, *Geometry of Quantum Theory*, (Van Nostrand, New York, 1970).

V2. S. R. S. Varadhan, 'On the Behavior of the Fundamental Solution of the Heat Equation', *Comm. Pure Appl. Math.* **20** (1967), 431–455; 659–685.

V3. A. B. Venkov, 'Expansion in Automorphic Eigenfunctions of the Laplace Beltrami Operator in Classical Symmetric Spaces of Rank One and the Selberg Trace Formula', *Proc. Steklov Inst. Math.* **125** (1973), 1–48.

V4. A. B. Venkov, V. L. Kalinin, and L. D. Faddeev, 'Nonarithmetic Deduction of the Selberg Trace Formula for the Upper Half Plane', *J. Sov. Math.* **8** (1977), # 2.

V5. A. B. Venkov, 'Selberg's Trace Formula', *Math. USSR Izv.* **12** (1978), 448–462.

V6. M. Vergne, 'La structure de Poisson sur l'algebre symetrique d'une algebre de Lie nilpotent', *Bull. Soc. Math. Fr.* **100** (1972), 301–335.

V6a. I. E. Viney, 'Asymptotic Expansions of the Expressions for the Partition Function', *Proc. Comb. Phil. Soc.* **29** (1933), 142–148.

V7. A. M. Vinogradov and B. A. Kupershmidt, 'The Structures of Hamiltonian Mechanics', *Russ. Math. Sur.* **32** (1977), 177–243.

V8. N. J. Vilenkin, *Special Functions and the Theory of Group Representations*, (Amer. Math. Soc., Providence. R. I., 1968).

V9. A. Voros, 'Semi-Classical Approximations', *Ann. Inst. H. Poincaré* **A24** (1973), 31–90.

V9a. A. Voros, 'The Zeta Function of the Quartic Oscillator', *Nucl. Phys.* **B165** (1980), 209–236.

W1. N. R. Wallach, 'On the Selberg Trace Formula in the Case of Compact Quotient', *BAMS* **82** (1976), 171–195.

W2. N. R. Wallach, 'An Asymptotic Formula of Gelfand and Gangolli for the Spectrum of $\Gamma\backslash G$', *J. Diff. Geom.* **11** (1976), 91–101.

W3. N. R. Wallach, *Symplectic Geometry and Fourier Analysis*, (Math. Sci. Press, Brookline, Mass., 1977).

W4. H. C. Wang, 'Closed Manifolds with Homogeneous Complex Structures', *Amer. J. Math.* **76** (1954), 1–32.

W5. G. Warner, *Harmonic Analysis on Semisimple Lie Groups*, (Springer-Verlag, New York, 1972).

W6. G. Warner, 'Selberg Trace Formula for Nonuniform Lattices, the *R-Rank One Case*', *Adv. Math.* **6** (1979), 1–142.

W7. A. Wawrzynczyk, 'Reciprocity Theorems and the Theory of Representations of Groups and Algebras', *Diss. Math.* **126** (175).

W8. A. Weinstein, *Lectures in Symplectic Manifolds*, (CBMS, # 29, Amer. Math. Soc., Providence, 1977).

W9. H. Weyl, 'Das asymptotische Verteilungsgesetz der Eigenschwingungen eines beliebig gestalteten elastischen Korpers', *Rend. Circ. Mat. Palermo* **39** (1915), 1–50.

W10. H. Weyl, 'Das asymptotische Verteilungs geseta der Eigenwerte linearer partieller Differential gleichungen (mit einer Anwendung auf die Theorie der Hohlraum straglung)', *Math. Ann.* **71** (1911), 441–469.

W11. H. Weyl, 'Ramifications, Old and New, of the Eigenvalue Problem', *BAMS* **56** (1950), 115–139.

W12. H. Weyl, *The Theory of Groups and Quantum Mechanics*, (Dover, New York, 1931).

W12a. E. P. Wigner, *Phys. Rev.* **40** (1932), 749.

W13. J. Wolf, *Spaces of Constant Curvature*, (McGraw-Hill, New York, 1967).

14. J. A. Wolf, 'A Contact Structure for Odd Dimensional Spherical Space Forms', *Proc. Amer. Math. Soc.* **19** (1968), 196.

W15. J. A. Wolf, 'Complex Manifolds and Unitary Representations', *Lecture Notes in Mathematics* **185** (1971), 242–287.

W16. J. A. Wolf, 'Unitary Representation on Partially Holomorphic Cohomology Spaces', *Mem. Amer. Math. Soc.* **138** (1974).

W17. J. A. Wolf, 'The Action of a Real Semisimple Group on a Complex Flag Manifold', *Bull. Amer. Math. Soc.* **75** (1969), 1121–1237.

W18. J. A. Wolf, 'Representations Associated to Minimal Coadjoint Orbits', in *Lecture Notes in Math.* **676** (1978), 328–349.

W19. J. A. Wolf, 'Remark on Nilpotent Orbits', *Proc. Amer. Math. Soc.* **51** (1975), 213–216.

W20. J. A. Wolf, 'Unitary Representations of Maximal Parabolic Subgroups of the Classical Groups', *Mem. AMS* **180** (1976).

W21. J. A. Wolf, 'Complex Manifolds and Unitary Representations', in *Lecture Notes in Math.* **185** (1971), 242–287.

W22. J. A. Wolf, 'Conformal Group, Quantization and the Kepler Problem', *Lecture Notes in Physics* **50** (1976), 217–222.

W23. L. S. Wollenberg, *Proc. Amer. Math. Soc.* **20** (1969), 315.

Y1. K. Yosida, *Lectures on Semigroup Theory and its Applications to Cauchy's Problem in Partial Differential Equations*, (Tata Inst. Fund. Res., Bombay, 1957).

Z1. V. E. Zakharov and L. D. Faddeev, 'KdV Equation: A Completely Integrable Hamiltonian System', *Func. Anal. and App.* **5** (1971), 280–288.

INDEX

Mathematics and Its Applications

1. Willem Kuyk, *Complementarity in Mathematics, A First Introduction to the Foundations of Mathematics and Its History*. 1977.
2. Peter H. Sellars, *Combinatorial Complexes, A Mathematical Theory of Algorithms*. 1979.
3. Jacques Chaillou, *Hyperbolic Differential Polynomials and Their Singular Perturbations*. 1979.
4. Svtopluk Fučik, *Solvability of Nonlinear Equations and Boundary Value Problems*. 1980.
5. Willard L. Miranker, *Numerical Methods for Stiff Equations and Singular Perturbation Problems*. 1980.
6. P. M. Cohn, *Universal Algebra*. 1981.
7. Vasile I. Istrăteşcu, *Fixed Point Theory, An Introduction*. 1981.
8. Norman E. Hurt, *Geometric Quantization in Action*. 1982.
9. Peter M. Alberti and Armin Uhlmann, *Stochasticity and Partial Order. Doubly Stochastic Maps and Unitary Mixing*. 1982.
10. F. Langouche, D. Roekaerts, and E. Tirapegui, *Functional Integration and Semiclassical Expansions*. 1982.